Tableau Certified Data Analyst Certification Guide

Ace the Tableau Data Analyst certification exam with expert guidance and practice material

Harry Cooney

Daisy Jones

‹packt›

Tableau Certified Data Analyst Certification Guide

Authors: Harry Cooney and Daisy Jones

Reviewers: Jess Hancock and Mahendra Singh

Publishing Product Manager: Sneha Shinde

Senior-Development Editor: Ketan Giri

Development Editor: Kalyani S.

Presentation Designer: Salma Patel

Editorial Board: Vijin Boricha, Megan Carlisle, Simon Cox, Ketan Giri, Saurabh Kadave, Alex Mazonowicz, Gandhali Raut, and Ankita Thakur

First Published: June 2024

Production Reference: 1240624

Published by Packt Publishing Ltd.

Grosvenor House

11 St Paul's Square

Birmingham

B3 1RB

ISBN: 978-1-80324-346-7

www.packtpub.com

Contributors

About the Authors

Harry Cooney, is a Senior Data Consultant who has worked with Tableau for the past five years, transforming business questions into analytical dashboards that enable users to generate insights about their business as well as mentoring Tableau users of all skill levels, both individually and through corporate training. In addition to this professional experience, Harry has even had his Tableau work featured in the Tableau Conference Virtual Gallery and made the shortlist for the 2023 World Data Visualization Prize.

Outside of work, Harry is a fan of football and enjoys incorporating this interest into Tableau by creating reports to compare team and player stats.

LinkedIn profile: `https://www.linkedin.com/in/harrycooney/`

Daisy Jones is a Senior Data Consultant and Chief Party Officer at The Information Lab Ireland. After completing a degree in Chinese Studies, Daisy discovered The Information Lab Ireland's training program where she gained the skills in Tableau and Alteryx that would spark a passion for creative design with Tableau dashboards. Four years later, she is now working with clients to help them utilize best data practices and create data driven answers for their business.

Away from the office, Daisy is normally found in the gym, crocheting, or playing computer games.

LinkedIn profile: `https://www.linkedin.com/in/daisy-jones1995/`

About the Reviewers

Jess Hancock has worked closely with a range of clients across the finance, aerospace, higher education, and non-profit sectors. For Jess, there's nothing more satisfying than bridging the gap between developers and decision-makers to turn raw data into useful information.

In her spare time, Jess enjoys painting, birdwatching, and being out in nature.

LinkedIn profile: `https://www.linkedin.com/in/jess-h/`

Mahendra Singh is a seasoned data analyst, consultant, and data scientist with over 10 years of experience in the industry. He has been using Tableau Desktop and Server applications for seven years and has worked with several large-scale US firms to build and maintain Tableau visuals and installations. Mahendra completed his Bachelor's degree in Information Technology with a focus on Software Development in 2015. He has worked in various industries, including healthcare, food and beverage, and banking and finance. Mahendra's areas of expertise and interest include business intelligence, big data and analytics, Spark, and machine learning.

Table of Contents

3

Calculations   107

4

Grouping and Filtering   169

5

Charts   223

6

Dashboards   295

7

8

9

Preface

The *Tableau Certified Data Analyst Certification Guide* is created to help you ace your *Tableau Certified Data Analyst* exam!

Data has been referred to as the new oil, and Tableau has consistently been at the top of the pack when it comes to data analytics software. For an analyst looking to excel in their job, or someone looking to move into the data industry, Tableau is the ideal tool. This book provides a detailed introduction to using Tableau and provides the necessary skills to pass the Tableau Certified Data Analyst exam. It will further set anyone up to use Tableau to enhance their career.

The Tableau Certified Data Analyst certification validates the fundamental knowledge required to explore, analyze, and present data and further your career in data analytics. This book is a best-in-class study guide that fully covers the Tableau Certified Data Analyst exam objectives and will help you pass the exam the very first time.

Complete with clear explanations, chapter review questions, realistic mock exams, and detailed solutions, this guide will help you master the core exam concepts and build the understanding you need to go into the exam with the skills and confidence to get the best result.

With the help of relevant examples, you will learn fundamental Tableau concepts such as transforming data and building dashboards. As you progress, you will delve into the important domains of the exam, including relationships, table calculations, and forecasting data.

This book contains a wide range of content and realistic scenarios that will give you everything you need to both pass the exam and utilize Tableau in your career.

The key features of the textbook include the following:

- All topics relevant to the exam are covered in detail

- Additional content is included to aid your understanding of Tableau

- Step-by-step practical exercises within each topic are discussed to reinforce learning

- Practice questions and exam preparation tips are included to ensure you are confident with the types of questions that will come up during the exam

Whether you are studying independently or as part of a structured course, this textbook is designed to support your learning journey and help you achieve success in the Tableau Certified Data Analyst exam. We encourage you to engage actively with the material, practice regularly, and leverage the resources provided to maximize your understanding and proficiency in Tableau.

We wish you the best of luck in your Tableau certification journey and hope that this textbook serves as a valuable companion along the way.

Who This Book Is For

This book is for anyone interested in using Tableau to effectively explore and present data. You could be an IT professional looking to upskill or someone looking for a career in data.

There are no prerequisites for this book. This book will assist those with no prior knowledge of Tableau to prepare for the exam. Analysts with prior experience in exploring and presenting data will be able to utilize this book to improve and solidify their use of Tableau.

What This Book Covers

Chapter 1, Connecting to Data, introduces you to data source connections in Tableau. This includes the types of data sources that can be connected to and the configuration options when doing so.

Chapter 2, Transforming Data, takes you over the complete suite of tools and methods for transforming data with Tableau. Both **Tableau Desktop** and **Tableau Prep** data transformation methodologies are covered, as well as the data types available in Tableau. You will come out of the chapter with an understanding of how to clean and prepare data sources so that they are ready for analysis.

Chapter 3, Calculations, provides a detailed walk-through of the calculated field logic available in Tableau. Basic calculations are broken up by the data type they relate to. All available table calculations are described and fixed level of detail calculations are explained in terms of functionality.

Chapter 4, Grouping and Filtering, teaches you how to structure and filter data with Tableau. Tableau's set, bin, group, and hierarchy functionalities are covered along with filtering methods and how to improve interactivity using parameters.

Chapter 5, Charts, covers how data can be visualized into charts, which is Tableau's primary functionality. All relevant chart types are listed with explanations for the required setup along with additional chart-related functionality and analytical features.

Chapter 6, Dashboards, provides you with a detailed breakdown of how to combine multiple charts into a single piece of analysis primarily through dashboards but also using Tableau's story feature. How to combine multiple charts in a dashboard is covered along with the types of objects that can be included on a dashboard and the available interactivity options.

Chapter 7, Formatting, walks you through how to apply formatting at a workbook level, including adding custom color palettes and shapes, formatting options available for individual charts, and how to customize the look of dashboards.

Chapter 8, Publishing and Managing Content, provides you with an introduction to publishing and managing content on Tableau Cloud. Once data has been presented in a visual format via charts, dashboards, or stories, the analysis needs to be shared with end users. This chapter covers how to share content as well as how to keep data fresh for end users.

Online Practice Resources

With this book, you will unlock unlimited access to our online exam-prep platform (*Figure 0.1*). This is your place to practice everything you learn in the book. How to access the resources. To learn how to access the online resources, refer to *Chapter 9*, Accessing the Online Practice Resources at the end of this book.

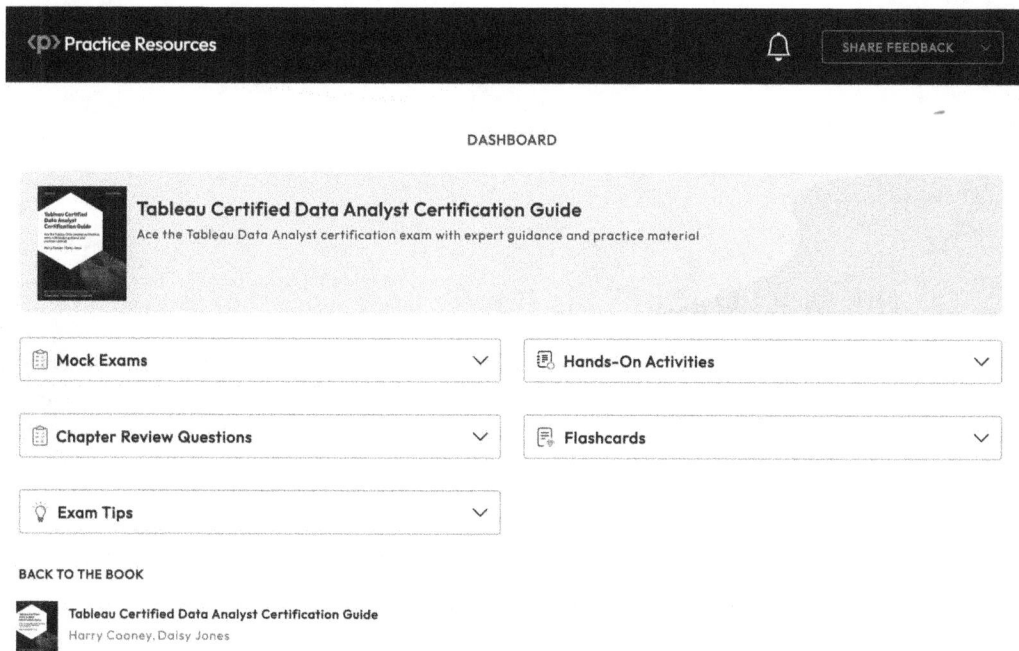

Figure 0.1: Online exam-prep platform on a desktop device

Sharpen your knowledge of Terraform concepts with multiple sets of mock exams, interactive flashcards, and exam tips accessible from all modern web browsers.

Download the Color Images

We also provide a PDF file that has color images of the screenshots/diagrams used in this book. You can download it here: `https://packt.link/swloG`.

Conventions Used

`Code in text`: Indicates code words in text, database table names, folder names, filenames, file extensions, pathnames, dummy URLs, user input, and X (formerly Twitter) handles. Here is an example: "The `COUNT` function takes a single input of any data type and counts the number of items input (excluding `null` values). It is formatted as `COUNT(value)` and the example field with the values 1, `null`, and 3 would be formatted as `COUNT([Field])` and would return 2."

Bold: This indicates a definition or an important word or words that you see on screen. For instance, words in menus or dialog boxes appear in bold. Here is an example: "Scroll down on the release page to the **Download files** section and click either the **Windows** or **Mac** download link to start the installation application download."

A block of code is set as follows:

```
IF test1 THEN output1
[ELSEIF test THEN outputn]
[ELSE defaultoutput]
END
```

To Get the Most Out of This Book

In this book, Tableau-specific terms will be used to refer to areas of the **Tableau Desktop interface**. The following is a list of numbered terms that will be used throughout this textbook and can be identified in the following screenshot. A brief description of the functionality of each part of the interface is also provided:

- **Data pane**: This is where the tables and fields from your data source can be found. These can be dragged onto the other areas of the interface to create charts.

- **Canvas**: This is where charts are displayed visually. Fields can be dropped directly onto the canvas positionally, and Tableau will usually incorporate the field into the chart accordingly.

- **Columns and rows**: Fields can be dragged onto columns and rows to create x and y axes or tables positioned as rows or as columns.

- **Marks card**: This is used to add detail to charts and visuals via color, text elements, tooltips, and a general level of detail for a visual. The specific mark type for the visual can be selected here as well (for example, pie chart, bar chart, or line chart).

- **Filters shelf**: Fields placed here can be configured to filter the chart to specific data points.

- **Data Source tab**: Selecting this tab will navigate to the data source interface where data sources can be configured.

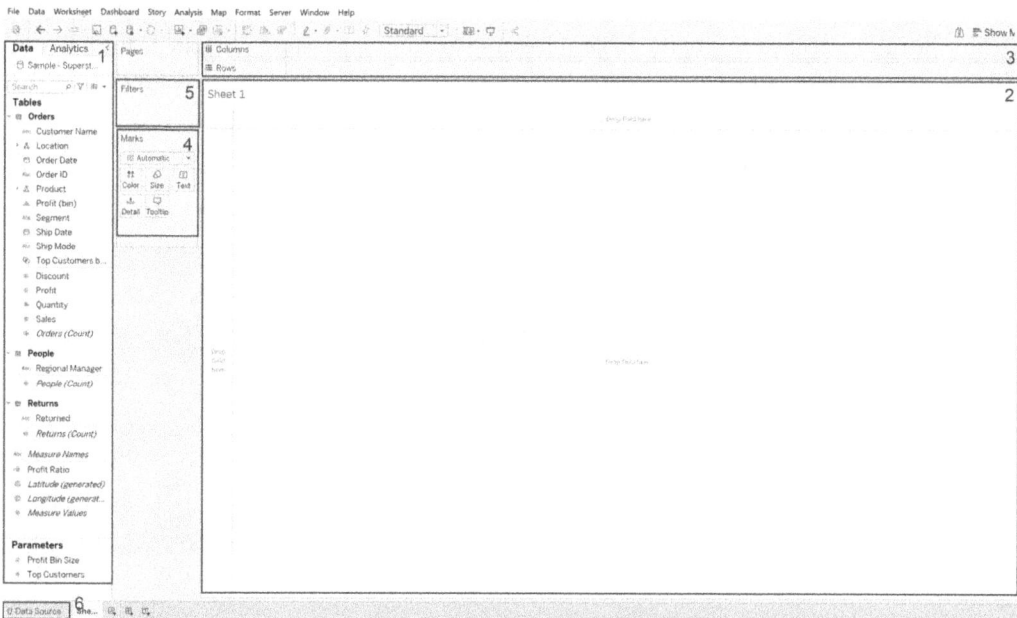

Figure 0.2: Tableau Desktop Interface

Setting up Tableau

To complete the exercises in this textbook, you will need to download and install both **Tableau Desktop** and **Tableau Prep Builder**. If you do not have a license for Tableau, a two-week trial can be utilized. You will also need to register for a trial with **Tableau Cloud**.

This chapter will walk through the necessary steps for downloading and installing Tableau Desktop and Tableau Prep. It will also outline the process for signing up for a 14-day Tableau Desktop, Tableau Prep, and Tableau Cloud free trial.

Download and Install Tableau Desktop

Tableau Desktop is the Tableau tool that is used for analytics and data visualization. The bulk of the Tableau Certified Data Analyst exam consists of functionality within Tableau Desktop. There is a browser-based version of Tableau Desktop called **Web Authoring**, but there are some differences in functionality between the two tools, with Tableau Desktop having a wider range of features. It is recommended that you download and install Tableau Desktop and use it to complete the exercises contained within the textbook. If you already have Tableau Desktop installed can skip this section.

It is free to download and install Tableau Desktop, but a license is required to use it. If a license cannot be purchased, then a 14-day free trial can be started after installation.

To find the most recent version of the Tableau Desktop installer application, navigate to the following URL: `https://www.tableau.com/support/releases`. The page should look similar to the following screenshot (although the versions shown may be different).

Figure 0.3: The Tableau Desktop released versions page

Click on the large blue button at the center of the screen that says **VIEW THE CURRENT VERSION**. This will open up a sign-in page. If you already have a Tableau account, then log in here, otherwise, fill in the form and create an account. Once you have created an account and logged in, return to the downloads page and click the blue **VIEW THE CURRENT VERSION** button again.

The download page for the most up-to-date version of Tableau will now be open. The following screenshot was taken when version 2023.3 was the most current, but this will likely have changed.

Figure 0.4: Most recent Tableau version download page

Scroll down on the release page to the **Download files** section and click either the **Windows** or **Mac** download link to start the installation application download. The `.exe` file will begin to download and will likely be saved in the `Downloads` folder.

Download files

Windows

• TableauDesktop-64bit-2023-3-0.exe (573 MB)

Mac

• TableauDesktop-2023-3-0.dmg (707 MB)

Figure 0.5: The Windows and Mac download links

Once the application installer has been downloaded, the installation process can begin. To install the application, the user must be signed in to the machine as an administrator. If the machine is Windows, then navigate to the downloaded `.exe` file and double-click to run it. To start the installation on a Mac, open the `.dmg` file and then select the `.pkg` installer.

The Tableau installer will now start and the installation steps can be walked through. Read through the terms and conditions and then tick to confirm they have been read. Decide whether to send product usage data to Tableau. The installation can be customized in terms of the location where Tableau Desktop will be installed, whether to include shortcuts, and whether to automatically check for Tableau product updates. When customization is complete, the **Install** button can be clicked followed by **Yes** to confirm and begin the installation.

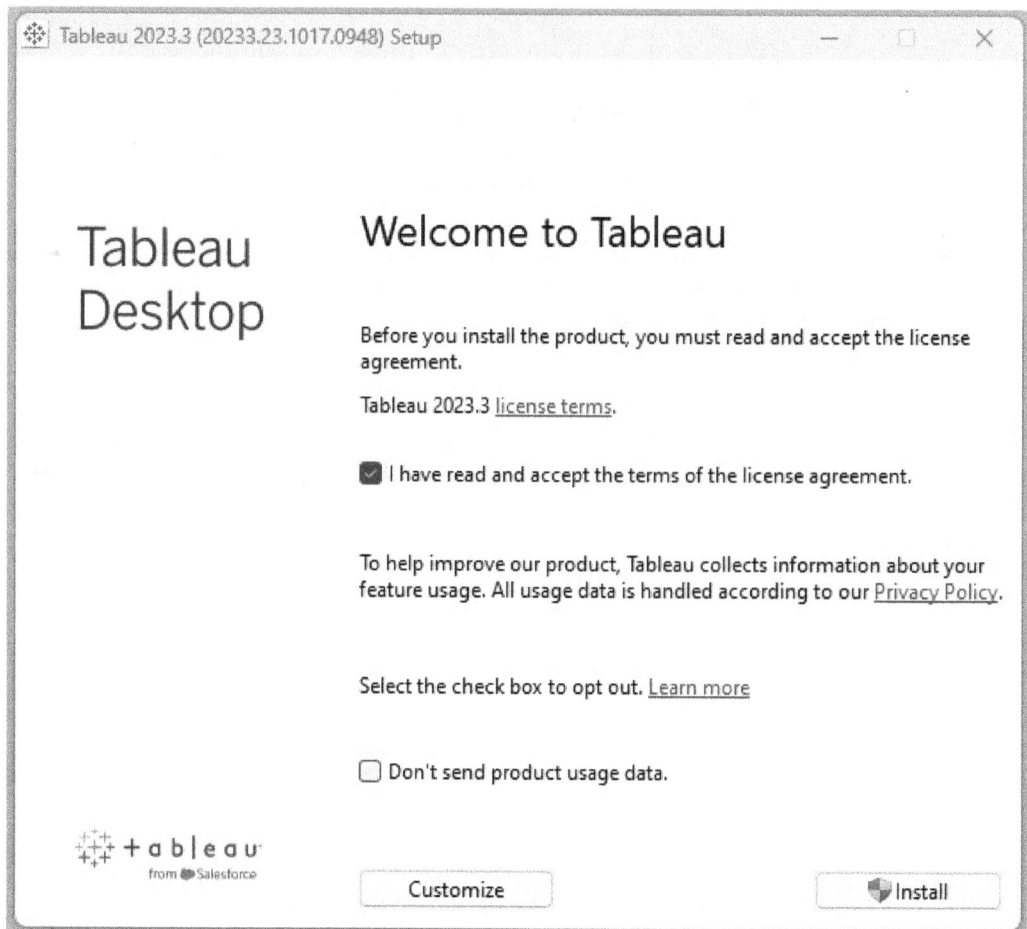

Figure 0.6: The Tableau Desktop Windows installation wizard

Once the installation process is complete, Tableau will open automatically. The Tableau registration interface prevents the usage of Tableau Desktop. If a Tableau license has been purchased, then the **Activate Tableau** link can be clicked and the corresponding license key or Tableau Server or Cloud credentials can be put in. If opting for the 14-day free trial, the form can be filled in, and the **Start trial now** button pressed.

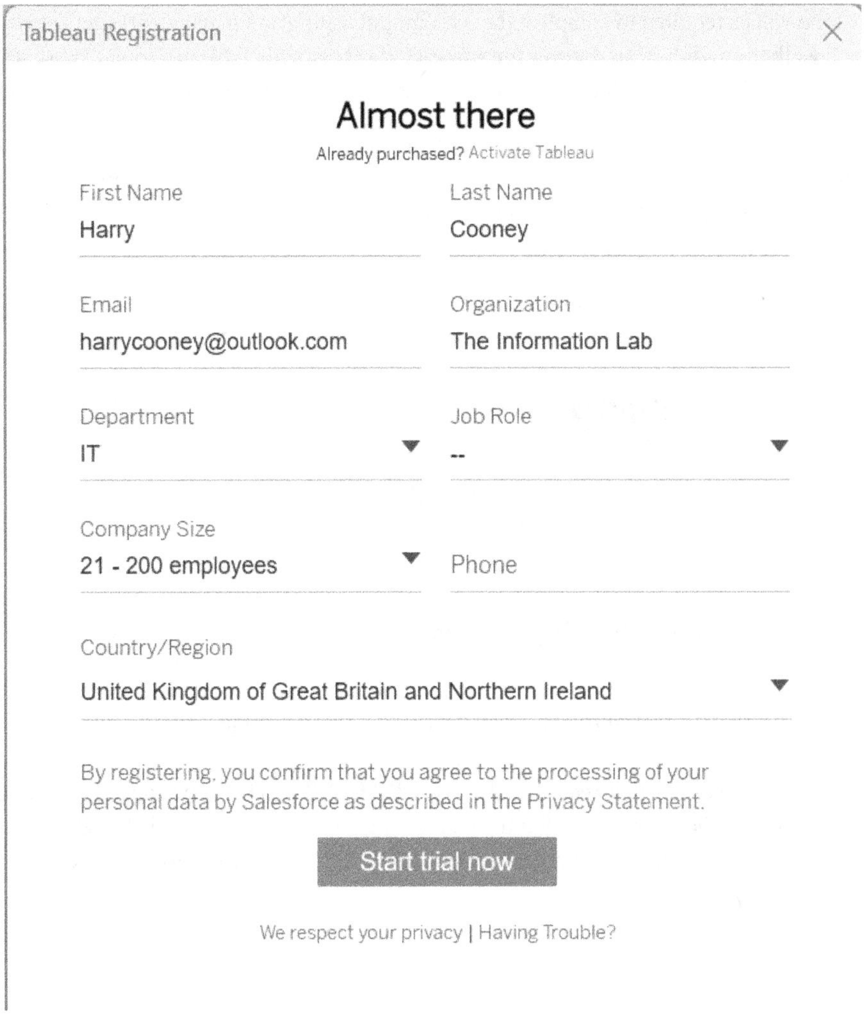

Tableau Registration   ×

Almost there

Already purchased? Activate Tableau

First Name

Harry

Last Name

Cooney

Email

harrycooney@outlook.com

Organization

The Information Lab

Department

IT   ▼

Job Role

--   ▼

Company Size

21 - 200 employees   ▼

Phone

Country/Region

United Kingdom of Great Britain and Northern Ireland   ▼

By registering, you confirm that you agree to the processing of your personal data by Salesforce as described in the Privacy Statement.

Start trial now

We respect your privacy | Having Trouble?

Figure 0.7: Tableau Desktop product activation

Tableau Server and Tableau Cloud

Tableau Server and Tableau Cloud are web-based interfaces used to house and share Tableau content. Tableau Server is hosted by the end user, whereas Tableau Cloud is hosted by Tableau. Publishing and managing content on a Tableau Server or Tableau Cloud is a significant part of the Tableau Certified Data Analyst exam and will be covered in the final chapter of the textbook. Access to Tableau Server or Tableau Cloud will be required to complete the relevant publishing and managing content exercises. This section will walk through how to register for a free 14-day trial with Tableau Cloud. Those with access to a Tableau Server or Tableau Cloud that have at least a Creator license can skip the following section.

To start a free 14-day Tableau Cloud trial, log in to `https://www.tableau.com/` and then navigate to `https://www.tableau.com/en-gb/products/online/request-trial`. The page will look similar to the following screenshot:

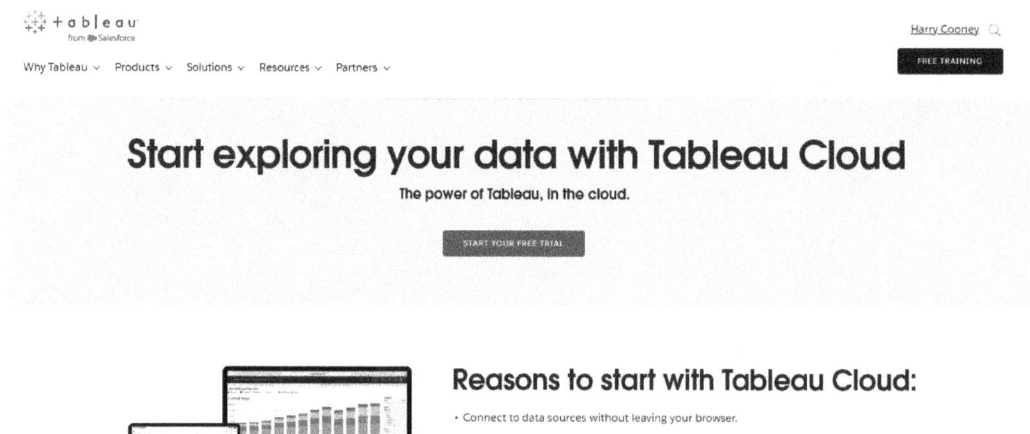

Figure 0.8: The Tableau Cloud free trial page

The blue button in the center that reads **START YOUR FREE TRIAL** can be clicked, which will scroll the page down to a registration form. Fill in the registration form and then click **Get free trial**. This will open a landing page where the blue **START A FREE TRIAL** button can be clicked.

Why Tableau
Products
Solutions
Resources
Partners
Harry Cooney
FREE TRAINING
Success! It looks like you're ready to start a trial.
START A FREE TRIAL

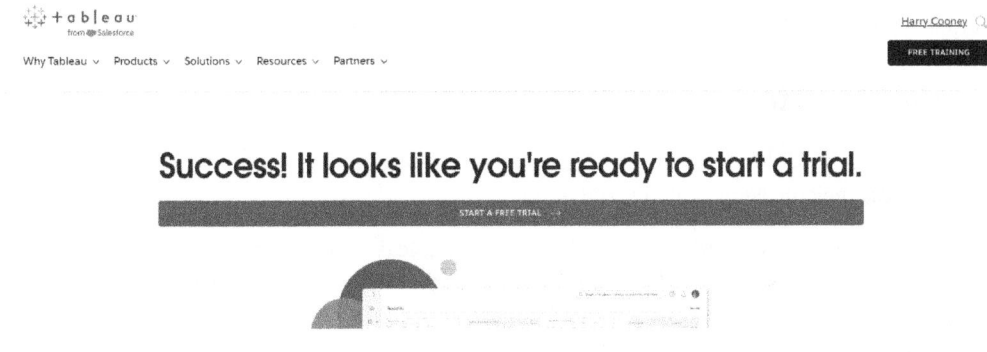

Figure 0.9: Post registration form landing page

Finally, a site name can be typed in and a location for where the site should be hosted can be selected. Once the terms and conditions have been read and accepted, the **ACTIVATE MY TRIAL** button can be clicked to begin the free 14-day trial.

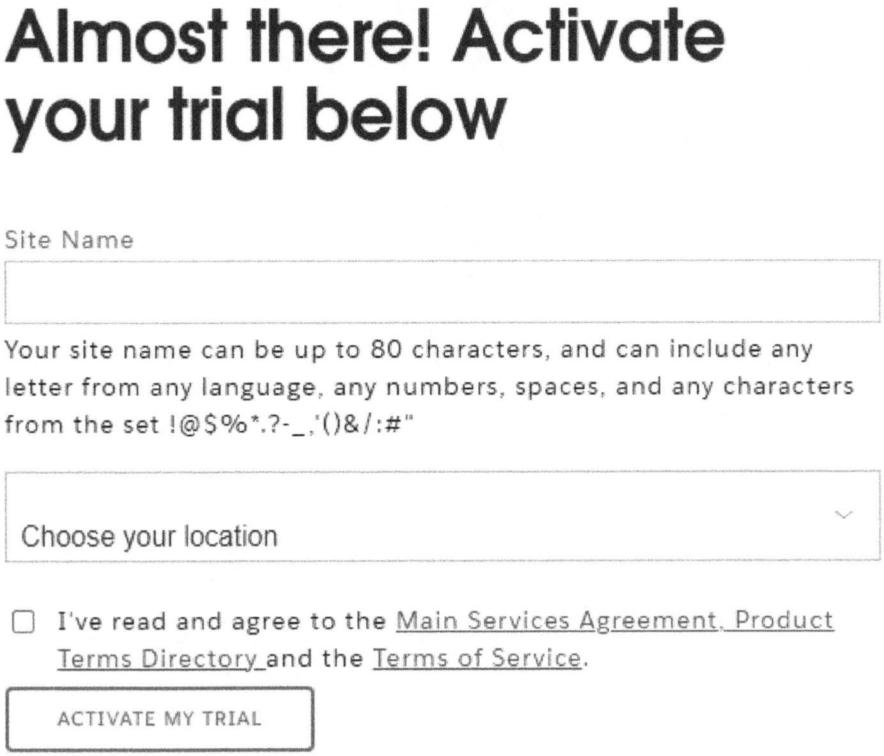

Almost there! Activate your trial below
Site Name
Your site name can be up to 80 characters, and can include any letter from any language, any numbers, spaces, and any characters from the set !@$%*.?-_,'()&/:#"
Choose your location
I've read and agree to the Main Services Agreement, Product Terms Directory and the Terms of Service.
ACTIVATE MY TRIAL

Figure 0.10: Tableau Cloud site name, location, and activation

It may take a few minutes, but the Tableau Cloud site will be created and then navigated to.

Tableau Desktop Practical Exercises

At the end of each section in each chapter, there will be practical exercises that will be a great way for you to reinforce the concepts learned throughout.

When it comes to Tableau Desktop, there is a default Tableau training data source that is readily available with each Tableau install. This is the data source that will be used for each exercise. The data source is called **Sample – Superstore** and can be selected from the bottom of the blue **Connect** pane on the left-hand side when Tableau Desktop is first opened.

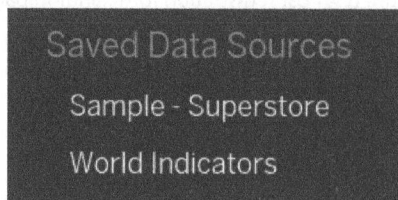

Figure 0.11: The Sample – Superstore data source connection used for practical exercises

Information about the Exam

The Tableau Certified Data Analyst examination is a test of knowledge of and skills in the data visualization tool Tableau Desktop, the data cleaning and preparation tool Tableau Prep, and the content hosting product Tableau Server and Tableau Online.

There is no prerequisite for taking the Tableau Data Analyst exam and this textbook will walk through the content required to pass. Upon passing the exam, the title **Tableau Certified Data Analyst** is awarded with a digital badge confirmation on the certification website, **Credly**. The title lasts for two years, at which point the exam will have to be taken again to renew the title.

Direct experience with Tableau Prep and Tableau Desktop is considered essential for passing the *Tableau Certified Data Analyst* examination as it is geared toward aspiring data analysts looking to prove a base level of technical skill in these platforms. Theoretical knowledge of Tableau Cloud or Server is beneficial, as the examination covers them briefly in the text-based questions.

General advice from Tableau is for users to have six months of practical experience, but this is merely a benchmark, as the quality of experience can vary enormously. Users should therefore meet the following criteria:

Be comfortable navigating the user interfaces of Tableau Prep and Tableau Desktop

- Understand the key features and functions covered in this guide, and be able to confidently apply them to an intermediate level of difficulty

- Have some experience creating basic charts and compiling these into simple interactive dashboards

Note that this study guide contains a host of practical questions that can form the baseline of a student's practical experience.

The following section outlines the four domains covered in the Tableau Certified Data Analyst exam, followed by a description of the examination format and some useful tips for passing the exam.

Examination Domains

This section will describe the focus of each of the four examination domains, including what proportion of the examination they account for. The sections within each domain will be described along with the chapter the corresponding content can be found in within this textbook.

Connect to and Transform Data

The first examination domain is *Connect to and Transform Data*. This section accounts for 24% of the examination content, and is focused on the following:

- Connecting to data from Tableau Desktop or Prep

- How to clean, transform, and combine the data within those tools

- Customizing the final data source to be ready for analysis in Tableau Desktop

These topics are covered across four subsections:

- Connect to Data Sources

- Prepare Data for Analysis

- Perform Data Transformation in Tableau Prep

- Customize Fields

The first subsection, *Connect to Data Sources*, covers the different data source types that Tableau Desktop and Tableau Prep can connect to. This includes Tableau-specific files such as .hyper or .tde files, data sources that are published on Tableau Server/Online, spreadsheet files, and both direct connections to tables/views and custom SQL queries to relational databases. Knowledge of the advantages and disadvantages of the different data sources and when to choose one type over the other is also important. Similarly, an understanding of the benefits of live data source connections compared to data source extracts is required. Specific Tableau Desktop skills when it comes to how to replace data sources for existing content are also covered. This content is explored in *Chapter 1, Connecting to Data*.

The second subsection, *Prepare Data for Analysis*, covers basic data cleansing operations in Tableau Desktop. Knowledge of how to prepare data using Tableau's Data Interpreter as well as how to pivot and split columns is required. General cleaning operations and organization of columns into folders are covered, as well as how to assess columns for key measures of data quality such as completeness, consistency, and accuracy.

You will need to know how to combine data sources within Tableau Desktop via relationships, joins, unions, and blending. It is important to understand how the data structure is impacted by these data combinations, as well as the benefits of one type over the other. Pre-filtering data using extract filters is another skill specific to Tableau Desktop that is required. *Chapter 2, Transforming Data*, covers the content in the subsection.

The third subsection, *Perform Data Transformation in Tableau Prep*, requires an understanding of how to clean, filter, transform, and combine data in Tableau Prep, including knowing which transformation is most suitable given a specific business scenario. How to perform unions, joins, aggregation, and pivoting is covered, and it is important to understand how each of these transformations will impact the granularity and structure of the data. *Chapter 2, Transforming Data*, also covers the content contained within this subsection.

The final subsection within the *Connect and Transform Data* domain is *Customize Fields*. This subsection is specific to Tableau Desktop and covers the final customization of columns in preparation for data visualization. Required knowledge includes how to rename columns, default property customization, such as changing the field type and sort order, and aliasing names. It is important to fully understand the distinction between **discrete** and **continuous** fields, as well as **dimensions** and **measures**, particularly the implications of the combination of these field types. The content for this subsection is covered in *Chapter 1, Connecting to Data*.

Explore and Analyze Data

Explore and Analyze Data is the largest examination domain, accounting for 41% of the examination's content. The domain covers the bulk of Tableau Desktop's features for exploring and analyzing data, including custom logic using calculated fields, the creation of dynamic view-based table calculations, the various methodologies for filtering data, and how to use Tableau's parameters to increase interactivity. The options available for structuring data is required knowledge and so is how to create charts using geographic field types. Finally, Tableau's analytics features are all included – from reference lines to predictive models.

The *Create Calculated Fields* subsection requires you to know how to create custom logic in Tableau Desktop using Tableau's own calculated field language. Simple logic (such as converting data types) is required as well as knowledge of how to aggregate measures and write string and number functions. Basic logical expressions (such as `if` and `case` statements) must be understood, as well as the various date functions available. The most complex type of calculation needed for the examination is the **Level of detail** calculations that allow Tableau users to fix data at a specific level of aggregation. The required calculations will be covered in *Chapter 3, Calculations*.

The second subsection covers the creation of quick table calculations. Quick table calculations are logical calculations that work on aggregated data in the view and are created via the interface selections. The examination can include questions on creating quick table calculations to show a moving average, a percent of the total, a running total, a difference or percent of a difference, a percentile, and a compound growth rate. It is important to understand what these table calculations show in the view and how to customize them to work in different directions or based on different levels of aggregation. Quick table calculations will be covered in *Chapter 3, Calculations*.

The *Create Custom Table Calculations* subsection builds on the quick table calculations by requiring an understanding of how to use table calculations to implement either date-based or ranking logic. Date-based logic includes how to use table calculations to dynamically display year-to-date, month-to-date, and year-over-year values. Ranking logic includes adding IDs or rank values to the view using either index, ranking, first, or last table calculations. Custom table calculations will be covered in *Chapter 3, Calculations*.

How to create and use filters is a subsection focusing on limiting the data shown in the view using Tableau Desktop's various filtering methods. The difference between filtering dimensions and measures must be understood and it is also important to understand the types of filtering available for dimensions, such as top and bottom N, include, exclude, wildcard, and conditional. The order in which different types of filters occur is covered, including the impact of adding filters to context. You will also need to know how to apply a filter across multiple sheets and data sources. This content will all be covered in *Chapter 4, Grouping and Filtering*.

Parameters are user-created static variables that are available in Tableau Desktop. Knowledge of how to apply parameters in calculations, filters, and reference lines is needed for the examination and this will be covered in *Chapter 4, Grouping and Filtering*.

The next subsection within *Explore and Analyze Data* focuses on how to structure the data. This section includes the various methodologies Tableau Desktop users can implement to create groupings within their data. Fields can be grouped into hierarchies where some field values represent a higher or lower-level grouping of the values of other fields. Field values can also be manually grouped together, in the case of dimensions, using Tableau's groups or sets. Dynamic grouping can also be set up for dimensions using sets and can be set up for measures using bins. The functionality available for all the mentioned methods of data structuring must be understood, and these will be covered in *Chapter 4, Grouping and Filtering*.

Geographic data can be mapped in Tableau via symbol, heat, density, and choropleth (filled) maps. The difference between these as well as an understanding of how to create each is required for the examination. Geographic chart creation will be covered in *Chapter 5, Charts*.

The final subsection in the *Explore and Analyze Data* domain involves summarizing, modeling and customizing data by using the **analytics** feature. You are required to know how to use the **analytics** feature to add subtotals and totals, reference lines and bands, average lines, trend lines, and distribution bands. Knowledge of advanced analytics features such as default and customized forecasting and the creation of predictive models is also required. *Chapter 5, Charts*, ends with an explanation of each Analytics feature.

Create Content

The third examination domain is again specific to Tableau Desktop only and covers 26% of the examination content. It focuses on the creation of content, namely how to create the basic chart types available in Tableau and then how to combine them into dashboards and stories. Dashboard development in terms of the interactivity options available as well as formatting capabilities are also covered.

The *Create Charts* subsection is comprehensive, requiring users to know how to create bar charts, line charts, pie charts, highlight tables, scatter plots, histograms, tree maps, bubbles, data tables, Gantt charts, box plots, area charts, dual axis charts, and combo charts. You need to know how to create these charts from scratch as opposed to using Tableau's **Show Me** functionality. This means you will need to understand the key areas of the Tableau Desktop interface, including the rows and columns shelves as well as the marks card, and how placing different field types on each of these sections will result in different types of charts being created. Sorting of the data within the charts must also be understood, including custom sorting. Chart creation and sorting is covered in *Chapter 5, Charts*.

The creation of dashboards to combine worksheets (charts) into a consolidated piece of analysis is also covered. This means you will need to know how the various layout options available in Tableau Desktop work, and more specifically how layout containers work. You also need to know what the different dashboard objects are and how to bring them onto a dashboard. The combination of dashboards into stories must also be understood. Dashboard and story creation are covered in *Chapter 6, Dashboards*.

Knowledge of the interactivity options available when creating dashboards is another requirement for the examination. This includes how to apply filters to a view, and how to add interactivity to charts in the form of hover, click, and menu options. You will need to know what types of actions are available, including filtering, highlighting, and URL actions, as well as how to configure them and provide guiding sentences if required. You need to know how to implement navigation via buttons as well as how to implement sheet swapping using parameters or sheet selectors. The functionality for this subsection is covered in *Chapter 6, Dashboards*.

The final subsection of the domain is *Format Dashboards*. This section requires you to be able to use all the formatting options available to make charts look more presentable and focus the analysis on key insights. This includes how to customize color, fonts, shapes, and styling using the marks card as well as how to add custom tooltips. Cleaning up charts by removing gridlines and table formatting must also be understood. You must also know how to add annotations to charts to point out key insights. At the dashboard level, you will need to know how to add padding to dashboards to create whitespace and avoid a cluttered look, and you must also know how to customize device-specific layouts for dashboards. Tableau has standard shapes and color schemes available to all users but an understanding of how to add custom shapes and color schemes to Tableau Desktop is also required. The content for this subsection is covered in *Chapter 7, Formatting*.

Publish and Manage Content on Tableau Server and Tableau Cloud

The final examination domain, *Publish and Manage Content on Tableau Server and Tableau Cloud*, is mainly focused on the sharing and managing of content in Tableau Server/Online but also requires knowledge of uploading content from both Tableau Desktop and Tableau Prep. This is the smallest domain, accounting for 9% of examination content. The domain covers publishing content from Tableau Desktop and Tableau Prep, scheduling data refreshes and Tableau Prep flows on Tableau Server/Online, and managing alerts and subscriptions on and to published workbooks.

The first subsection, *Publish Content*, requires an understanding of how to publish workbooks and data sources to Tableau Server/Online from Tableau Desktop and Tableau Prep. How to print and export from Tableau Desktop and Tableau Server/Online must also be understood, along with the print and export format types available. This content is covered in *Chapter 8, Publishing and Managing Content*.

Scheduling data updates must also be understood in terms of how to schedule both an extract refresh and a Tableau Prep Workflow on Tableau Server/Online. Tableau Server and Online scheduling will be covered in *Chapter 8, Publishing and Managing Content*.

Tableau user alerts on data points within workbooks and subscriptions to workbooks must be understood in terms of how to set up, configure, and manage them. This will also be covered in *Chapter 8, Publishing and Managing Content*.

Examination Format

The following applies to both the test center and remote/online examination format unless otherwise stated.

Prerequisites

This examination does not require any pre-existing qualifications. A Tableau account is required to book and take the examination. If you do not already have a Tableau account, you can do so by following official Tableau instructions at `https://mkt.tableau.com/files/CertificationAccountCreation.pdf`. As part of this process, the Tableau/Salesforce Terms of Service must also be agreed to.

Once created, your Tableau account should be used to log in to the **Pearson VUE Certification** portal using the following link: `https://cp.certmetrics.com/tableau/en/home/`. Once preliminary information has been entered, and terms agreed upon, the user can schedule directly with Pearson VUE, who proctor the examination on behalf of Tableau. A legal, photographic form of identification must be shown to the proctor on examination day before the examination begins to confirm your identity. Further information on ID requirements may be found under **View ID requirements** here: `https://home.pearsonvue.com/tableau/onvue`. It is important to ensure the first and last names used for your Tableau account are identical to the name on your chosen form of identification, or examination access may be denied without recompense.

The online examination is accessible only when a number of minimum technical requirements are met. Crucial amongst these are a computer with video capabilities, an internet connection of consistently high quality, and a screen resolution of at least 1,024 x 768. A comprehensive list is available at `https://home.pearsonvue.com/op/OnVUE-technical-requirements`.

These standards and checks are pre-established when the examination is taken at a Pearson VUE testing center.

When beginning the examination, participants must also agree to the terms and conditions of testing, which includes a strict non-disclosure agreement.

Examination Length

The examination is 2 hours long in total, 93 minutes of which are allocated to the examination itself. Approximately 8 minutes are permitted for the administrative tasks of signing the non-disclosure agreement and completing the tutorial; the tutorial describes the question types the user can expect and familiarizes the participant with the user interface.

The examination environment will open 30 minutes prior to the scheduled examination time. Participants are advised to join as soon as possible after this time to conduct environment, system, and ID checks, which occur live with a proctor.

Cost

There is a USD 250 base fee, excluding any taxes, such as VAT – this can vary based on location.

> **Note**
> Should the examination need rescheduling, a fee of USD 25 (plus required taxes) applies.

Assessment Limitations

The Tableau Certified Data Analyst examination is a "closed book" exam. No resources of any nature, be it digital or hard copy, are permitted to be present or used during the examination.

Language and Accessibility

The examination is most commonly conducted online, but it may also be taken at one of Pearson VUE's examination centers. There is no material difference between taking the examination online or at a test center, but if your machine is at risk of failing system checks, it is recommended to take the examination at a center. The system checks can be run from the following page: `https://system-test.onvue.com/system_test?customer=pearson_vue&clientcode=CHECKPOINT&locale=en_US`. The examination is currently available in English and Japanese only.

Should participants have specific accessibility requirements, adjustments may be requested by completing the following form: `https://app.smartsheet.com/b/form/2d4c22dfbbd74aee9aea5d07c223f676`. Please note that terms are to be agreed upon prior to scheduling an exam, and students may be required to submit appropriate evidence as part of their appeal.

Question Structure

The Tableau Certified Data Analyst certification contains 55 questions, split across three sections. Questions from any of the four domains discussed previously may appear in these sections, in any order. The majority of these test theoretical knowledge; expect approximately 45 static, text-based questions. The format of these may be any of the following:

- **Multi-select**: Four potential answers are presented, in which one or more may be correct
- **Multi-choice**: Four potential answers are presented, with a single correct answer
- **Active screen**: An image is presented with interactive regions that the user must select in response to the question

There are also approximately 10 practical questions, which are administered in Tableau itself. (Tableau is launched in a controlled manner directly inside the examination environment; this set-up process occurs automatically and requires no expertise on behalf of the examination participant.)

The examination is split into three sections: the first set is static (theoretical); the second practical, completed within an instance of Tableau Desktop; and the third is once again static. The static questions can be marked for review, which makes them easy to return to at the end of the section. All prior questions can be revisited and updated until the participant moves to the next section, at which time sections are closed and answers locked. The domains covered here are spread roughly evenly across each of the three sections.

Pass Criteria and Results

A passing grade of 750 is required. (Note that this score is not the true score achieved, but a mathematically calculated one to standardize difficulties across examinations).

Tableau policy is to send examination results to the email address associated with the participant's account within 48 hours of the examination finishing. (Note that there may be a short delay before the certification is registered on the participant's account and/or issued at Credly.)

Share Your Thoughts

Once you've read *Tableau Certified Data Analyst Certification Guide*, we'd love to hear your thoughts! Scan the QR code below to go straight to the Amazon review page for this book and share your feedback.

https://packt.link/r/1803243465

Your review is important to us and the tech community and will help us make sure we're delivering excellent quality content.

Download a Free PDF Copy of This Book

Thanks for purchasing this book!

Do you like to read on the go but are unable to carry your print books everywhere?

Is your eBook purchase not compatible with the device of your choice?

Don't worry, now with every Packt book you get a DRM-free PDF version of that book at no cost.

Read anywhere, any place, on any device. Search, copy, and paste code from your favorite technical books directly into your application.

The perks don't stop there, you can get exclusive access to discounts, newsletters, and great free content in your inbox daily.

Follow these simple steps to get the benefits:

1. Scan the QR code or visit the link below:

https://packt.link/free-ebook/9781803243467

2. Submit your proof of purchase.
3. That's it! We'll send your free PDF and other benefits to your email directly.

1

Connecting to Data

Introduction to Data

You are likely to have a basic grasp of the concept of data – it is difficult to avoid in a world increasingly driven by information, in which this information is both easier to gather and more valuable than ever. However, it is useful to begin with a basic definition.

Take, for example, a coffee chain store. Every day for the past five years, there has been an abundance of data collected for the number of sales, the number of cups of coffee sold, the different types of coffee sold, and even the stocks of coffee. A regional store manager may want to be able to see how well their sales are doing, which stores are performing best, and whether their specialty drinks are selling better than the classics. Five years of data can be hard to analyze in spreadsheets, and this is where Tableau can be the ideal tool for use. Tableau can connect to the data source and build dynamic dashboards that can help answer these questions. It can even define any outliers that exist, allowing the coffee chain to make data-driven decisions to help improve its business.

The following topics will be discussed in this chapter:

- Connecting to sources
- Data structure
- Choosing an appropriate data source type
- Connecting to Tableau Server or Tableau Cloud
- Connection management

This chapter is designed to inform you how to connect to data in Tableau with different data sources and servers that are available.

Making the Most Out of This Book – Your Certification and Beyond

This book and its accompanying online resources are designed to be a complete preparation tool for your **Tableau Certified Data Analyst Certification Guide**.

The book is written in a way that you can apply everything you've learned here even after your certification. The online practice resources that come with this book (*Figure 1.1*) are designed to improve your test-taking skills. They are loaded with timed mock exams, interactive flashcards, and exam tips to help you work on your exam readiness from now till your test day.

> **Before You Proceed**
>
> To learn how to access these resources, head over to *Chapter 9, Accessing the Online Practice Resources*, at the end of the book.

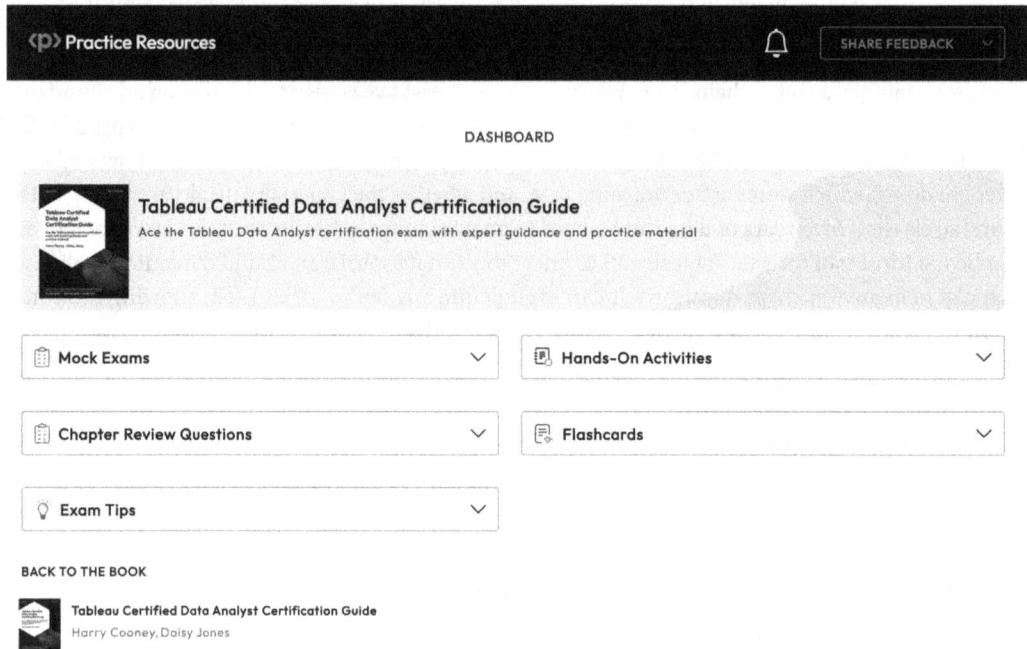

Figure 1.1: Dashboard interface of the online practice resources

Here are some tips on how to make the most out of this book so that you can clear your certification and retain your knowledge beyond your exam:

1. Read each section thoroughly.

2. **Make ample notes**: You can use your favorite online note-taking tool or use a physical notebook. The free online resources also give you access to an online version of this book. Click the BACK TO THE BOOK link from the Dashboard to access the book in **Packt Reader**. You can highlight specific sections of the book there.

3. **Chapter Review Questions**: At the end of this chapter, you'll find a link to review questions for this chapter. These are designed to test your knowledge of the chapter. Aim to score at least **75%** before moving on to the next chapter. You'll find detailed instructions on how to make the most of these questions at the end of this chapter in the *Exam Readiness Drill - Chapter Review Questions* section. That way, you're improving your exam-taking skills after each chapter, rather than at the end.

4. **Flashcards**: After you've gone through the book and scored **75%** more in each of the chapter review questions, start reviewing the online flashcards. They will help you memorize key concepts.

5. **Mock Exams**: Solve the mock exams that come with the book till your exam day. If you get some answers wrong, go back to the book and revisit the concepts you're weak in.

6. **Exam Tips**: Review these from time to time to improve your exam readiness even further.

This chapter covers the following main topics:

*   The benefits of cloud computing
*   Cloud deployment models
*   Cloud service models
*   The core concepts of Azure

Connecting to Sources

The first step in virtually any data analytics process is to connect to data. Tableau (and machines in general) cannot natively know where, how, or when to look for data. This **link** must be initially set up by the user.

Tableau offers many options for data connectivity, from simple spreadsheets on a local device to large online repositories accessible anywhere in the world. Naturally, with such varied means of storing data, the process for setting up these connections differs. Nonetheless, the location must always be specified in some way.

Data often contains sensitive information and is relied upon to represent real events. It is, therefore, important to maintain its integrity. Some kind of authentication is often required to prove user identity, especially for data stored on the cloud and intended to be accessed remotely; this most commonly involves entering credentials (such as a username or password).

It is important to note that none of the services in the Tableau Suite (such as Desktop, Prep, or Server) have the tools or permissions to change the underlying source data. Tableau can only edit duplications of the data contained within its own environment.

Organizations with a base level of data maturity, typically medium to large companies, often have established data sources and servers, such as Teradata, Snowflake, or even Tableau Server, which Tableau users are expected to utilize. However, it is certainly possible for data analysts and those in similar roles to find themselves establishing new sources. In any scenario, there are requirements and best practices for data that all users should be aware of.

Data Structure

Tableau is optimized for use with data in a tabular (table-based) structure. This is a structure that you may be familiar with through working with Microsoft Excel or similar software. Vertical columns store values for whatever the column (or field) represents, such as an item description or order date; horizontally, each row of values collectively forms a record (which can be thought of as an observation).

For example, a company may record transactions across its stores in a table such as *Table 1.1* (sampled to the first transaction – 000001 – in store number 677):

Store Code	Transaction ID	Item Code	Quantity	Purchase DateTime
677	000001	0000145-GRY	10	2024-01-03 09:15:32
677	000001	0000096-AAA	5	2024-01-03 09:16:01
677	000001	0000452-BLU	2	2024-01-03 09:16:23

Table 1.1: Example of a tabular structure

As each record represents an item bought, it is apparent that three distinct products were bought as part of this single transaction. Each key data point relating to these records (such as quantity, purchase date, or time) is stored neatly in a distinct column.

Relational databases such as Microsoft SQL Server and many common file types have this tabular structure as a default, or at least a simple alternative structure (such as comma-separated values) that Tableau can quickly convert into a table as it loads the data.

Choosing an Appropriate Data Source Type

Before any user or developer can start building a visualization, an appropriate data source must be defined. Without data, there will be no visuals to be built and no stories to be told. Multiple factors should inform decision-making when it comes to choosing a data source. In summary, these include the following.

Content and Quality

The following points are key for any data preparation for a developer to be able to analyze data properly. It is important for you to familiarize yourselves with these practices as you will be questioned about them in the exam.

Level of detail: Dimensions and Measures

Tableau should be approached not merely as a tool for data visualization. Charts should be created and used thoughtfully as a means of answering business questions. Therefore, data should be selected with a goal in mind: does it contain the **fields (columns)** and **records (rows)** required to answer the questions at hand?

When it comes to fields, data should contain the appropriate dimensions (to divide the view) and measures (for assessable metrics). It is impossible to review the relative performance of each salesperson, for example, without their name or other unique identifiers alongside a **Profit** field. And if those fields are incomplete – lacking all salespeople, or profits for certain months of the year – then accurate conclusions cannot be drawn.

Data Quality

The previous point touched on completeness as an important facet. This can be expanded further: any data source used should have an appropriate level of **completeness**, **accuracy**, and **consistency** for the resulting insights to be valuable. You need to make sure that the data used is complete, all field names are named appropriately, and the spellings are kept consistent. Please see *Chapter 2, Transforming Data*, for further details.

Technical Requirements

This section will look into the technical requirements that need to be considered when building a report. The purpose is to inform you about the different data connections that can be made and their performance.

Performance: Data Size and Structure

Tableau is capable of processing large volumes of data, but performance sits on a curve: the larger the dataset, the more computational power is required to access and process it. There is no fixed rule for when performance will meaningfully decrease, as this depends on a complex combination of factors, including the specification of the machine running the query (one with lots of resources, such as RAM, can handle greater quantities of data). It is fair to say that a data source with dozens of columns will be processed slower than one with a handful of them; similarly, a source with millions or even billions of records will be less performant than one with a few hundred.

There are stricter limitations for data sources hosted on Tableau Server or Tableau Cloud rather than a local machine; for example, joins and relationships cannot be established, only blends. These are covered in more detail in *Chapter 8, Publishing and Managing Content*.

It is worth noting that Tableau generally prefers data that is **long** rather than **wide** in structure: that is, Tableau can handle more records better than it can handle more fields.

Data Format and Compatibility with Tableau

Users should be sure that a connector exists natively for the given data source type. This can be a type of file that exists locally on the computer such as an Excel file.

Users should consider whether data is accessed live or saved as an extract – that is, whether the data is a saved snapshot, such as an extract, or whether it would run on a real-time basis, such as a live data source.

The description and limitations of these connections will be explained further in this chapter.

Microsoft Excel (.xlsx) and Comma-Separated Values (.csv) Types

This section is written to differentiate between Excel and CSV files and discuss how they are connected to Tableau. Because of their simplicity and familiarity with users, these are the most likely files used to connect to Tableau.

Excel is a specialist spreadsheet format from Microsoft that has been popular for data storage for decades. The **comma-separated values (CSV)** file is named as such because values within it are distinguished (delimited) by commas. Each line in the file constitutes a distinct record. The CSV format is generic and not associated with a particular software or service, though files are often opened and used with Microsoft Excel or Google Sheets.

The **plain-text** format of CSVs is simple to create and straightforward for programs to read, making it easier to move data between systems or locations without complicated parsing steps. CSV files appear no different from spreadsheets when opened in Excel or Tableau – almost all software designed for use with data automatically displays the mass of comma-delimited text in a tabular structure of columns and rows.

As both file types are commonplace, there is a low barrier to entry – the formats are familiar to a wide range of computer users. However, as with most files, these types offer a **snapshot** of data in time; they are not automatically updated in the way that connections to live source systems are.

Note that CSV files typically cast values as plain text, even when all values are in a specific format – fields that are fully numerical, for example, may still be returned as strings. When first opened in Tableau Desktop, fields often need converting to appropriate data types such as **String**, **Integer**, or **Float**, so that there will be no errors when building calculations. Excel files have a hard limit of approximately one million records, determined by the maximum memory available in the Microsoft Excel software itself; note that this also applies to imported CSVs.

Relational Databases

While using Excel and CSV is a simple and easy way to connect to data in Tableau, these files can easily be changed by human error and are not dynamic. Most organizations have outgrown using Excel and CSV for the following reasons:

- They require data that can be found quickly when needed and is trusted to be reliable and accurate

- The solutions need to be able to comfortably handle the natural growth of data and the number of people wanting to access and manipulate it

- The files can often be duplicated and shared freely, risking unwarranted access

- Alternative solutions offer greater opportunities to connect from other locations, rather than a single local machine

Relational databases are often a reliable means of achieving these benefits. They are data storage systems that organize information in the familiar tabular structure, with rows and columns; when **databases** are discussed in a Tableau-specific context, users are usually referring to relational databases. Databases are often hosted on a server, which provides the resources required to run and manage the database; servers can often host multiple databases simultaneously, each with a distinct function.

Tables inside these data repositories are usually set up by developers to capture conceptually distinct types of information. For instance, a marketing center may have a **Telephone Enquiries** table with each record representing an outgoing call (with columns such as start time, duration, and operator), but store customer-level information (such as phone numbers, addresses, and last-contact dates) in a separate table called **Clients**.

Common elements allow tables to be related to each other for analytical purposes. This is usually done through **keys**. Primary keys are either a single field or multiple fields in combination that can be used to identify distinct records. To do this effectively, values in the primary key column(s) must be unique for each row, and primary key columns must be fully populated – that is, all records must have a value (with no missing values, known as `null` values). Tables typically have just one primary key. Primary keys are useful for identifying duplicate values, which reduce the reliability of the data and result in issues such as double counting.

Foreign keys are columns in a table that refer to the primary key in another table. They are used to link tables on a common identifier. To continue the preceding example, the **Clients** table might have a **Client ID** column as the primary key, which also appears as a foreign key in the **Telephone Enquiries** table. Analysts can match the numbers between tables and identify which client was called in each instance. For example, they could identify which clients have had the greatest volume of successful calls and are therefore worth investing in. This process maintains the original values in a single location – the confidential **Clients** table – to make the data easier to govern.

Relational databases need to be communicated with for records to be accessed, updated, added, or deleted. This is achieved using a programming language called **Structured Query Language (SQL)**. SQL is discussed further later, in the *Custom SQL Query* section.

Relational databases are popular as they often enforce rules to maintain data consistency and accuracy; for example, rules may be built to only allow values with a certain range when adding new records. In the **Clients** table, a **Telephone Number** field may require a 10-digit format with a country code prefix for a new record to be accepted in the table.

Popular relational database management systems include PostgreSQL, MySQL, and Oracle.

Initial Data Connections

There are multiple ways to connect to data when Tableau Desktop is first opened. The initial screen appears as shown in *Figure 1.2*:

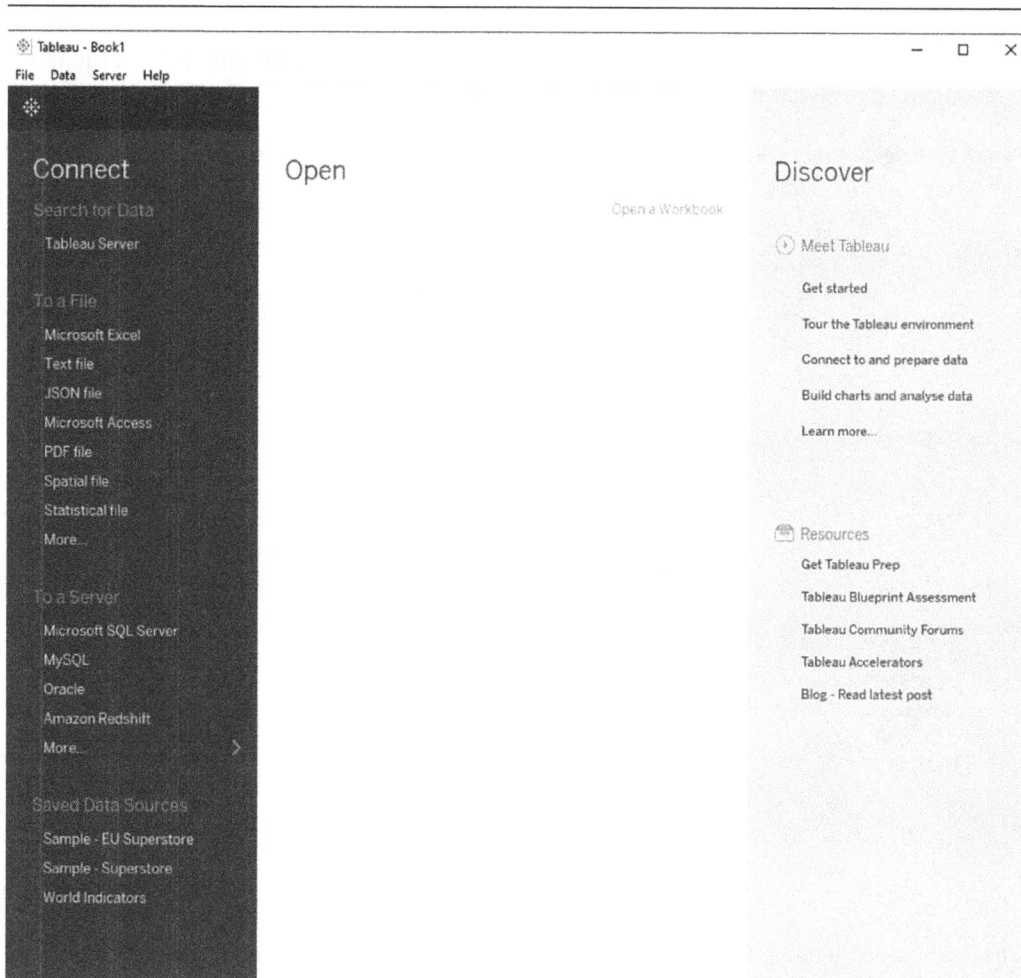

Figure 1.2: Tableau Desktop start page

Users can connect to data primarily through the blue **Connect** data pane on the left-hand side of the window.

Some popular connection types are listed here for quick access:

- **Tableau Server** (including Tableau Cloud) will be described briefly soon or in more detail in *Chapter 8, Publishing and Managing Content*.

- **Files** on the user's local machine, or a cloud location, mapped and accessible by the current machine.

- A **server** (containing databases) whose available options were discussed later.

- The **Data** tab at the top of the window is largely useful when data has already been connected to. At this early stage, the only available option is to allow the pasting of data from the clipboard.

When pasting data onto Tableau from the clipboard, the values are spun into a table with an arbitrary name, as shown in *Figure 1.3*:

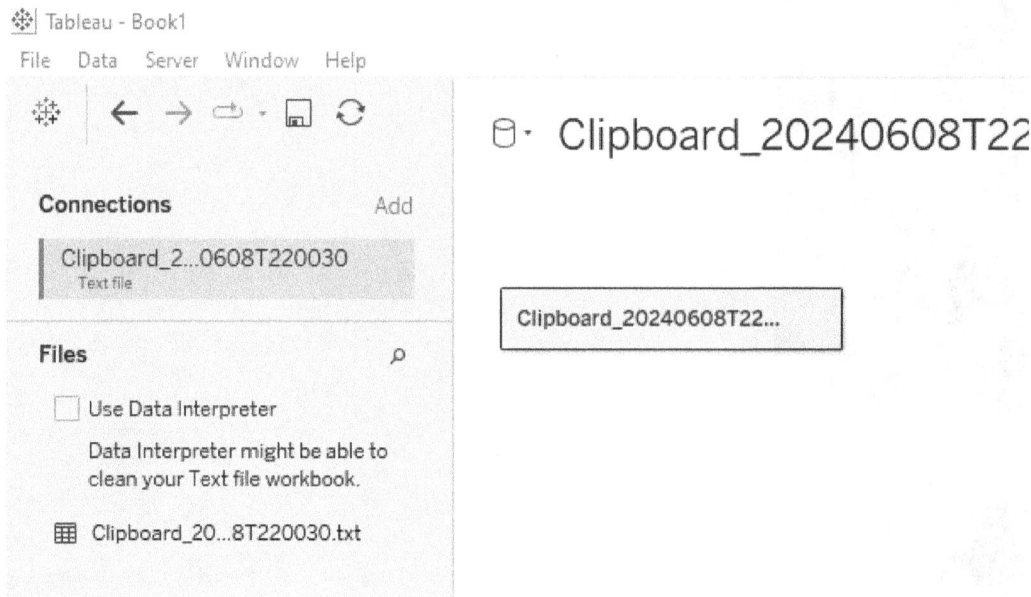

Figure 1.3: Data connector pane

The data is saved automatically by Tableau as a temporary file of the plain-text type in a predetermined location:

Figure 1.4: File directory

Though this option can be useful in certain cases – such as for mocking up quick examples in a teaching environment or troubleshooting data issues – it is not recommended for serious data work. Firstly, it relies on the user to copy data accurately. The temporary file location is also not intuitive to access for maintenance purposes and may be cleared depending on user settings and behavior.

The window that appears when a data source is first connected to is called the **Data Source** page. It is accessible at any time from the bottom of the Tableau Desktop window, should a data source need checking, adding to, or otherwise modifying while visualizations are being developed.

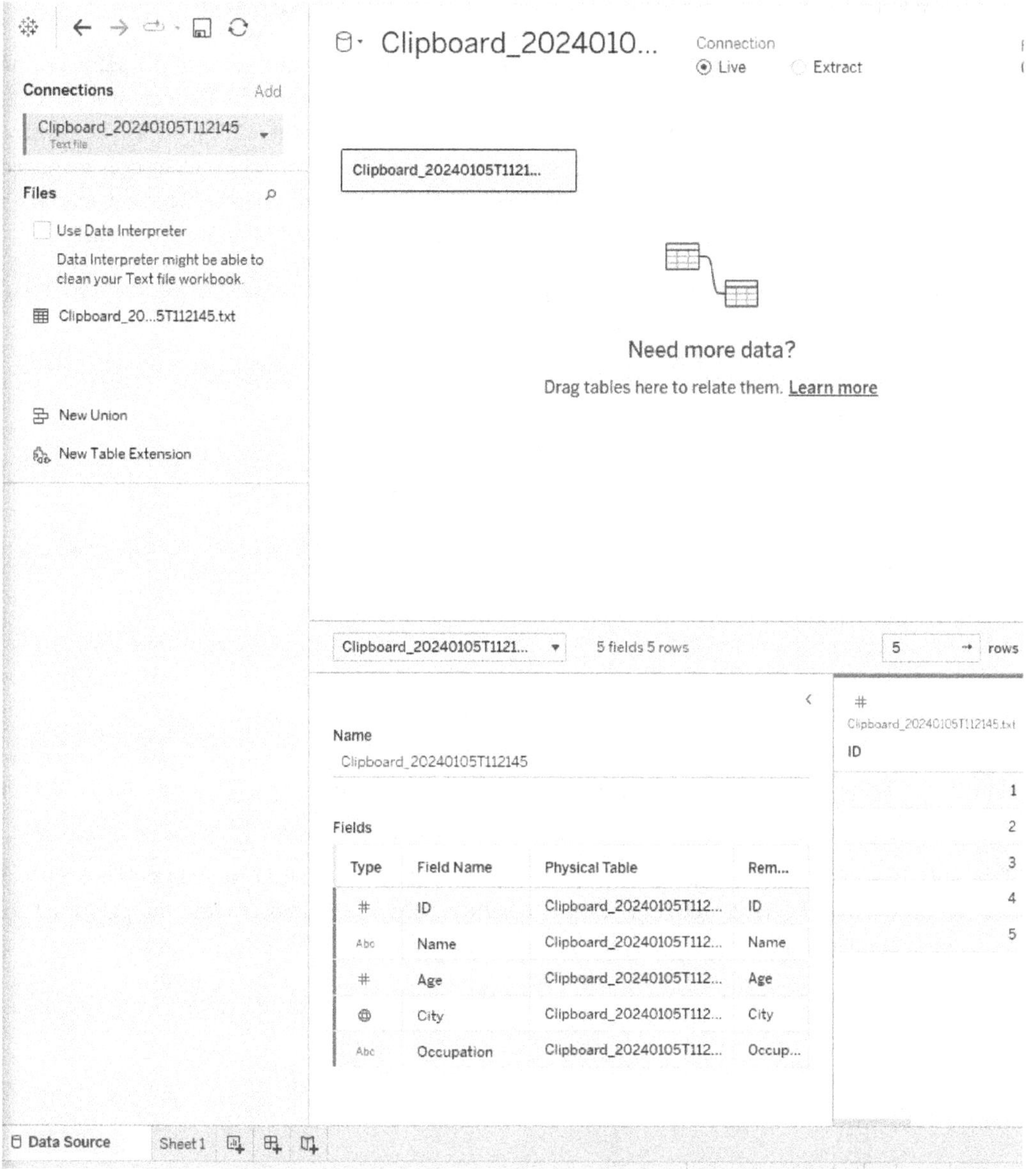

Figure 1.5: The Data Source tab

If multiple connections are present in a workbook, the primary one used in the sheet being navigated away from will appear in this window.

Connecting to Tableau Server or Tableau Cloud

Both Tableau Server and Tableau Cloud can be thought of as different flavors of the same online service: they enable organizations to securely host, organize, share, and edit Tableau-specific data sources and visualizations, usually through a web browser. Tableau Server and Cloud will be covered in more detail in *Chapter 8, Publishing and Managing Content*. In short, Tableau Server is managed independently by an organization with the skills and resources to do so, whereas Tableau Cloud is an out-of-the-box solution hosted and maintained by Tableau, and is therefore more accessible.

Both services can be accessed on Tableau Desktop through the **Tableau Server** button at the top of the **Data** pane, or through the **Server** tab at the top of the window.

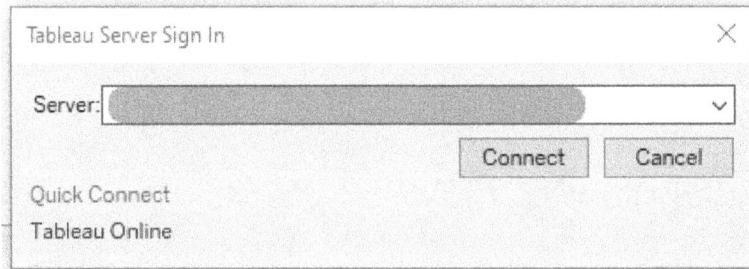

Figure 1.6: The Tableau Server Sign In window

Note that if **Tableau Server** is selected, you will be prompted to enter the URL for a Tableau Server or Cloud instance.

If the URL is recognized, the following window will prompt you to authenticate. Tableau Desktop will open a new tab in the default web browser with a login screen. If a valid username and password are provided, a connection to Tableau Server will be established.

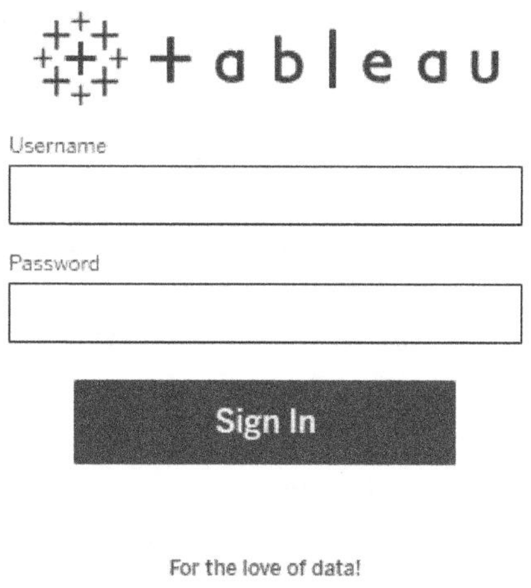

Figure 1.7: Tableau Server login window

When connecting to Tableau Server, the user will be prompted to choose a site to connect to. Sites are described fully in *Chapter 8*, *Publishing and Managing Content*, but in brief, they can be thought of as a folder; they are the highest level and the strictest way to separate and organize content, as sites cannot communicate with each other.

Figure 1.8: Tableau Server site selection

Note that this screen does not appear for Tableau Cloud instances, as Tableau Cloud only permits a single site on which all content must be stored. An entirely new instance of Tableau Cloud must be spun up if content requires strict separation.

Connecting to a File

Connecting to a file will require a file location on a local machine or a network that the machine can access, such as SharePoint or GDrive. These kinds of files have different ways in which they are brought into Tableau. This section will go into detail on how to connect to each file type.

CSV and Excel Files

When Tableau Desktop is open and is asking to connect to data, it is likely that the developer will choose to use an Excel or CSV file. Note that `.csv` is a text file, so it is accessible under the **Text File** button under **Connect > To A File**.

In this selection, the **Data** pane will be populated with tables that exist in the files. From this point, the developer will need to identify the main table and then bring it into the main view. Then, the developer can establish the relationships of other tables and create their data model for analysis.

Tableau is intelligent enough to be able to automatically identify the data type of the fields for Excel files, but it is likely that for CSV, the developer will need to assign the data types for the fields.

.hyper or .tde Files

One type of file to connect to is called a `.hyper` file, the successor to the `.tde` file that was deprecated in 2023. This kind of file is a Tableau extract that is a local copy of the data that can be shared and has an improved performance.

This extract is very simple to connect to. Select the **More…**option in the **To A File** option and locate the extract in the file directory.

These two files are important as they are considered the main type of files to create extracts. This is likely to come up in the exam.

JSON

A JSON file is a formatted text file that stores data in pairs or arrays. To use this kind of file requires the user to select a schema level, in which Tableau then flattens this information for the schema.

The following is an example of connecting to a JSON file and the window that opens that allows the user to select the schema.

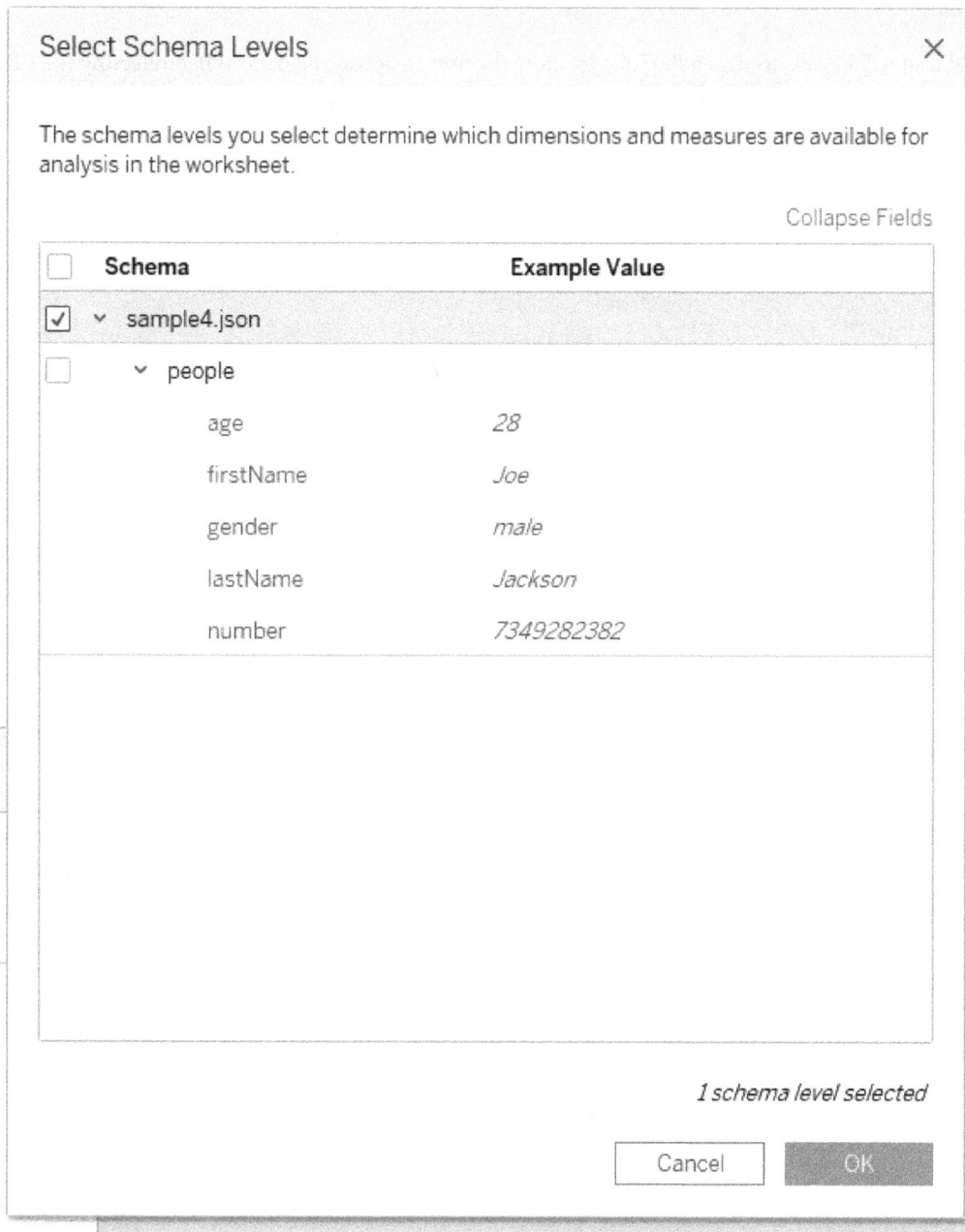

Schema	Example Value
☑ ⌄ sample4.json	
☐ ⌄ people	
age	28
firstName	Joe
gender	male
lastName	Jackson
number	7349282382

Select Schema Levels   ✕

The schema levels you select determine which dimensions and measures are available for analysis in the worksheet.

Collapse Fields

1 schema level selected

Cancel   OK

Figure 1.9: Example of a JSON file connection

Once the schema has been selected, the data will be opened for preview and then the user can prepare the data for analysis.

PDF

Tableau will be able to scan a PDF file to allow the user to access the data. When selecting the PDF, the user will be prompted to select what areas Tableau can scan through.

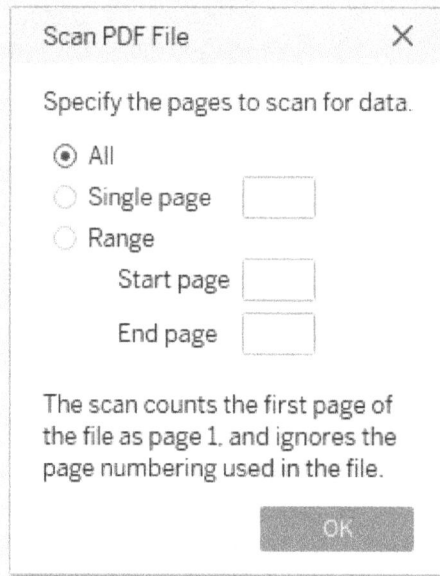

Figure 1.10: Scan PDF File selector

Depending on how the PDF is set up, a user can simply scan the whole file, or find the specific page on which the table is located.

Once the file has been established, the **data connection** pane will be populated with tables that exist in the sheets. The user can then union the tables together to get a complete set.

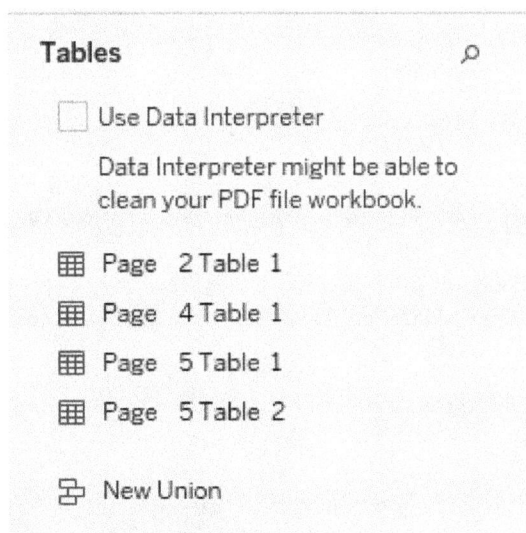

Figure 1.11: Data connection pane

Note that there is a chance the data will not be clean and may need some preparation, which can be done in Tableau Prep.

Spatial Files

Spatial files can be used to create maps or plot out geographical areas such as the counties in the United Kingdom, but there are several considerations that need to be taken into account before connecting to one. In the local file directory where the spatial file is saved, it is important to note that each type of spatial file will require supporting files for it to work in Tableau.

For KML, GeoJSON, and TopoJSON files, there only needs to be one file. For the other types, Esri, Esri geodatabases, and MapInfo tables, the folder should include other files to allow the tables to work.

In the **London Datastore**, there are spatial files of London boroughs that can be downloaded for use.

Once the file has been located and established, there will be a dataset that includes a **Geometry** column.

Figure 1.12: Example of a spatial file and a Geometry column

Note that the fields use **Polygon**, which is what created the shape of the border for each London borough. When creating a new sheet, double-clicking or dragging the **Geometry** field will create a singular shape of the map.

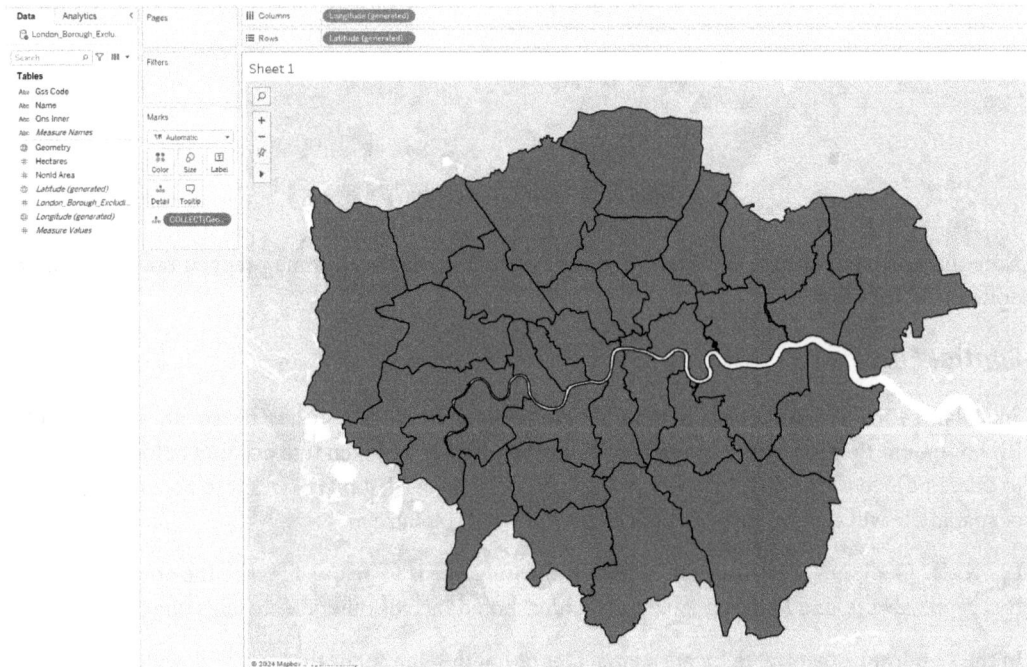

Figure 1.13: Tableau Geometry field showing a map of London boroughs

Notice that the whole map has been highlighted as one. To be able to get each separate borough, the developer will need to drag in the detail that will pinpoint which space belongs to the borough.

Statistical Files

For the use of statistical files, Tableau can connect to **Statistical Analysis System (SAS)**, **Statistical Package for the Social Sciences (SPSS)**, and R data files. These kinds of files are straightforward to connect to. There are only a few items to note:

- The statistical file does not accept values as label headers
- These connections only allow for one table per file
- If more than one table is present in an R file, then Tableau will take the first one

Now that the types of files have been described, take some time to find examples of these files and connect to them. Explore how the different types connect and what is required to do so. There will be a Tableau data extract and an Excel file available in the **Datasource** folder located in the Tableau repository folder. The other files can be created or found online.

Connecting to Databases and Cloud Services

So far, this chapter has covered files and relational databases that Tableau can connect to. However, large corporations with complex and abundant amounts of data are likely to use databases and cloud services that hold their data. This section will go through the different types of databases and cloud connectors and discuss how to connect to the data.

Default Server Connectors

Tableau has a vast number of default servers readily available for use. Go to the **Connect** pane and, under where it says **To a Server**, hover over **More…** to see all the types of server connections that can be made.

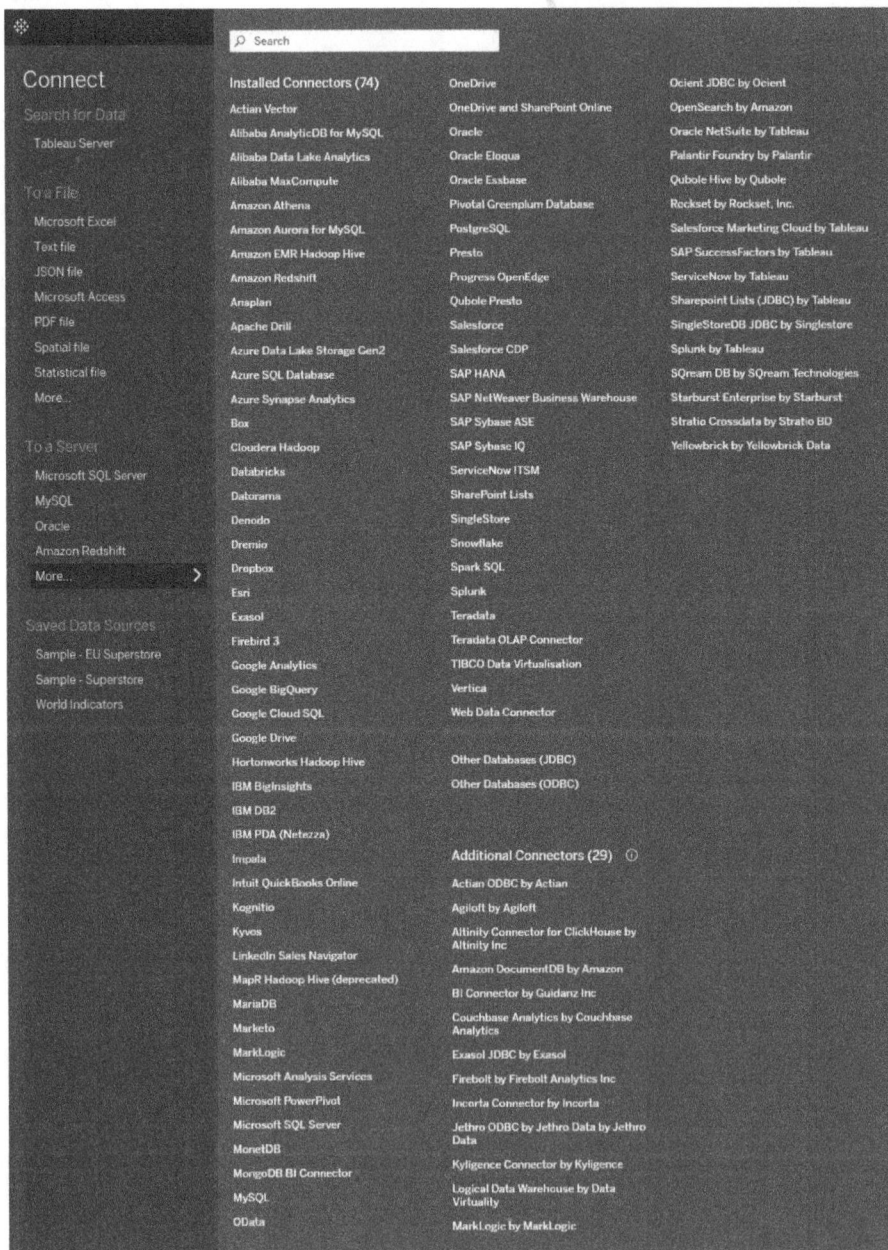

Figure 1.14: Tableau Server connection options

Servers is a broad term, but in the context of Tableau, largely refers to the host of relational databases (or other technically different but functionally similar services, such as Microsoft OneDrive, which allows business users to share and collaborate on the web).

As visible in *Figure 1.14*, the list of connectors is comprehensive, even for Tableau version 2022.3. You are not expected to memorize the full list of connectors, and Tableau is continually adding new connectors to this list. However, it is useful to be aware of their general types.

Relational databases (introduced earlier in this chapter) are communicated with and queried using SQL, which is compiled by Tableau behind the scenes. These queries are written to be highly efficient, so queries can return results as quickly as possible.

When a server is selected, the resulting window will ask the user for connection and authentication details. (Though the exam is unlikely to pose questions about this window specifically, it is important to understand what is required to establish a connection that works reliably.)

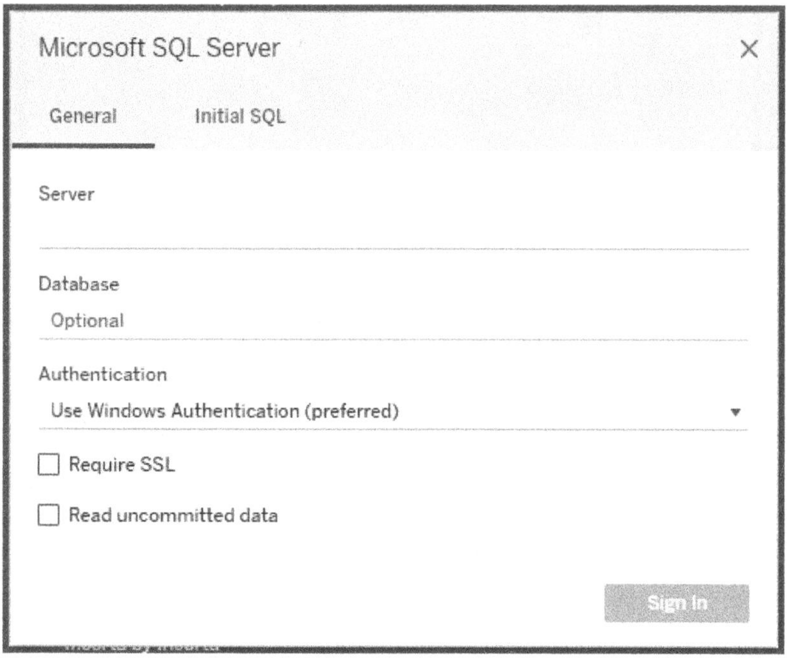

Figure 1.15: Credentials section to SQL Server

Note that for certain services, such as Microsoft OneDrive and SharePoint Online, **single sign-on** (**SSO**) may be the authentication type configured by the network administrator. This will be someone who has access to permissions for others in the organization.

When this is the case, users are usually directed to their default browser to sign in with their organizational account.

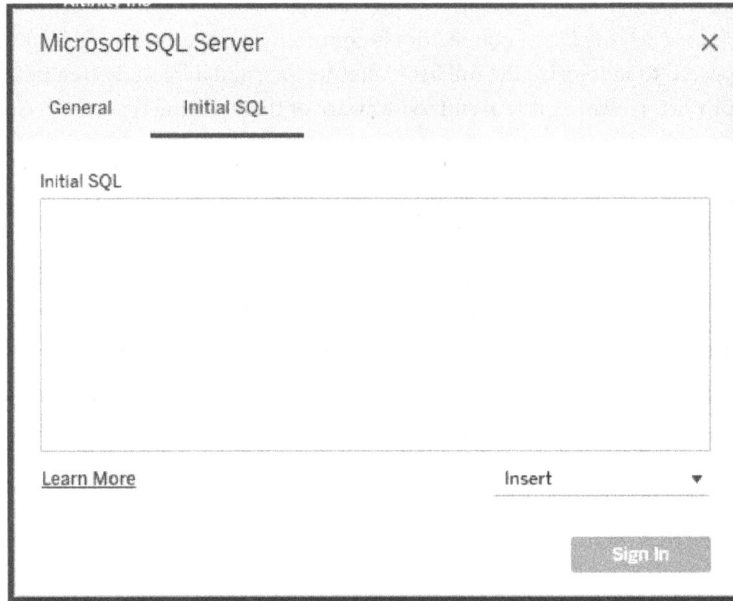

Figure 1.16: SQL query builder

Initial SQL is an advanced option in Tableau Desktop that developers can use if they wish to recall a specific subset of data (for example, population data for a particular region only). It runs when the workbook is first opened and generates a cache (or temporary table) for the duration of the session. Any changes to the data will not be reflected until the workbook is reopened.

Initial SQL can result in noticeable performance improvements when used alongside custom SQL. See the *Custom SQL Connections* section later in the chapter.

Once connected, the database tables available to the user will appear on the left-hand side, in the **Data** pane. At this stage, a connection has been established, but Tableau cannot predict which data the user is interested in. The table(s) of interest must be dragged onto the relationship canvas, as shown here:

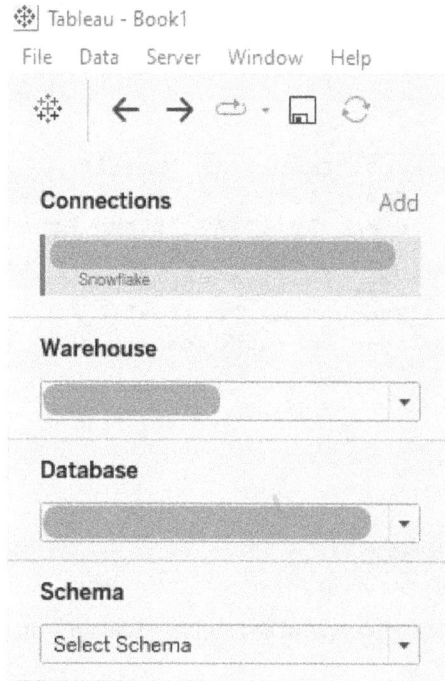

Figure 1.17: Connection pane and tables to be connected

To preview the data for the given table in the **Results** pane, select **Update Now**:

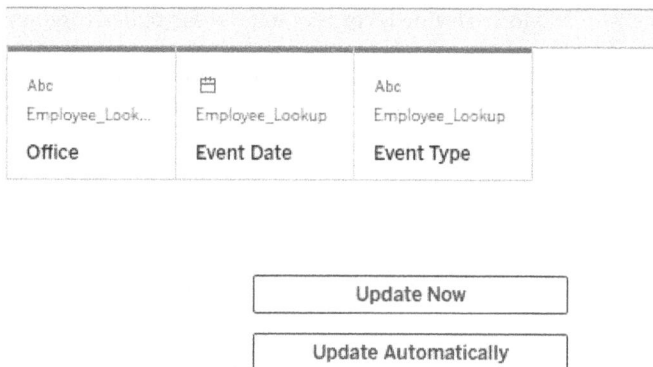

Figure 1.18: The Update Now button to preview the data

Custom SQL Queries

A more technically advanced but powerful option for connecting to database tables is **Custom SQL**. This allows a developer to specify the exact fields and even filter the data from the source to create an efficient dataset to work with in Tableau. The following section will look into SQL queries and how to connect to them in Tableau.

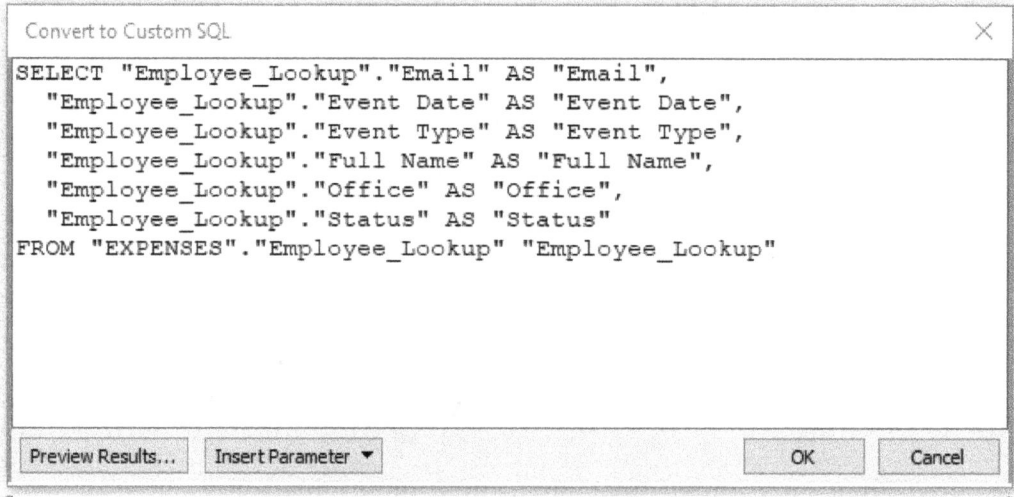

Figure 1.19: Example of an Employee Lookup Custom SQL

Introduction to SQL and Custom Queries in Tableau

SQL (pronounced **sequel**) is a programming language commonly used to communicate with relational databases. It can be used for a variety of tasks, such as inserting, updating, or deleting records. However, in Tableau, it is largely limited to retrieving records of interest for analysis and reporting purposes.

Though some SQL queries may be long and complex, the keywords used by SQL use natural language, and clauses (pieces of logic) are often structured in a logical manner. Take the following query, which is pulling through the latest 1,000 records from a PostgreSQL database table:

```
SELECT
*                                              ←  Select
all records ('*' is a wildcard character)
FROM "Database"."Schema"."QA_CHECKS"    ←  From the database, schema,
table specified
ORDER BY LAST_MODIFIED DESC                ←  Sort records descending
by date field given
LIMIT
1000;                                          ←  Return
the first 1,000 records only
```

In the preceding example, PostgreSQL is specified as a specific database management system. This is because SQL is a generic term; there are multiple different **flavors** of the language, the same way human languages often have different dialects depending on which region the speaker is from. Each system has its own grammatical structure (known as **syntax**), which needs to be exact to run correctly. For example, another system may use single quotes (') rather than double quotes (") to identify the **Database**, **schema**, and **table** objects.

Custom SQL is useful in the following cases:

- The user is only interested in specific subsets of data that cannot be achieved using the interface (no-code) tools.

- Custom SQL can be helpful if the user wishes to select records or combine data from different tables, using more complex logic that the interface does not permit.

- The dataset is extremely large either in width (number of columns) or length (number of records).

- It is best practice to only bring strictly necessary data into Tableau to avoid wasting valuable computing resources (and therefore time). However, Custom SQL can be slow (as we will discuss soon). Therefore, it can be a case of trial and error to discover whether custom SQL can improve performance. Where possible, simpler joins and unions are best set up using the user interface – see *Chapter 2, Transforming Data*, for further details.

- The analysis is one-off or exploratory – we will discuss performance concerns shortly.

- The database itself does not permit certain functionalities using the user interface.

- There are times when certain operations, such unions or pivots, are not supported by the physical layer.

A major downside to using custom SQL is that it is slow. Queries are almost always slower to run than those automatically generated by Tableau with basic connections that do not use custom SQL. This is aggravated when many complex clauses are used or when you attempt to select large amounts of data – for example, with the SELECT * (select all records) statement. If a custom SQL query is proving valuable, it is almost always worth formalizing as a database table or view that can be connected to directly by the user, especially if Tableau workbooks are intended to be reliable, performant, and for mass consumption (production workbooks).

Custom SQL queries are powerful, but this can cause errors if the data has not been prepped properly for analysis. Custom SQL queries should only be used when the following apply:

- The user has a clear idea of the data required to write efficient queries that answer the business questions at hand

- The user understands the structure of the database and is comfortable writing SQL queries, to avoid pulling incorrect, excess, or duplicate records

Live versus Extract Connections

Almost all connections can be configured to be either live or extract. These types of connections can determine performance and how the data is accessed, so making sure to choose the appropriate type is important for any data analyst. This section will look into the two types and their benefits and limitations.

Live Connections

Live data connections are the simplest to conceptualize. Here, the source data is pinged directly with every query, producing an almost real-time accuracy of data. Little needs to be done to maintain live data connections by their direct nature as they should be self-sufficient. There is no need to set up a schedule for refresh.

Note that a live connection does technically use caching technology. **Caching** is the temporary gathering and storing of information in Tableau for use in future queries.

This means that the fetching of data from a database only triggers if the charts are interacted with in some way; a visualization left open for a long time may need refreshing or clicking on to show the most up-to-date information. Minute delays are also inevitable as the machine sends and receives queries from the external source; it takes time, though often only milliseconds, to complete this process.

Pros of Live Data

Take a look at some advantages of live data connections:

- Updates are virtually in real time. It can be reassuring for stakeholders to know that the data being visualized is up to date whenever it is accessed, and live connections can be more convenient to use if an organization does not have access to refresh schedules.
- The data is updated automatically without the user needing to access the source.

Cons of Live Data

It is less performant. This kind of connection can take longer to retrieve data as it would be expected to run every time there is an action performed on the data. This can be detrimental for large companies with huge quantities of data, especially if there are complex joins and unions involved.

Now that live connections have been described, it is time to look into data extracts and how they perform as well as their pros and cons.

Data Extracts

An extract is a duplicate of the source data in a format optimized for Tableau use. The data can be limited to a smaller set, or represent the source data fully and be considered a kind of **snapshot** in time.

For some of the best practices, it is useful to filter the data to only the data you need, such as a **Datasource filter**. This creates a smaller sample for the data that will improve efficiency and eliminate unnecessary queries or the querying of irrelevant data.

For example, the user could be asked to generate visualizations to analyze trends in transactional data from a particular store. If the visualization in development is purely concerned with this store independently, and not looking to compare it to others, taking an extract will allow the conditional filtering in the **Store** field at the data-source level. Now, whenever the visualization queries the data – when initially loading charts, for example – queries will not be wasted on miscellaneous stores.

Users of Tableau Desktop can update any extract they own or have permission to edit. This can be done manually or by setting a schedule.

The initial refresh occurs when publishing an extract for the first time (either directly, by publishing the data source itself, or by publishing a workbook with the data source embedded). The user will then be prompted to save the extract in the directory that can be accessed locally. This option appears once the user has selected the **Extract** option in the top-right corner of the **Datasource** page.

Figure 1.20: Selection of the Extract type

Any other manual refreshes will require the original datasource to have changed and then for the developer to refresh the data in the workbook. This can be done by selecting **Data** > **Refresh Data Source** or by clicking on the circular **Refresh** button.

Figure 1.21: Refresh symbol

There are two types of refresh that can be set when using a data extract:

- **Full**: The entire dataset is recalled and the extract is completely regenerated – no records from the previous extract are retained.

- **Incremental**: New records only are returned and appended (added to) the existing dataset. The user can specify the field used to identify new records, but note that this must be of the numeric or date type; Tableau will look for records that are greater than the maximum value in that given field.

These refreshes can be set in the configuration window for **Extract Data**. This box opens when the user selects **Edit** next to the radio options in the top-right corner of the **Datasource** page.

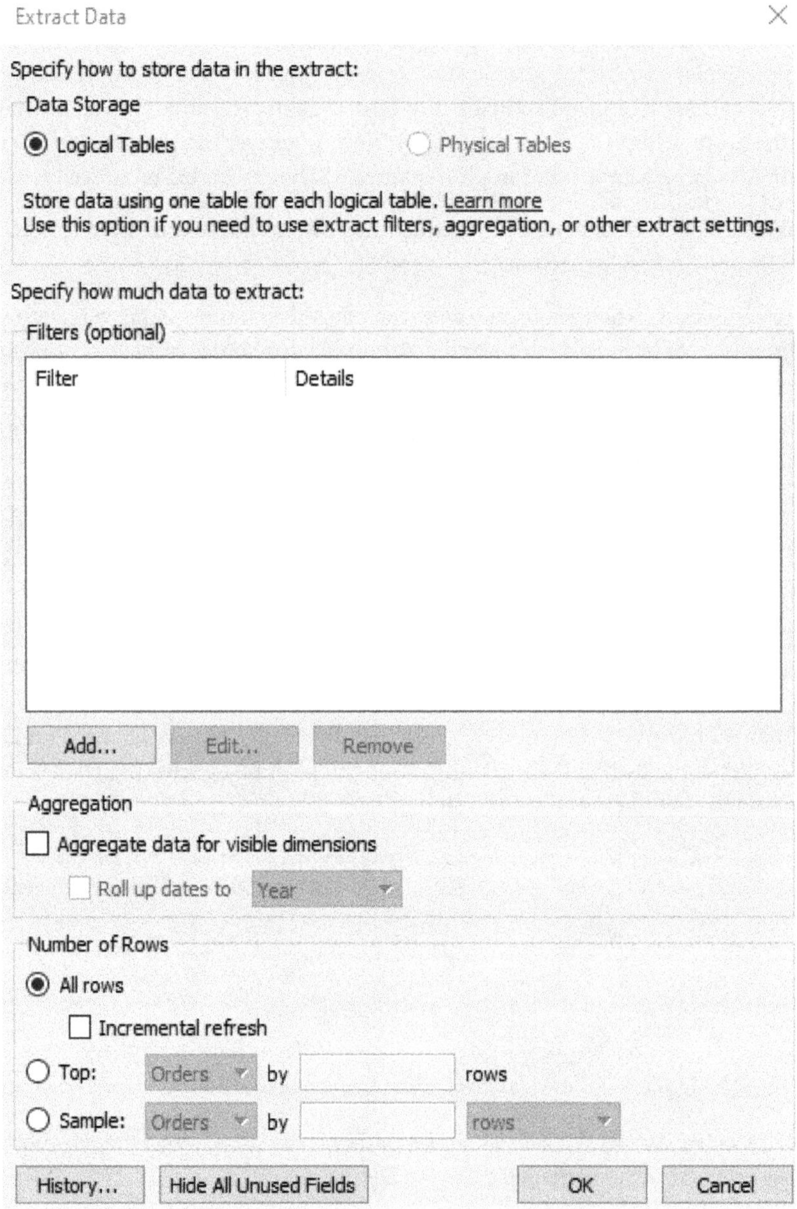

Figure 1.22: Window to add extract filters or incremental refresh

The manual refreshing of data in these ways is possible even when the data is running on a schedule. As mentioned previously, data is often analyzed for historical periods and rarely needs to be live to each second. Generally, there is an **Extract Refresh** schedule suitable for most needs: data can currently be set to refresh each hour, day, week, or month.

As refresh schedules can only be set through Tableau Server or Cloud, data sources must be hosted there to be eligible, either as standalone data sources or contained within specific workbooks.

The scheduling options available to Tableau users are set by Tableau Server administrators, a senior role intended for the management of the server, and may vary from instance to instance. As these processes are more strictly in the realm of Tableau Server/Cloud, they are explored further in *Chapter 8, Publishing and Managing Content*.

A full extract refresh reflects the source data most accurately at the time of running. However, it can be slow to complete. It can also use a lot of computational resources in doing so. The impact of these depends on multiple factors. For example, large sets of data, with wide (large number of columns) and long (number of rows) tables, can take longer for the data to fetch. It could also depend on whether the database is optimally structured for Tableau queries; the details of this are out of this exam's scope.

An advantage of incremental refresh is that the field used to identify unique records can be specified by the user. However, the logic for identifying new records is limited to a **single field** only. Tableau cannot consider the values of multiple fields in combination. It is also restricted to a field of either the numeric, date, or datetime type or a simple greater-than logic.

Tableau does not permit custom logic for returning records; it automatically returns records with values greater than the maximum value in the current extract. For example, if a date field is chosen and the latest date is 2023-12-01, any record found at the source with dates beyond this value (such as 2024-01-01) is appended. Numeric types work the same way. This may be an issue if, for example, a record identifier field is not stored as a numeric field (and therefore cannot be selected), or if new records are not given consecutive numbers.

The filtering mentioned here is just one type of filtering possible in Tableau (at the highest level: limiting the data itself). For further filtering options, see *Chapter 4, Grouping and Filtering*.

Other Factors Affecting Performance

Here are some general factors that may impact performance. These factors are important to keep in mind for any developer so that the performance of any future workbooks that are being built can be improved:

- A developer will need to be aware of the location of the files – whether they are on a local computer or a network location. A network location could slow the connection and also be at risk as many people may have access to the source and could even accidentally delete it.

- If there are multiple tables and complex joins, this could impact the performance of the workbook.

- Tableau works best with tables that are long, which means fewer columns and more rows. If a table is wide, this can impact performance.

Connection Management: the Datasource Pane and Data Tab

The **Datasource** pane is important for a user to be able to create reports on Tableau. This section looks at the **Datasource** pane as well as the tab that will likely appear in the exam. There are several options in these tabs that can change the data connection type and even modify the data coming into the report for analysis, such as adjusting the data type to an appropriate one.

The Datasource Window

This window allows users to view, edit, add, remove, and combine data sources present in a workbook. The window is generally structured as follows:

- Metadata grid
- Results pane

The number of records will be displayed along the top of the **Results** pane. If data contains fewer than 1,000 records, all records will be displayed; otherwise, it is limited to a sample of 1,000 rows.

Some key features are worth identifying in this window, as listed after the figure:

Figure 1.23: Tableau Data Source page

Refresh button: ⟳

Add connection: Add

Relate/join tables: Drag tables here to relate them. L

Union tables: New Union

Extend tables: New Table Extension

Connection type: Connection ⦿ Live   Extract

Data source filters: Filters 0 | Add

Results pane settings cog: ⚙

The Data Tab

When at least one data connection is set within a workbook, the **Data** tab at the top of the screen contains multiple options for managing sources. (This tab is always accessible, regardless of which tab or sheet in the workbook the user is currently.)

Options in this tab are as follows:

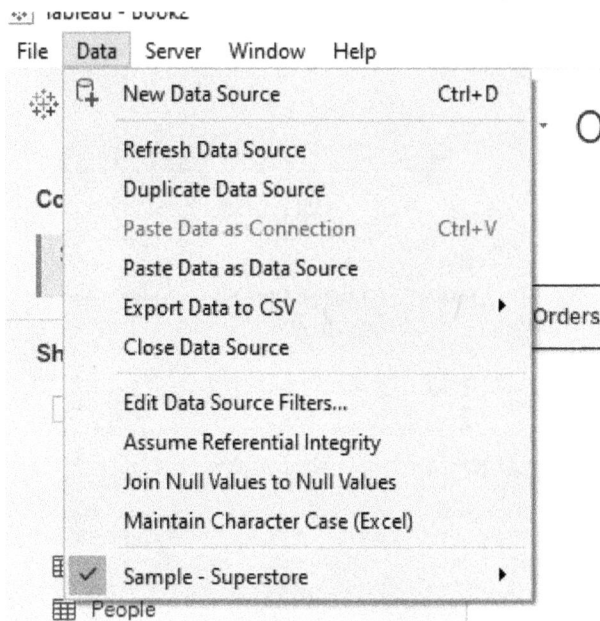

Figure 1.24: Data tab options

Take some time to go over these options and explore what happens. It is good to familiarize yourself with what these options mean and how they work in a workbook setting.

Making New Connections in Existing Workbooks

Various data sources can be combined as required in a single Tableau workbook to suit specific analytic requirements. They can be used separately or combined together through methods such as joins, unions, relationships, and blends.

Joins use a key identifier or linking field to join two tables together, while unions stack data tables on top of each other.

Relationships are a new model and a flexible way to connect two or more tables together. Blends allow the user to establish a connection between a detailed table and an aggregated table.

These connections will be explained later in this book.

Replacing Data Sources

There might come an occasion when the original data source is outdated or has been improved and in a new location. If a specific visualization has been created with a data source, a user can easily replace the data source with a new one by adding the new data source.

Once this has been done, right-click on the old data source and select **Replace Data Source**.

There will be an option to select the current data and then the replacement data. Once this is done, the data should automatically update with the new data source. If the data is kept the same as before, maybe a new column is included, then the replacement should go smoothly and all charts will be able to take on the new fields.

However, it is important to note that if the field has changed or a specific field that was used is no longer in the data, this can potentially break the workbook.

Notes, Caveats, and Unsupported Data Sources

Uncommon connectors may require a few additional steps to establish a connection.

In some cases, a specific driver needs downloading that does not come pre-packaged with Tableau Desktop; for example, for a connection to **SAP HANA** (a database), a user will need to have a driver installed, as shown here:

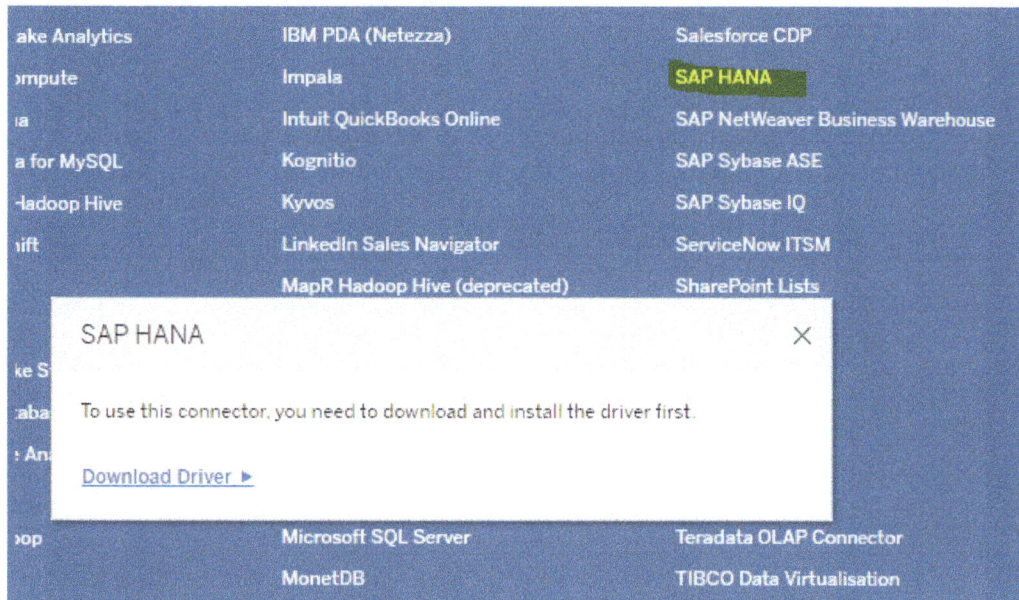

Figure 1.25: SAP HANA connection error

Drivers convert information between each end of the connection so applications can speak to each other; they can be thought of as translators at each end of the connection.

For caveats on multidimensional sources such as cubes, please see the previous section/s regarding database connections.

Web-Based Data Access: Drivers and APIs

Servers such as the previous ones are made accessible through drivers; as discussed previously, some of these come pre-packaged with the Tableau software when it is downloaded, and others require downloading by the user. Tableau can also connect directly to data on the internet that is accessible by **Application Programming Interfaces** (**APIs**) rather than ODBC or JDBC drivers, where a user can access a database system using SQL; this is done using Web Data Connectors. The details of Web Data Connectors are not required for the exam.

Multidimensional Systems

There are also **multidimensional systems** that are structured and queried with different technologies to relational databases. The technical details of these data source types are not expected for the exam, but it is useful to have an awareness of them for comparison against relational sources.

Data in multidimensional systems is pre-aggregated with particular business questions in mind. For example, the developer of an **OLAP cube** (a common type of multidimensional system) may create a summary view of profits at a country level, rather than having a user query billions of records at the transaction level to calculate this higher-level value outside of the source. As the user can query these aggregations directly with algorithms optimized to do so, multidimensional databases are much quicker to query. However, they are far less flexible to use than relational sources in Tableau: useful features such as data source filters, level of detail calculations and forecasting are not available when using cube sources.

Only the following sources are supported in Tableau at the time of writing:

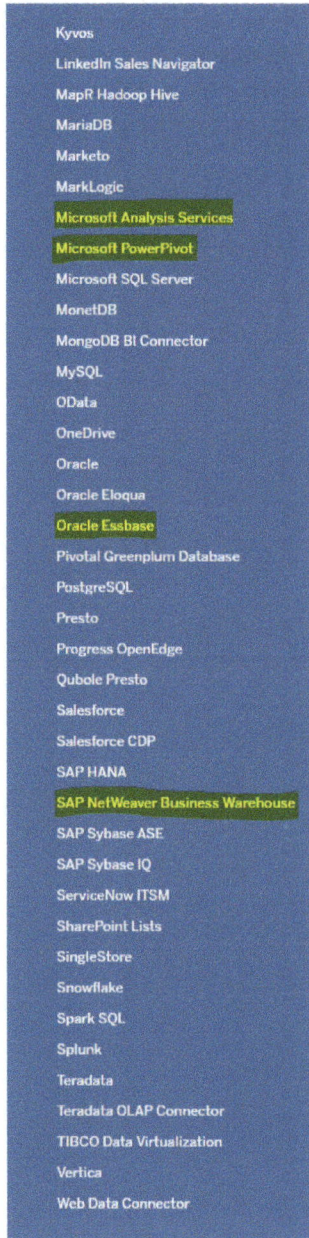

Figure 1.26: Supported sources available in Tableau

Summary

This chapter went over the different types of data connections and how a user can connect to databases to get their data. After this chapter, you should have confidence in connecting to data and should be able to move on to the next chapter to learn how to transform data for analysis. It is recommended that you take some time to practice connecting to the different data types and experiencing the connections.

Exam Readiness Drill – Chapter Review Questions

Apart from a solid understanding of key concepts, being able to think quickly under time pressure is a skill that will help you ace your certification exam. That is why working on these skills early on in your learning journey is key.

Chapter review questions are designed to improve your test-taking skills progressively with each chapter you learn and review your understanding of key concepts in the chapter at the same time. You'll find these at the end of each chapter.

> **How to Access these Resources**
>
> To learn how to access these resources, head over to the chapter titled *Chapter 9, Accessing the Online Practice Resources*.

To open the Chapter Review Questions for this chapter, perform the following steps:

1. Click the link – `https://packt.link/TDA_CH01`.

    Alternatively, you can scan the following **QR code** (*Figure 1.27*):

Figure 1.27: QR code that opens Chapter Review Questions for logged-in users

2.   Once you log in, you'll see a page similar to the one shown in *Figure 1.28*:

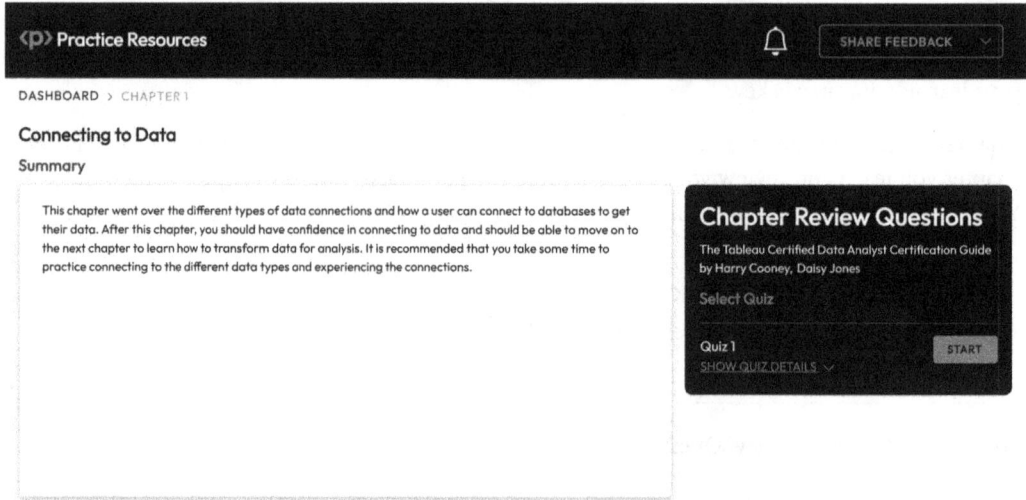

Figure 1.28: Chapter Review Questions for Chapter 1

3.   Once ready, start the following practice drills, re-attempting the quiz multiple times.

Exam Readiness Drill

For the first three attempts, don't worry about the time limit.

ATTEMPT 1

The first time, aim for at least **40%**. Look at the answers you got wrong and read the relevant sections in the chapter again to fix your learning gaps.

ATTEMPT 2

The second time, aim for at least **60%**. Look at the answers you got wrong and read the relevant sections in the chapter again to fix any remaining learning gaps.

ATTEMPT 3

The third time, aim for at least **75%**. Once you score 75% or more, you start working on your timing.

Tip

You may take more than **three** attempts to reach 75%. That's okay. Just review the relevant sections in the chapter till you get there.

Working On Timing

Target: Your aim is to keep the score the same while trying to answer these questions as quickly as possible. Here's an example of how your next attempts should look like:

Attempt	Score	Time Taken
Attempt 5	77%	21 mins 30 seconds
Attempt 6	78%	18 mins 34 seconds
Attempt 7	76%	14 mins 44 seconds

Table 1.2: Sample timing practice drills on the online platform

---

Note

The time limits shown in the above table are just examples. Set your own time limits with each attempt based on the time limit of the quiz on the website.

With each new attempt, your score should stay above **75%** while your "time taken" to complete should "decrease". Repeat as many attempts as you want till you feel confident dealing with the time pressure.

2

Transforming Data

Introduction

This chapter will start by covering the data types available in Tableau and how they interact. You will look into ways to clean data once it has been brought into both Tableau Desktop and Tableau Prep, followed by the data transformation options available in both. Finally, the methods available to combine data sources in both Tableau Desktop and Tableau Prep will be covered.

When importing data into Tableau for analysis, there are a number of considerations to make first. For example, a company wanting to analyze quarterly sales data in Tableau would first need to consider whether the data is complete and include sales for the whole quarter. The product names in the quarterly sales data may need to be cleaned first and some transformations may be required, such as calculating total sales from the unit price and the quantity sold. To make these decisions, the user needs to understand the different data types in Tableau so that they can be configured correctly, as well as the data cleansing and transformation abilities available in both Tableau Desktop and Prep.

Sometimes, a single data source is not enough for a complete analysis, and multiple sources of data will need to be combined. There are multiple methods to combine data in both Tableau Desktop and Tableau Prep, each with different benefits and drawbacks, depending on the situation.

The following topics will be covered in this chapter:

- The types of data fields in Tableau
- Cleaning data
- Transforming data
- Combining data

Types of Fields in Tableau

This section will walk you through the data types available in Tableau and the ways in which the default behaviors of different field types can be set, as well as their implications.

Data Types

Tableau data types are the methods by which Tableau understands the information stored within a field. Fields can be either the text or string type, numeric, date, date and time, Boolean, or geographic.

Numeric data types can be specified as either whole numbers or decimal numbers. Both have the same symbol in Tableau, and there is no significant difference in functionality. If it is not known whether a numeric field contains details down to the decimal level, it is recommended to set the field to decimal anyway just in case.

By default, Tableau will assign a data type to a field, depending on what is most appropriate. For example, a field with only 1 and 0 values will be assigned as Boolean, whereas a field with values in a format such as 01/01/2024 will be read in as a date field.

A field's data type can be identified in Tableau by the symbol shown in the **Data** pane. Here is a table listing the symbols for each data type.

Tableau data type	Symbol
Text/string	Abc
Numeric	#
Date	🗓
Date and time	🗓🕐
Boolean	T|F
Geographic	⊕

Table 2.1: Tableau data type symbolic icons

Data types can be updated by clicking the icon in the view and then selecting from the popup the desired updated data type.

It is important to get data types correct because they have different default behaviors in the view. For example, date fields have the functionality to drill from a higher-level date, such as a year, down to lower-level dates, such as a month or day, by clicking a plus icon in the view. Similarly, geographic fields can generate longitude and latitude and plot points on a map.

Data Type Conversion

This exercise will help you familiarize yourself with the techniques that allow a user to change a field's data type in Tableau. It will also demonstrate the difference between the date and string data types.

1. On a new sheet, drag the **Order Date** field to the rows shelf. As the **Order Date** field has the date data type, it has been placed on the view aggregated to the year level. This is the default behavior of date fields.

2. Hover over the **Year of Order Date** table header and click the + symbol that shows up. This drills into the quarterly level, another functionality available to date data type fields.

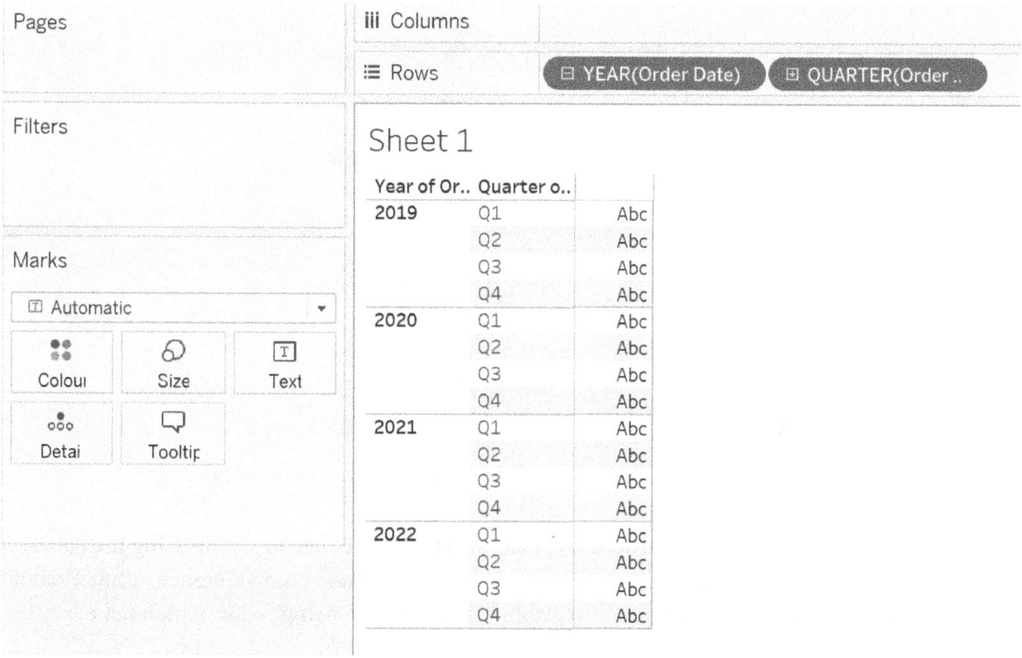

Figure 2.1: The Order Date field displaying date functionality

3. Remove the **YEAR** and **QUARTER**-aggregated **Order Date** fields from the rows shelf to return a blank canvas.

4. Now, click on the calendar icon to the left of the **Order Date** field in the **Data** pane.

5.   Use the popup to switch the data type from **Date** to **String**.

6.   Drag the **Order Date** field to the rows shelf and select **Add all members**. The difference in functionality between the date and string data type fields can now be observed. Now that the **Order Date** field is a string data type, every date in the dataset is listed as a piece of text in the format it was provided from the data source.

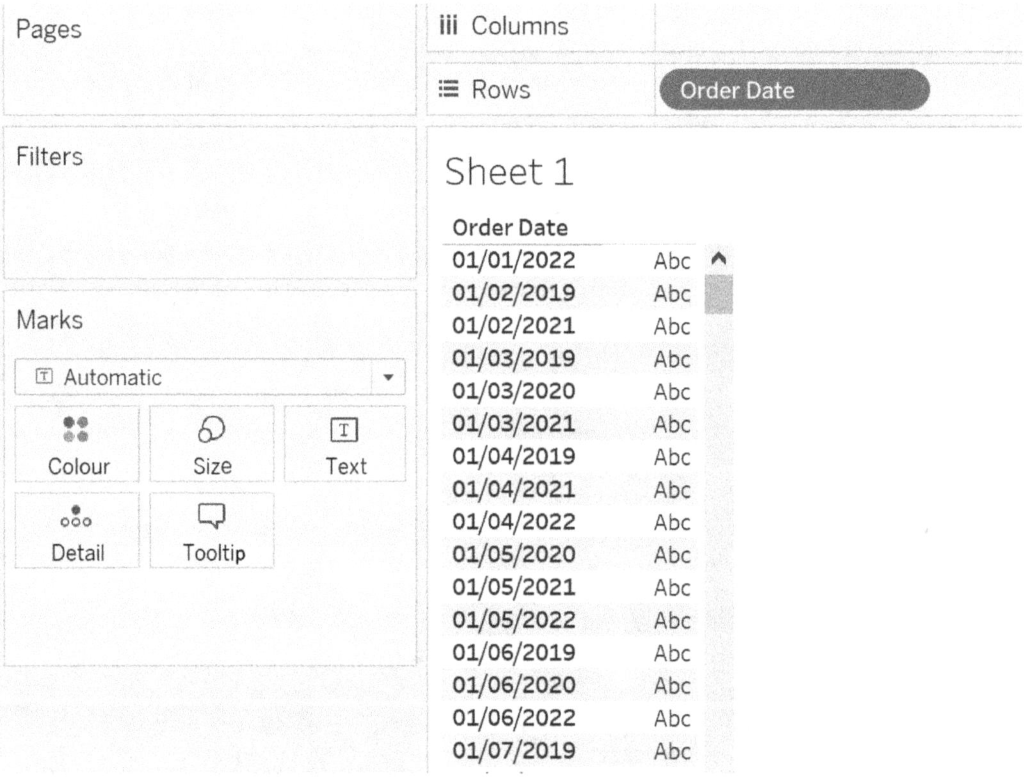

Pages		iii Columns	
		☰ Rows	Order Date

Filters

Sheet 1

Order Date

Order Date	
01/01/2022	Abc
01/02/2019	Abc
01/02/2021	Abc
01/03/2019	Abc
01/03/2020	Abc
01/03/2021	Abc
01/04/2019	Abc
01/04/2021	Abc
01/04/2022	Abc
01/05/2020	Abc
01/05/2021	Abc
01/05/2022	Abc
01/06/2019	Abc
01/06/2020	Abc
01/06/2022	Abc
01/07/2019	Abc

Marks

☐ Automatic ▾

Colour   Size   Text

Detail   Tooltip

Figure 2.2: The Order Date field displaying string functionality

You have successfully updated the **Order Date** field data type from date to string, using the data type icon present on the field in the **Data** pane. You have also demonstrated the difference in functionality between date fields, which default into a drillable hierarchy, and string fields, which list every text value in the field.

Dimensions and Measures

Fields in Tableau, in addition to having a data type, are considered either a dimension or a measure. Dimension fields comprise distinct qualitative values that break up and categorize a view. Some examples of dimensions in Tableau could be country, product name, and customer ID. Measures are made up of quantitative values that are aggregated to measure the relevant metric. Some examples of typical measure fields are sales, profit, and quantity. By default, dimensions are blue and listed first in the **Data** pane. Measures default to green and are listed below dimensions in the **Data** pane, with a line demarcating the split.

Fields can be switched between dimensions and measures by right-clicking the field in the **Data** pane and selecting **Convert to Dimension** or **Convert to Measure**, respectively. String fields are dimensions by default, but when they are converted to a measure, they become a distinct count of the values within that field, as measures have to be aggregated. For example, a **Customer Name** field is a string field that would default to a dimension, but if it were converted to a measure, it would be aggregated as a count of customer names. **Geographic** and **Boolean** fields are also dimensions by default, and geographic fields act the same as string fields when converted to measures. For example, a **Country** field would be a dimension, but if it were converted to a measure, it would return a count of countries in the data source. Boolean fields cannot be converted to measures.

Numeric fields are measures by default, but occasionally, it may make more sense for a numeric field to be a dimension as it will not be used as a metric, but rather as a level of detail in the view. A good example of a numeric field that should be a dimension would be an **ID** field, as it would most likely be used as a list, a level of detail in a scatter plot, or some other type of distinct breakout as opposed to an aggregated value (a sum of an **ID** field would not make sense).

Date and **Date and Time** fields are dimensions by default, and when placed on a view, they default to a high level of detail (year) from which they can then be further broken down into lower levels of detail, such as month or day. **Date** fields cannot be converted to measures in the **Data** pane but can be aggregated on a view.

Dimensions and Measure Field Conversion

In this exercise, you will learn how to convert a dimension to a measure field and convert a measure to a dimension field. This is useful, as it is important to ensure each field in a data source is the correct type of field, as measures and dimensions behave differently in the view.

On a new sheet, right-click the **Order ID** field and select **Convert to Measure**. Observe how the **Order ID** field has turned green and moved down to the measures section. The field also has **Count(Distinct)** as a suffix, indicating that the measure defaults to a count of each unique order ID in the dataset:

1. Drag the **Order ID** field onto the columns shelf, creating a single bar chart that shows the total count of order IDs in the dataset.

2. Right-click the **Order ID** field on the **Data** pane and select **Convert to Dimension** to turn the field back into a dimension. The field has turned blue and moved back up to the dimensions section in the **Data** pane.

3. Drag the **Order ID** field onto the rows shelf and observe the difference in behavior between the field as a measure and the field as a dimension. The dimension field has created a list of each order ID, and the distinct count of order IDs now shows one for each.

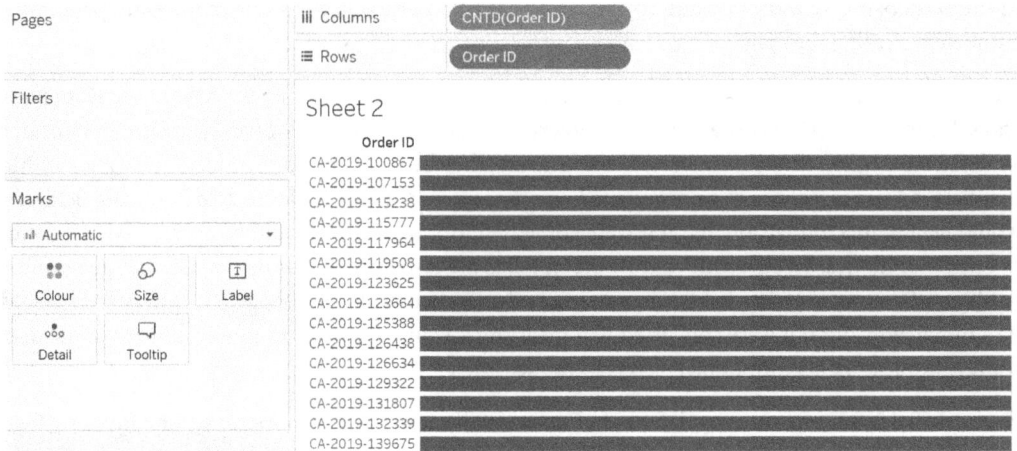

Figure 2.3: Order ID as both a measure and dimension in the view

You have successfully converted a field from a dimension to a measure, and then back from a measure to a dimension. The difference in behavior between dimensions and measures has been observed, with the measure creating a bar showing the total orders and the dimension creating a list of orders.

Discrete and Continuous

Tableau fields would appear to be colored green or blue based on whether they are a dimension or a measure, respectively, but this is not actually the case. Fields in Tableau are colored blue if they are discrete and green if they are continuous, and similar to dimensions and measures, each field must be either one or the other. The blue and green differentiation can be seen both in the fields icon in the **Data** pane and the pill color when a field has been dragged onto either the rows, columns, marks, or filter shelf. Whether a field is discrete or continuous determines how it will function in the view.

Discrete fields are fields that have a finite number of individually distinct values. Continuous fields contain a continuous list of values that could theoretically continue infinitely. Examples of discrete fields would be categories such as regions, whereas continuous fields are values such as sales. Tableau by default takes fields with numeric values that can be aggregated as continuous measures, whereas any field containing values that are not numeric is taken as a discrete dimension.

While discrete fields seem similar to dimensions and continuous fields similar to measures, it is possible for a dimension to be a continuous field, and it is possible for a measure to be discrete. For example, a date field is always a dimension but can be converted from the default discrete field setting to become a continuous field. Doing so will not move the date field from the dimensions section of the **Data** pane, but it will change the field icon color to green. Similarly, a price field may by default be a continuous measure field, but this can be updated to a discrete measure field if the user is more interested in price values as discrete categories. Note that string and Boolean fields can only be discrete dimensions.

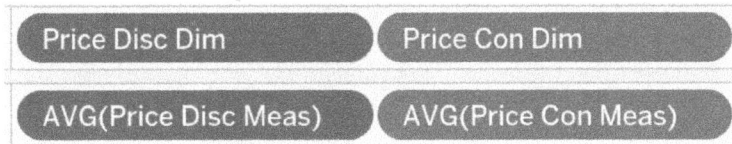

Figure 2.4: Each Price field pill shows a different possible combination. The top left shows price as a discrete dimension, and the top right shows price as a continuous dimension. The bottom two pills show price as measures, with the left discrete and the right continuous

Discrete and continuous fields behave differently when placed on the view in Tableau. Placing a discrete field on either the rows or columns shelf will result in the values of the field being presented as row or column table headers. Placing a continuous field on rows or columns will create a vertical or horizontal axis for that field, showing values between a minimum or maximum.

Similar to rows and columns, placing a discrete field on a color on the marks card differs from placing a continuous field. A discrete field will result in a legend with a separate color for each value in the discrete field, whereas a continuous field will result in a legend with a continuous range of colors from a minimum to a maximum value.

Dimension, Measure, Discrete, and Continuous Combinations

In this exercise, you will switch a field between continuous and discrete and observe the change in behavior. You will also observe how a dimension can be continuous and how a measure can be discrete. The interaction of these four field types underpins the creation of all chart types in Tableau:

1. On a new sheet, drag the **Sales** field onto **Rows**. The **Sales** field is numeric so defaults to a continuous measure. When placed on **Columns**, **Sales** is aggregated (sum) because the field is a measure, and an axis is created because it is continuous.

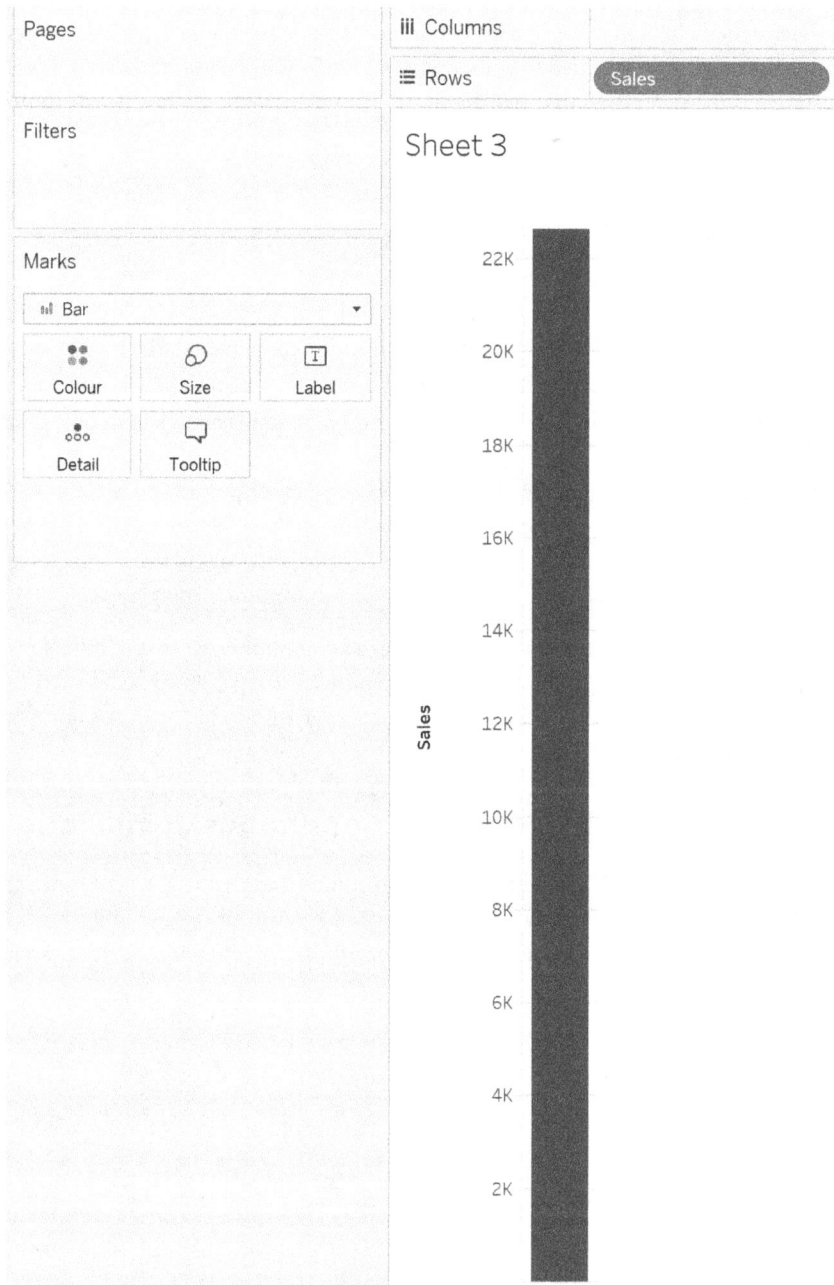

Figure 2.5: Sales as a continuous measure

2. Right-click the **Sales** pill on **Rows** and select **Discrete**. This changes the field to behave as a discrete field in the view only, while the field in the dataset remains continuous. This can be seen by the fact that the pill changed color to blue, indicating it is now discrete, but the number icon next to **Sales** in the **Data** pane remained green, indicating it is a measure. Note how the axis has now switched to a row header for the sum sales value. There is only one value because the field is still a measure and so is aggregated (sum of sales).

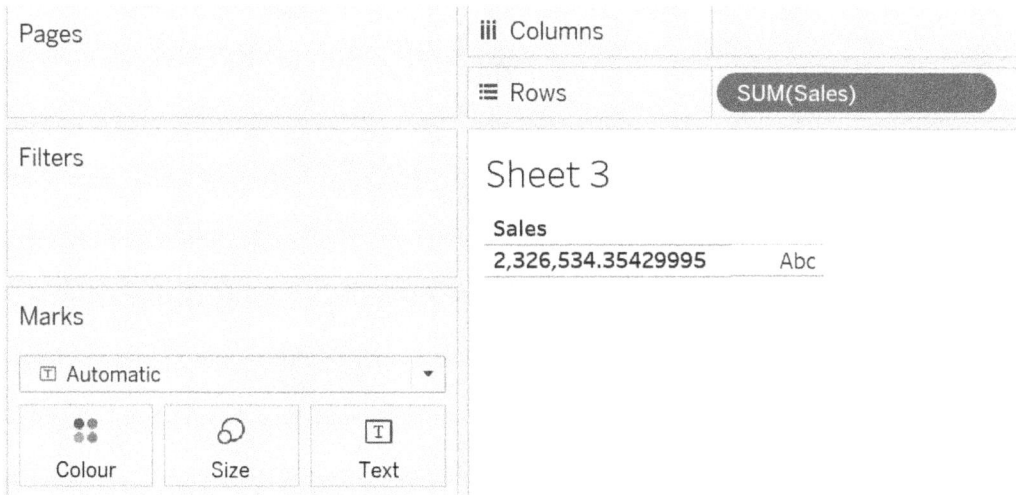

Pages		iii Columns	
		☰ Rows	SUM(Sales)
Filters		**Sheet 3**	
		Sales	
		2,326,534.35429995	Abc
Marks			
⊤ Automatic ▾			
⠿ Colour	⬡ Size	⊤ Text	

Figure 2.6: Sales as a discrete measure

3. To show all of the sale values in the dataset as row headers, the field needs to be a discrete dimension. Right-click the **Sales** pill on **Rows** and select **Dimension**. A large table is now created with a row header for every distinct sale value in the dataset.

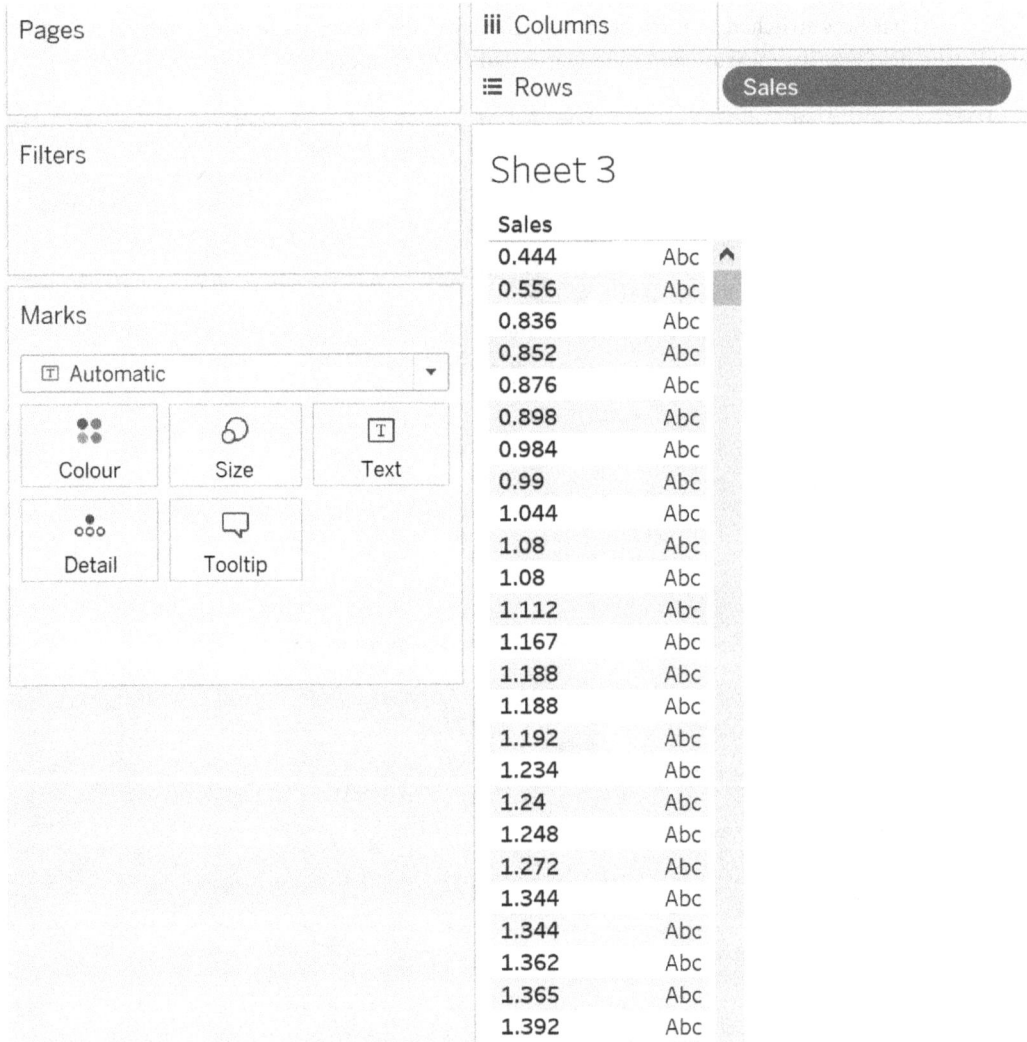

Pages		iii Columns	
		≣ Rows	Sales

Filters

Sheet 3

Marks

				Sales	
				0.444	Abc
				0.556	Abc
⊤ Automatic			▼	0.836	Abc
				0.852	Abc
				0.876	Abc
Colour	Size	Text		0.898	Abc
				0.984	Abc
				0.99	Abc
Detail	Tooltip			1.044	Abc
				1.08	Abc
				1.08	Abc
				1.112	Abc
				1.167	Abc
				1.188	Abc
				1.188	Abc
				1.192	Abc
				1.234	Abc
				1.24	Abc
				1.248	Abc
				1.272	Abc
				1.344	Abc
				1.344	Abc
				1.362	Abc
				1.365	Abc
				1.392	Abc

Figure 2.7: Sales as a discrete dimension

4. Finally, to view sales as a continuous dimension, right-click **Sales** on **Rows** and select **Continuous**. The table now switches to an axis, as the field is continuous. Instead of showing a single bar representing the total sales, it displays a Gantt line chart where each line represents a separate sales value in the dataset.

Pages		iii Columns	
		☰ Rows	Sales

Filters

Sheet 3

Marks

🔧 Automatic ▾

⠿ Colour	🔵 Size	T Label
⠿ Detail	🗩 Tooltip	

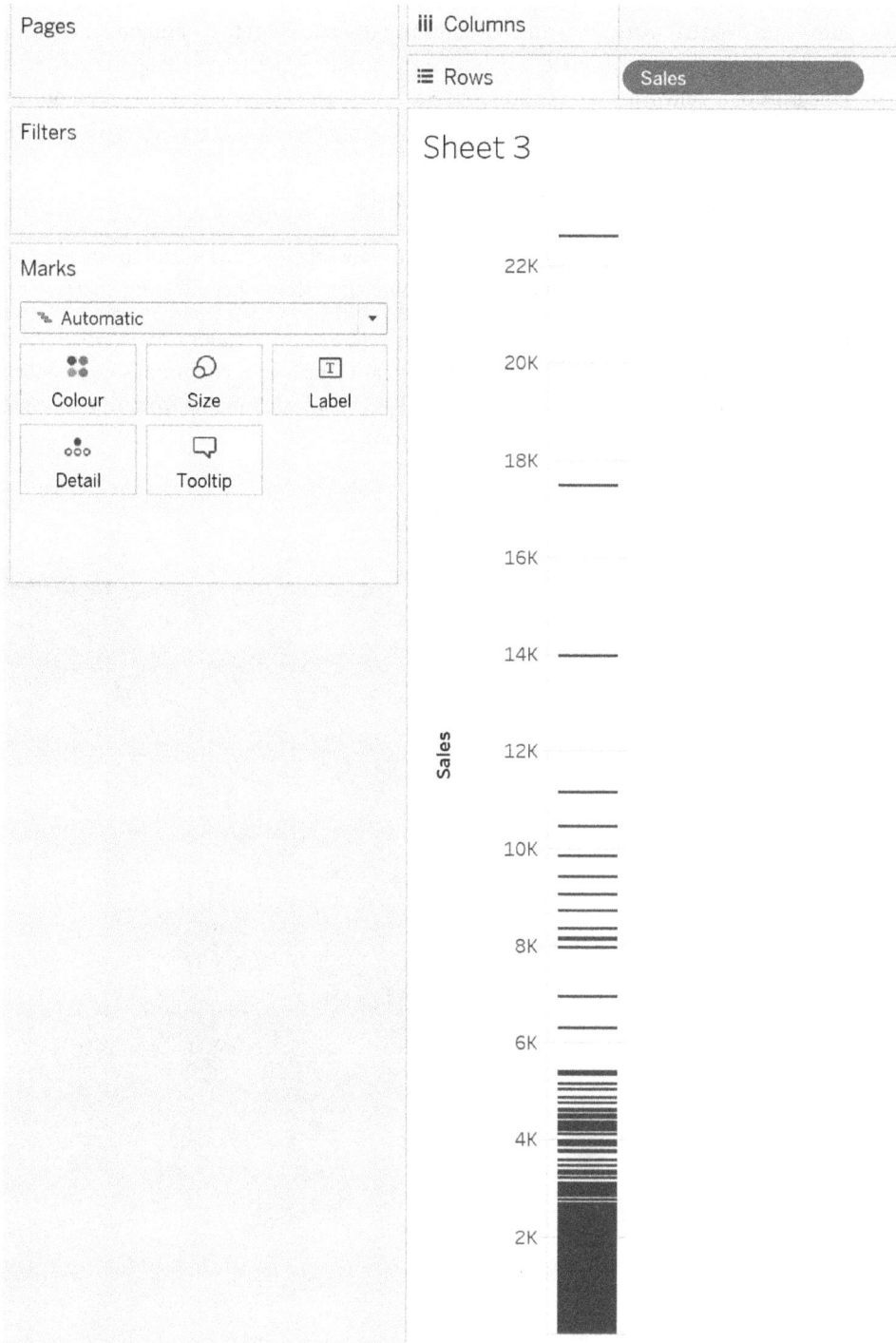

Figure 2.8: Sales as a continuous dimension

You have now converted a dimension (which defaults to a discrete field) to continuous and a measure (which defaults to continuous) to discrete. You have seen that discrete fields break up the view whether they are a measure or a dimension, and continuous fields create an axis whether they are a dimension or a measure.

Discrete and Continuous Color Legends

In this exercise, you will create continuous and discrete color legends. It is useful to understand the difference between these so that you can apply color more effectively when creating charts.

On the same sheet used for the *Dimension, Measure, Discrete, and Continuous Combinations* exercise, drag **Sales** from the **Data** pane onto **Color** on the **Marks** card. The field is continuous, so a continuous color legend is added to the data, coloring each **Gantt** line, on a scale ranging from the lowest to the highest sale value and getting darker as the values increase.

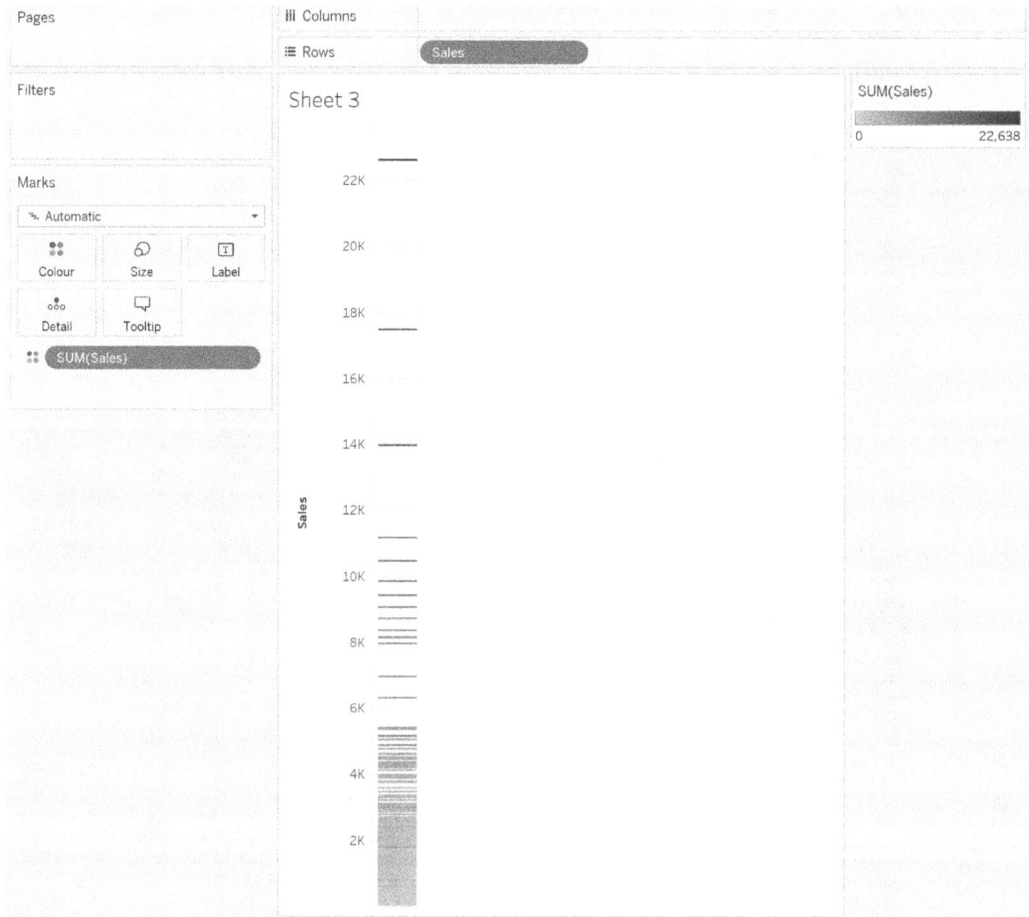

Figure 2.9: A continuous color legend added

To make the color legend discrete, right-click the **Sales** field on the **Marks** card and select **Discrete**. The color legend now updates to give each sales value a distinct color. In cases where a field has more discrete values than colors in the assigned color palette, the same colors will be used for multiple values.

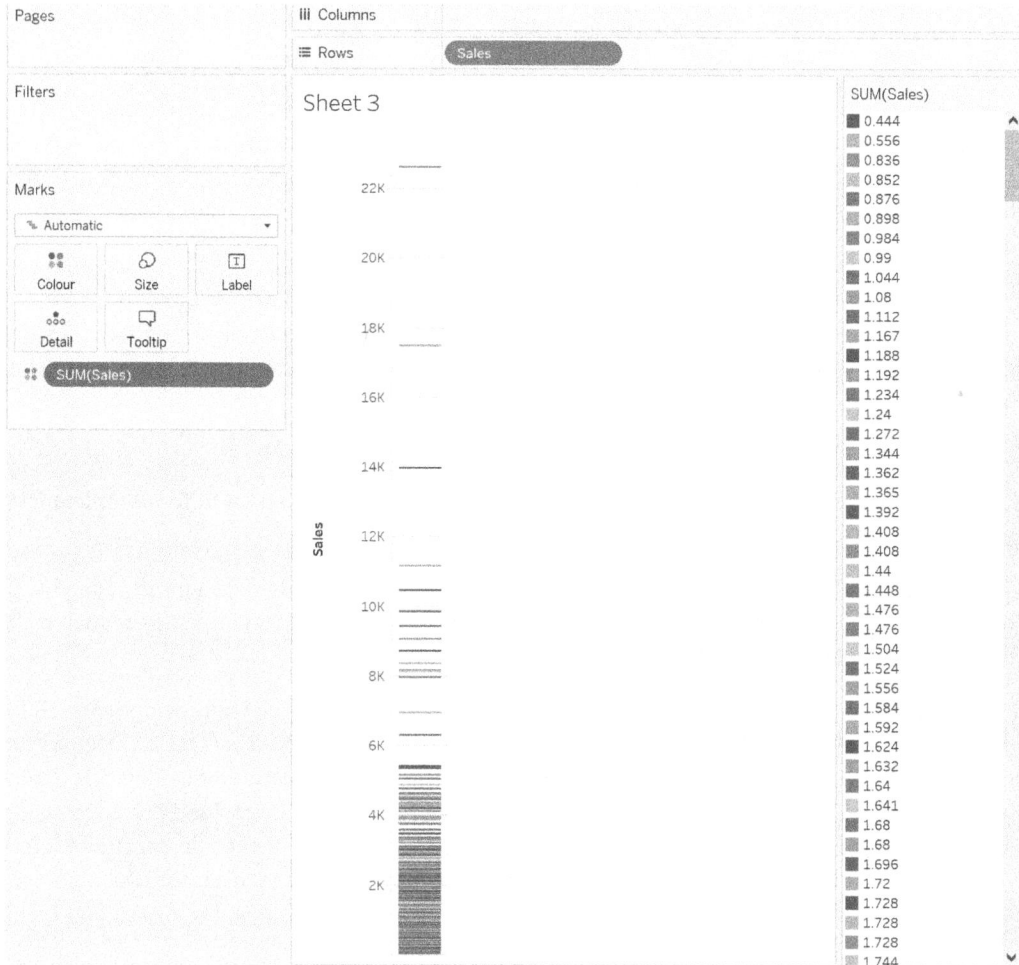

Figure 2.10: The color legend updated to Discrete

You have created a continuous color legend that is presented as a scale between two values, and also a discrete color legend that is presented with a distinct color per discrete item in a field.

Cleaning Data

This section will walk you through the Tableau Desktop and Tableau Prep cleaning functionalities, starting with what is important when assessing data quality. Cleaning in Tableau Desktop will be followed by using the data interpreter, using folders to organize data sources, and finally, cleaning in Tableau Prep.

Cleaning data refers to either removing unwanted data or fixing broken data. This can include processes such as removing duplicates and outliers or fixing incorrect values and formatting. For example, an invoice system may produce a data source with duplicate invoice records in error. For the data source to be usable in Tableau, the duplicate records would have to be removed to ensure that invoices are not double counted, resulting in inflated totals.

Ensuring data is clean before analysis is a key step, as the results of the analysis can only be trusted once the data source is confirmed as accurate and reliable. Every data source is different in terms of how it will need to be cleaned, but the principle of ensuring the data is accurate is always consistent.

Assessing Data Quality

To ensure a data source is sufficiently clean for Tableau, it should be assessed for completeness, consistency, and accuracy.

The completeness refers to whether the data source includes all the required data for the analysis. If, for example, a sales region was excluded from a company's quarterly sales dataset, then analysis of that dataset would not sufficiently explain all the sales for that quarter.

The consistency of a data source refers to ensuring that field values remain the same with no variation. Values within a field should be consistent – for example, a measure should not contain both percentage and total values; every value should refer to the same metric. Similarly, values in a dimension field should be in the same category – for example, it would not make sense to have countries and postcodes in the same column when they could be split into separate columns. It is also important to be consistent across data sources. A field of the same name should mean the same thing if used in another data source.

The accuracy of a data source refers to the fact that the data should reliably reflect true values. If different data sources result in differing totals, then these should be analyzed and a truly accurate value decided upon. The accuracy of the data is the most important piece of the puzzle, as any insight drawn from the analysis without this would be false.

To check a data source for completeness, consistency, and accuracy, it can be useful to start by removing duplicate and irrelevant values (a full dataset may not be necessary/relevant for the analysis being conducted). Similarly, the data can be checked for outliers, and if there is sufficient justification, these can also be removed. General cleaning to improve consistency is always useful. For example, if some values in a field are capitalized and others are not, this could be standardized. Similarly, some values can be grouped together. For example, both **UK** and **United Kingdom** in a country field refer to the same thing, so both of them should have the same name. Empty or null values should also be considered – should these be removed or retained? Could they be replaced with reasonably estimated values based on the complete data?

Cleaning data on Tableau Desktop

When it comes to cleaning data, Tableau Desktop has limited options, and if any serious cleaning needs to be done to ensure the completeness, consistency, and accuracy of a dataset, then Tableau Prep should be used. However, there is some functionality in Tableau Desktop for basic field updates. Tableau Desktop allows users to rename fields, change the data type (as mentioned earlier in the chapter), filter out unneeded records, update fields' default properties/behavior, and update field value aliases.

To rename a field in Tableau Desktop, right-click it in the **Data** pane and select **Rename**. This will highlight the field name and allow the user to change it. Every field name must be unique, and therefore, fields cannot be renamed to something that another field is already called.

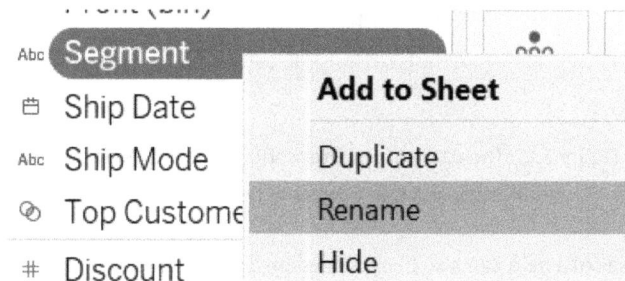

Figure 2.11: Renaming a field in Tableau

Some records may not be necessary for analysis. For example, if you are answering a question about sales in the **North** region, it does not make sense to retain data about sales in the **East**, **South**, or **West** regions, and therefore, these can be removed. It would also make sense to exclude specific values that have been identified as outliers. This can be done by adding a data source filter or an extract filter. Adding a data source filter removes data from the data source at the beginning of Tableau's order of operations. This can be done by right-clicking the data source name at the top left, selecting **Edit Data Source Filters**, and then adding the relevant filter. Extract filters are similar in that they filter data directly from the data source (during the creation of the extract), and these are added to the configuration screen when creating an extract.

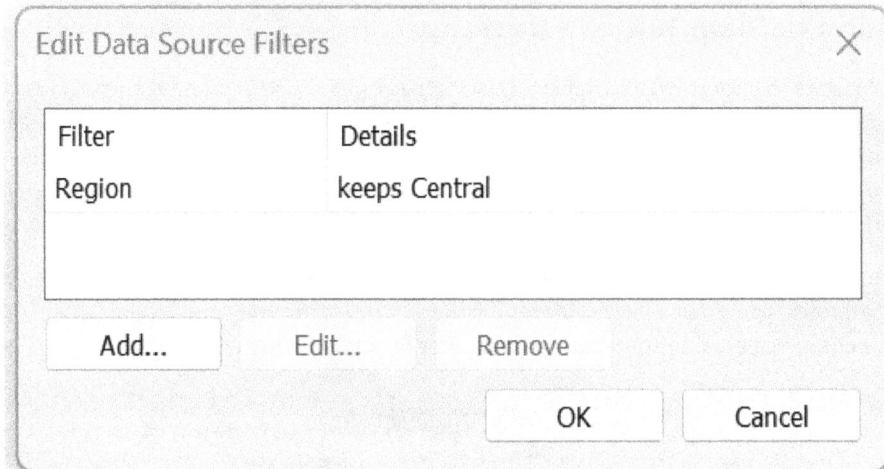

Figure 2.12: The data source filter configuration screen showing a
filter applied that keeps the Central region only

The default properties of a field can also be modified on Tableau Desktop to specify the behavior a field will initially display when placed on a view. To update a field's default properties, right-click the field in the **Data** pane and select **Default Properties**. In the **Default Properties** popup, there will be a list of default behaviors to update. Updating the color will allow the user to specify the colors that will be used on a legend when the field is dropped on **Color** on the **Marks** card. Numeric fields can have their number format specified – for example, setting a sales field to include the dollar symbol or setting a percentage field to include the percentage sign. Dimensions can have a default shape assigned to each field value and can also be given a default sorting methodology. Dimensions are, by default, sorted ascending alphabetically, but they can be given a manually assigned sort order too. Measures can have a default aggregation and total type set using the default properties – for example, price defaulting to show the average as opposed to the default sum, as well as showing the average when totaled.

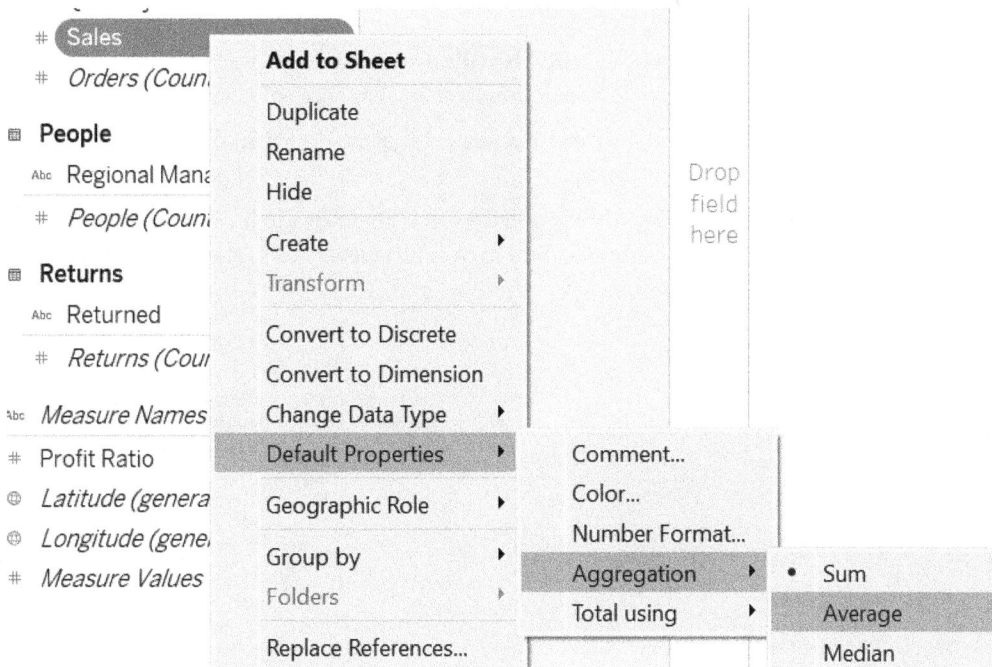

Figure 2.13: Updating the Sales field's default properties by setting its default aggregation as Average

Aliases can be used when a non-date dimension has an incorrect value that needs to be displayed differently. A good example of this would be an uncapitalized name where other names are capitalized. To update a field's alias, right-click the dimension and select **Aliases**. A configuration window will pop up that lists every member of a dimension, an asterisk for every member that has an alias, and the value/alias for each member, which can be manually overwritten. Aliases do not change the underlying data; they just change what is presented in a view.

Duplicating and Renaming a Field Updating the Aliases

In this exercise, you will duplicate a field, rename it, and then update the alias for a value in the field. It is useful to learn these data transformation steps in Tableau, as it will enable you to clean data sources within Tableau as and when you need to.

1. On a new sheet, right-click the **Customer Name** field in the **Data** pane and select **Duplicate**. This creates a new field in the dataset called **Customer Name (copy)**.

2. To rename the field, right-click it, select **Rename**, and then replace the **(copy)** part of the name with **With Updated Alias**. The field has now been successfully renamed.

3. To update the aliases of the customer names, right-click the **Customer Name With Updated Alias** field and select **Aliases**. The top value is **Aaron Bergman**.

4. Update the text in the **Value (Alias)** column to read **Aaron TEST**. Note how once the alias value no longer matches the member value, the **Has Alias** column is populated with an asterisk. This can be used to easily identify which members have an alias applied in a field:

5. Press **OK**, and then drag the **Customer Name** field to **Rows** to list all the customers. **Aaron Bergman** can be seen listed at the top.

6. Now, drag the **Customer Name With Updated Alias** field to **Rows** to the right of the **Customer Name** field. The alias can be observed in the view, with **Aaron TEST** showing on the first row as opposed to **Aaron Bergman**.

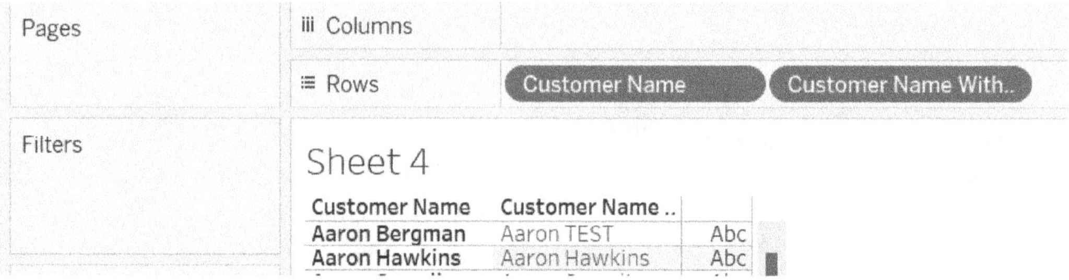

Figure 2.14: The Customer Name field is duplicated and the duplicate has been successfully renamed. An alias has been applied to one of the customer name values

You have successfully created a duplicate of the **Customer Name** field and renamed it to something that makes more sense in the context of the data. You then tested out the alias functionality, comparing a field value both with and without an alias.

Updating the Default Properties of Fields

In this exercise, you will learn how to update the default properties of a field. Becoming familiar with this feature can make you much more efficient in Tableau, as you no longer have to change field behavior every time it is brought into a view.

1. On the same sheet used for the first exercise of the chapter, right-click the **Customer Name** field and select **Default Properties**.

2. From the **Default Properties** menu, select **Sort** to set up a custom default sort for the field.

3. Use the sort by dropdown and select a **Manual** sort, and then select **Aaron Bergman** and move the value below **Aaron Hawkins**.

4. Close the configuration window and note how **Aaron Bergman** is no longer on the top row in the view and has been moved below **Aaron Hawkins**, due to the default sort set on the field.

5. Drag **Discount** onto **Columns** to show the total discount value for each customer as a bar chart.

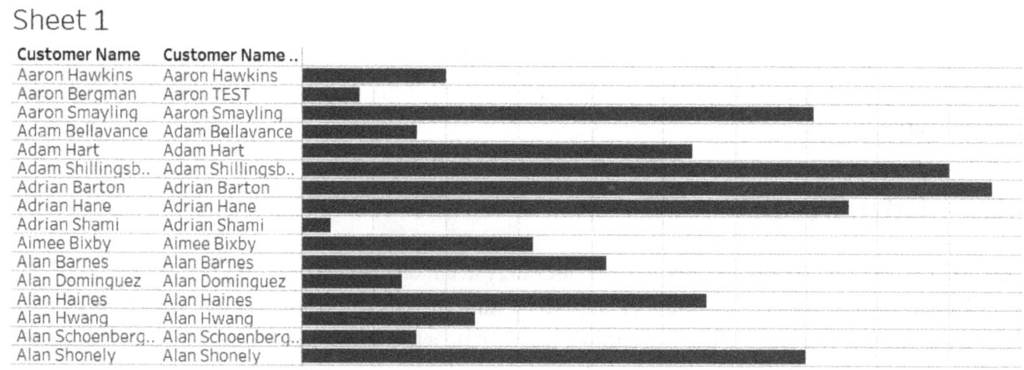

Sheet 1

Figure 2.15: Aaron Bergman has been sorted below Aaron Hawkins
and Discount has been added as a bar for each customer

6. Discount makes more sense displayed as a percentage, and the sum of every discount percentage a customer has received does not make sense. Right-click the **Discount** field, followed by **Default Properties**, and then **Number Format**.

7. In the configuration pop-up window, switch the default number format to **Percentage** with one decimal place, and then press **OK**. The axis has now been updated to show the discount values as a **Percentage**, but they are still being aggregated as a sum.

8. Right-click the **Discount** field, followed by **Default Properties**, then **Aggregation**, and then, finally, set the **Discount** field to **Average**.

9. This does not automatically update the aggregation set on fields in the view, so drag the **Discount** field from the **Data** pane on top of the **Discount** field on the columns shelf to replace it. The default aggregation is now the average discount value, and the view shows this broken down by customer.

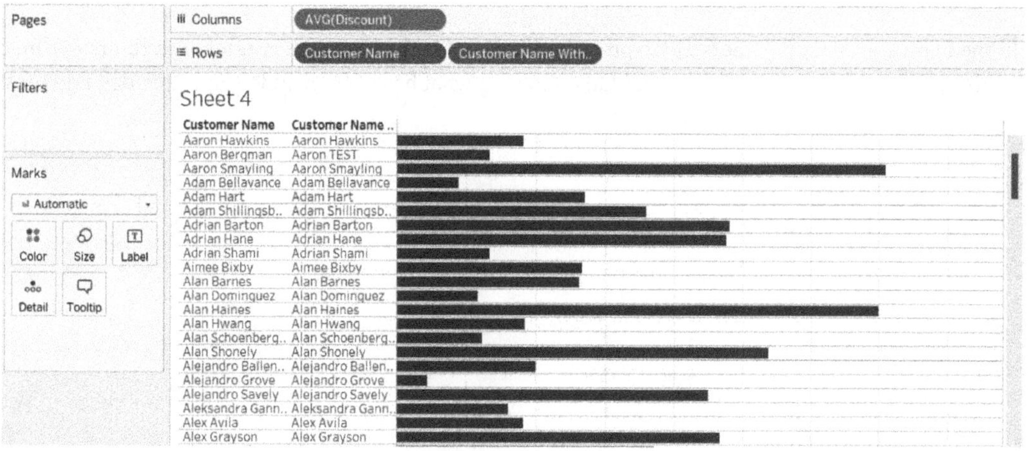

Figure 2.16: A custom sort, number format, and aggregations have been applied to the fields

You have successfully updated the default sort on the **Customer Name** field, resulting in a manual sort where the previously highest value now sorts into second place. You have also updated the default formatting and aggregation for the **Discount** field, ensuring it displays an average percentage whenever dropped onto the view.

Data Interpreter

Tableau Desktop's **Data Interpreter** is a methodology available to clean datasets from Excel, CSV, PDF, and Google Sheets files that are often created to be human-readable as opposed to machine-readable. The **Data Interpreter** helps clean a file into a usable dataset, getting rid of things such as titles, notes, and stacked headers. Additional tables and sub-tables within a file can also be identified and selected using Tableau's **Data Interpreter**. The data in the file will not be updated when using **Data Interpreter**, only Tableau's interpretation of the data.

To use **Data Interpreter**, go to the **Data Source** page using the tab in the bottom-left corner. If the data source is the correct file type, there will be an option in the **Sheets** pane to tick a box labeled **Use Data Interpreter**. Once this has been clicked, the label will switch to **Cleaned with Data Interpreter**, and there will be a **Review the results** link, which will take you to an Excel file that summarizes the **Data Interpreter** results.

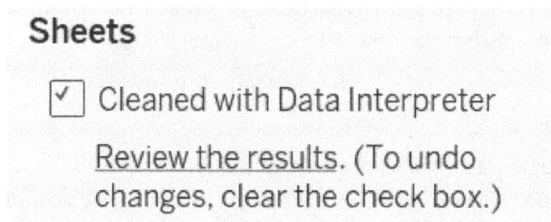

Figure 2.17: Check the box to confirm that Data Interpreter has been used

In the Excel summary file, the first tab provides the **Key for the Data Interpreter**. The colors outlined on the tab are utilized in the data source tabs, showing what has been included and excluded and what has been considered headers/field names.

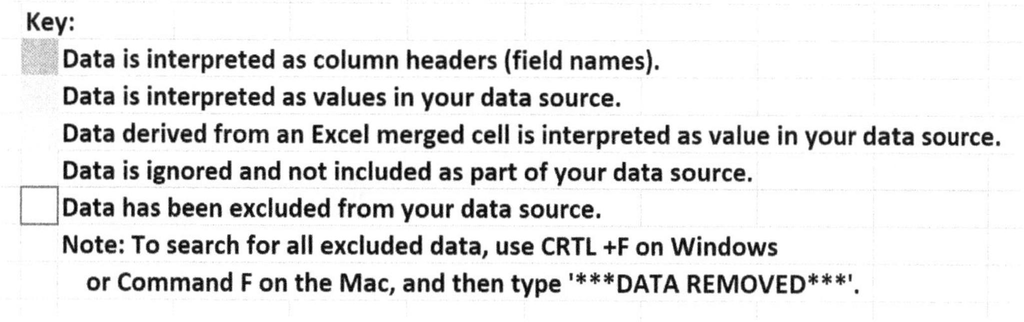

Key:

Data is interpreted as column headers (field names).

Data is interpreted as values in your data source.

Data derived from an Excel merged cell is interpreted as value in your data source.

Data is ignored and not included as part of your data source.

Data has been excluded from your data source.

Note: To search for all excluded data, use CRTL +F on Windows
  or Command F on the Mac, and then type '***DATA REMOVED***'.

Figure 2.18: The Data Interpreter key

If a sheet contains multiple tables, these will be identified in the **<Sheet name>_subtables** sheet, with black borders encasing each sub-table. The subsequent sheets will be each of the individual sub-tables separately. The sub-table sheets are named after the cell range within which the sub-table exists. The sub-table sheets use the key from the **Key for the Data Interpreter** tab.

Using Data Interpreter in Tableau Desktop

In this exercise, you will learn how to use **Data Interpreter** in Tableau Desktop. This is a useful tool for cleaning Excel, CSV, PDF, and Google Sheets files that do not follow a simplistic tabular structure, with column headers on the first row and the corresponding records below.

1. To create a new messy dataset for **Data Interpreter**, find the **Sample - Superstore** dataset used in Tableau on your machine. It can usually be found in **Documents\My Tableau Repository\ Datasources\2023.3\en_US-US** or a similar file path.

2. To find the definitive file path, go to the **Data Source** page in Tableau Desktop using the tab in the bottom left. From this page, the **Sample - Superstore** connection can be seen in the top-left **Connections** section.

3. Use the dropdown and select **Edit Connection…** to locate the Excel file.

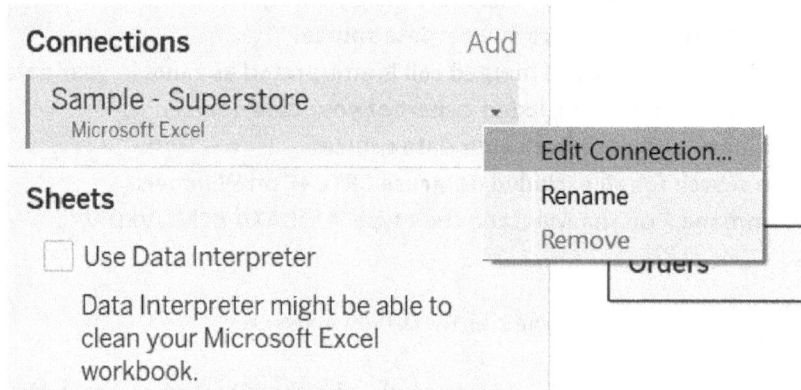

Figure 2.19: The Edit Connection… button leading to the Sample – Superstore file location

4. Create a copy of the **Sample - Superstore** Excel file and rename it **Sample - Superstore Messy**.

5. Open the file and delete the **Returns** sheet.

6. In the **Orders** sheet, add three rows to the top of the dataset by right-clicking row one's header and selecting **Insert**.

7. Add a title to cell A1, such as **Sample - Superstore**.

8. Now, copy the data from the **People** tab (cells A1–B5) and paste these onto the **Orders** sheet in cell Y10.

9. Finally, delete the **People** sheet and rename the **Orders** sheet **Messy** before saving the file and closing Excel.

10. Now that a messy dataset is available, open Tableau and add a new data source by selecting **Data** in the top menu, followed by **New Data Source**.

11. Navigate to the new **Sample - Superstore Messy** file and select it.

12. Tableau reads the file directly, and so the only header is the title added to A1, and the **Orders** and **People** sub-tables have not been identified as separate. Instead, three empty columns have been added between each table, and they have been combined as one large table.

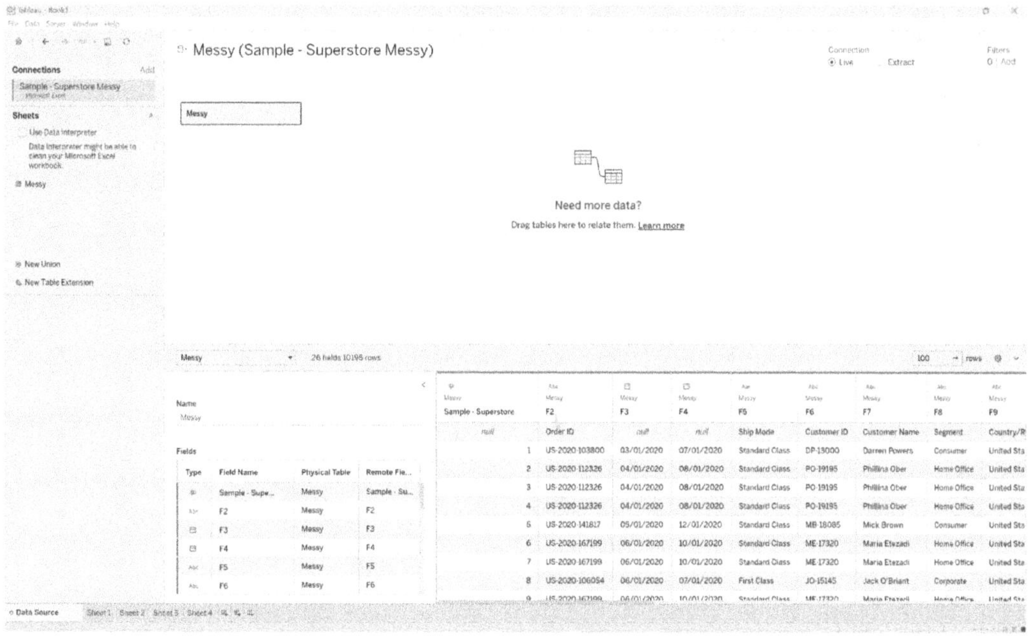

Figure 2.20: The messy data has been read in directly and has no column headers, other than the title from cell A1. The correct column headers are read in as column values on the first row

13. Check the **Use Data Interpreter** checkbox in the **Sheets** section, and then follow the **Review the results** hyperlink. Using the key, it can be observed in the **Messy** sheet that the correct headers are now identified for the orders table and the title and empty header rows have been excluded. The **Messy_subtables** sheet shows that each sub-table has been identified. The final two sheets are named the cell ranges of each sub-table, and the sub-table data can be seen within.

14. Returning to Tableau, the **Sheets** section shows that there are three options, the first and currently selected being the **Messy** sheet, which includes all the data but with the title and blank header rows removed.

15. Dragging **Messy A4:U10000** on top of the **Messy** sheet on the canvas replaces it as the active table and now correctly returns only the **Orders** sub-table data.

16. Doing the same with the **Messy Y10:Z14** sheet correctly returns the **People** sub-table data only.

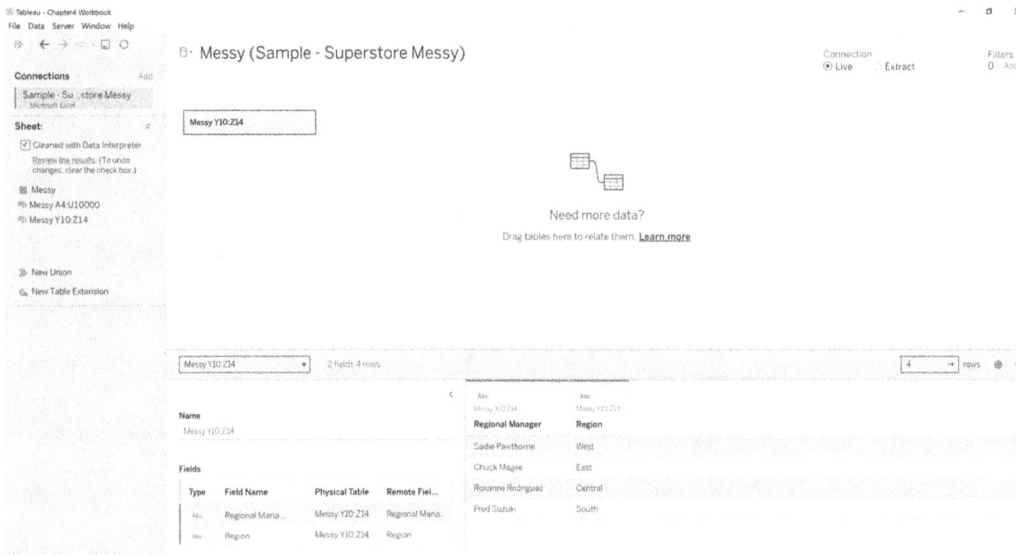

Figure 2.21: A subtable imported into the canvas

The messy data has been correctly interpreted using Tableau's **Data Interpreter**, with sub-tables being correctly identified and imported.

Organizing the Data Source

Once a data source has been cleaned and is ready for analysis, it can be organized in Tableau Desktop's **Data** pane. By default, a Tableau data source with multiple tables will group fields by the table they came from. Dimensions will be listed first, followed by measures for the first table, and then there will be the second table's list of fields, and so on.

To switch from table organization to grouping by folders, open the **Data** pane menu using the carrot in the top right and select **Group by Folder**. If using a multi-tabled data source, the **Data** pane will switch to a single list of fields, with dimensions above measures as opposed to table-specific lists. If the data source is a single table, then **Group by Folder** will be selected by default. Folder information is retained when switching between the **Group by Folder** and table views.

To create a folder, select the field that should be grouped within it, then right-click one of the fields, select **Folders**, and then **Create Folder**. Use the **Create Folder** configuration popup to give the folder a suitable name. A folder will be created at the top of the **Data** pane. Folders are ordered alphabetically, and dimensions and measures are split within each folder.

Fields can be added to or removed from folders by dragging and dropping the field in or out of the folder. Right-clicking a field in a folder and then selecting **Folder** and **Remove Field From Folder** will remove the field. Right-clicking a field that is not currently in a folder, then selecting **Folders**, and then **Add to Folders** will provide a list of available folders to add the field to.

Creating a Folder

In this exercise, you will create folders in Tableau that can be useful for organizing a data source. This comes in handy when a data source has many fields and when handing a workbook over to another user, who will be able to find fields more easily when they are grouped logically.

1. In the existing workbook from prior exercises, select the **Data** pane menu using the carrot in the top right and switch to **Group by Folder**. Fields are now displayed in one large list as opposed to being broken up by table.

2. Select the **Order Date** and **Ship Date** fields, then right-click any one of them, and select **Folders** and **Create Folder**.

3. Name the folder **Dates** and press **OK**. There is now a folder at the top of the **Data** pane with both date fields listed.

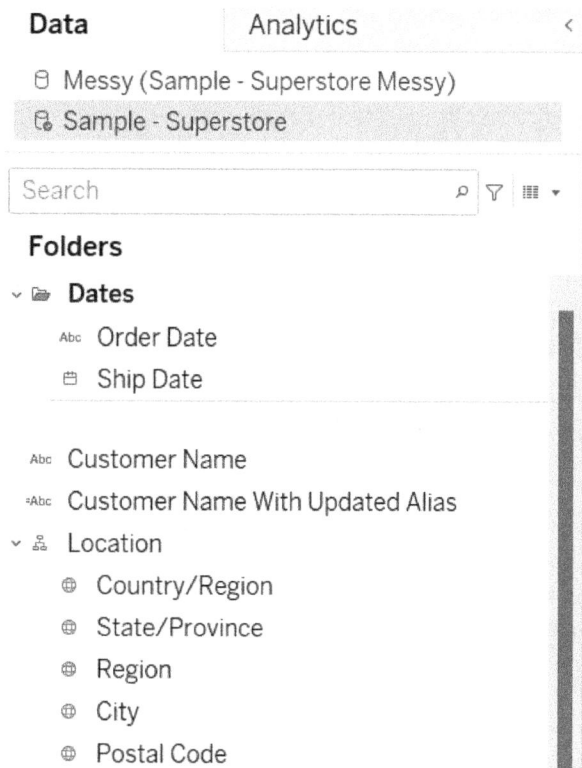

Figure 2.22: Date fields have been organized into their own folder

You have successfully updated the **Data** pane settings to allow for folder grouping and have created a folder grouping for date fields.

Cleaning Data in Tableau Prep

When a more comprehensive cleaning of a data source is required to ensure the analysis is accurate, the best tool to use is Tableau Prep. An example of this could be three distinct Sales data sources from offices in three different regions, each of them with their own methodology. Although the general metrics might be the same, the specific names of fields, the calculation logic applied, and the level of data aggregation may differ between each office. In this case, all three of the data sources can be cleaned in separate streams in Tableau Prep until they all have a common format, before combining them all together.

Tableau Prep offers a wide variety of data cleaning functionalities. Tableau Prep works by inputting data, which then goes through a series of logical steps. One of the logical steps available is called the clean step, and it provides all of the functionalities available for cleaning data in Tableau Prep.

To filter out outliers or unnecessary data in Tableau Prep using the clean step, there are three possible routes.

To filter on a specific value, the following steps should be followed:

1. Right-click the value to be filtered.
2. Select **Keep Only** or **Exclude** to filter to the value only or to remove the specific value.

Alternatively, a user could apply filter logic via the following steps:

1. Select the **More Options...** button in the top right of a field.
2. Select **Filter** and then the desired filter option. This can be a calculation using Tableau's calculated field logic, selected values from a list, `null` values to be included or excluded, a wildcard match if the field is text based, a range of values or dates if the field is numeric or a date, or a relative date if the field is a date field.

Finally, the **Filter Values** button in the clean step can also be used from the toolbar to create a specific filter calculation.

The **Group Values** functionality in the clean step allows users to group values either manually via the user's selection or automatically, based on either pronunciation, common characters, or spelling logic. This can be useful to ensure values are standardized in fields – for example, a **Region** field that includes the values **North** and **north** could have those values grouped into the single value **North**. To use this functionality, select **More Options…** for a field, then **Group Values**, and then select the desired methodology.

Some general data cleaning operations can be applied to a field by selecting **More Options...**, followed by **Clean**, and then the desired functionality. Options include making all text uppercase or lowercase, removing letters, numbers, and punctuation, trimming spaces, removing extra spaces, and removing all spaces. These functionalities help standardize fields and provide a cleaner dataset.

Often, when cleaning data, it can be useful to parse out specific date parts from date fields. Date fields can be converted to specific date parts in the clean step by selecting **More Options**, then **Convert Dates**, and then selecting the date part you would like to summarize the date to. Custom fiscal years can also be specified as part of the summarization.

Another scenario where a field needs to be cleaned is when the field needs to be split into multiple fields based on a delimiter. For example, a full name could be split into first name and last name fields by splitting on the space character. This can be done by selecting **More Options...**, followed by **Split Values**, and then selecting either **Custom Split** to provide a custom character or **Automatic Split** to have Tableau try and decipher where and how it is best to split the field.

When cleaning data, it can also be important to standardize field name formatting, fields can be renamed by either double-clicking the name of a field or selecting **More Options...**, followed by **Rename Field**. Each field name in the dataset must be unique.

Using Tableau Prep's Data Cleaning Functionality

In this exercise, you will use the clean step in Tableau Prep to parse out a date part, filter to a specific year, split a field based on a delimiter, remove punctuation from a field, and then group multiple values together. A clean data source is vital for analysis in Tableau, and this exercise will help you know how best to clean data in Tableau Prep.

To complete this exercise in Tableau Prep, connect to Tableau's readily available training data source, the **Sample - Superstore** data source, which can be located using the steps in the *Using Data Interpreter in Tableau Desktop* exercise in the *Data Interpreter* subsection of the *Cleaning Data* section. Add this connection to Tableau Prep, and drag the **Orders** and **People** datasets to the canvas.

1. Add a clean step to the **Orders** table input by hovering over it on the canvas, selecting the plus symbol, and then selecting the clean step.

2. Select the step to view the fields in the data source.

3. Select the **Order Date** field.

4. Select the **More Options...** button at the top right of the field, then **Convert Dates**, and then **Year Number**. The **Order Date** field has now switched from a date field to a numeric field, with every year in the dataset listed.

5. Right-click **2023** and select **Keep Only** to filter to 2023 data only.

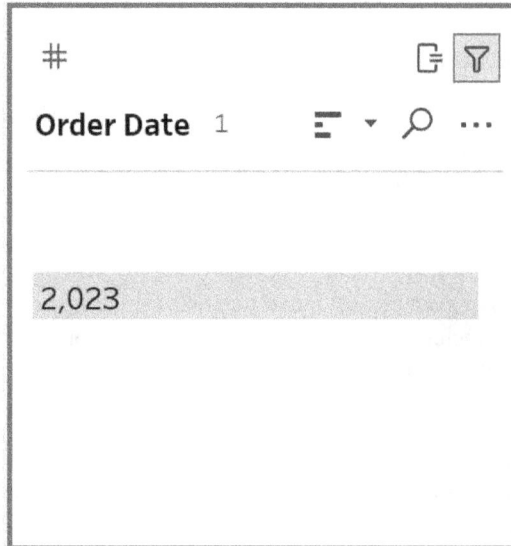

Figure 2.23: Order Date has been converted to a numeric field and filtered to 2023 data only

6. Using the **More Options…** button on the **Order ID** field, select **Split** and then **Automatic Split** to let Tableau split the **Order ID** field automatically. This will create **Order ID - Split 1** and **Order ID - Split 2** fields via splitting on the first found in the **Order ID** field.

7. **Order ID - Split 1** now contains the **Order ID** prefix, which indicates the order country. Rename **Order ID - Split 1** by double-clicking the name, and then call it **Order Country**.

8. **Order ID - Split 2** can be removed by selecting **More Options…** and **Remove**.

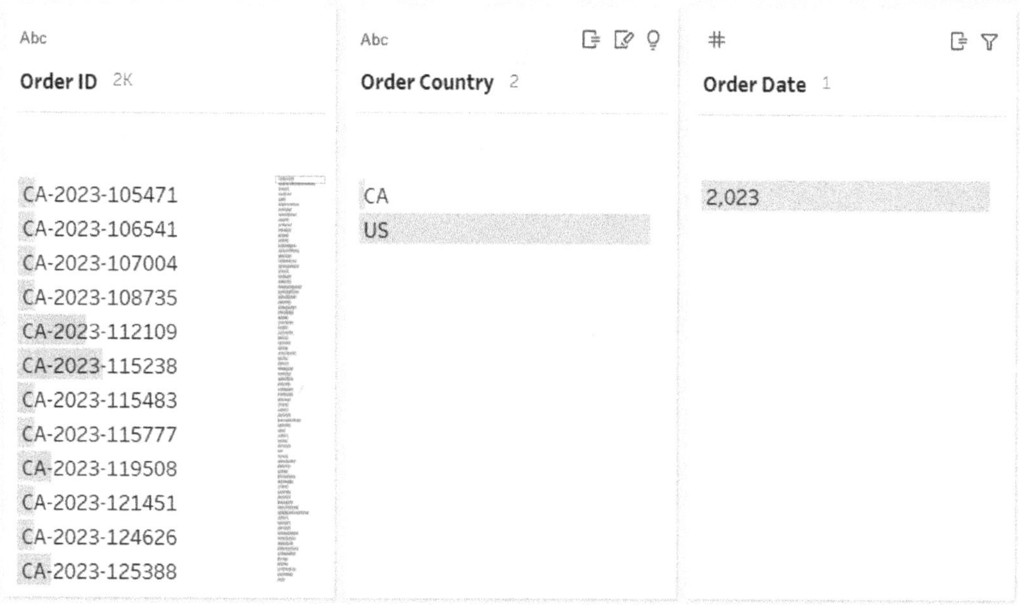

Abc			Abc			#	
Order ID 2K			**Order Country** 2			**Order Date** 1	

CA-2023-105471		CA		2,023	
CA-2023-106541		US			
CA-2023-107004					
CA-2023-108735					
CA-2023-112109					
CA-2023-115238					
CA-2023-115483					
CA-2023-115777					
CA-2023-119508					
CA-2023-121451					
CA-2023-124626					
CA-2023-125388					

Figure 2.24: Order Country has been created as a new field by splitting Order ID on the first instance of -

9. Navigate to the **Product Name** field, select **More Options…**, then **Clean**, and then **Remove Punctuation**. All punctuations have now been removed from the **Product Name** field.

10. Scroll down the **Product Name** field until products starting with **Acco** are visible.

11. Select the first **Acco** field, scroll down to the last **Acco** field, and then hold down Shift and select that field.

12. Right-click the selected values, then select **Group**, and name the newly created group **Acco Products**. All Acco products now have the product name **Acco Products**.

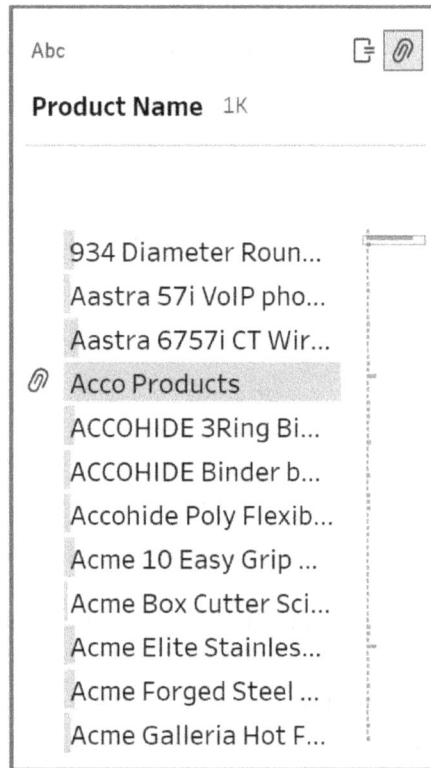

Figure 2.25: The Product Name field has been updated to remove
all punctuation and group Acco Products together

In this exercise, you used multiple data cleaning operations to parse out a year from a date field, and then filter to that year. You split a field on the - delimiter, renaming one field as a result and removing the other. You used the remove punctuation general cleaning operation and grouped multiple values into a single value.

Transforming Data

This section covers the data transformation functionality available in both Tableau Desktop and Tableau Prep. For basic transformations, such as splitting fields on a given character to create new fields or restructuring data by pivoting specific columns, Tableau Desktop can be used.

For more advanced transformations, such as aggregation of data and complex pivoting and unpivoting of data sources, Tableau Prep is best. Some of what could be considered data transformation was covered in the section on Tableau Prep's clean step. In general, data cleaning refers to removing incorrect data and fixing issues, whereas data transformation refers to modifying the structure of the data. This can include adding new fields, increasing or decreasing the level of detail, or flipping/pivoting rows and columns.

Tableau Desktop Transformations

Tableau Desktop allows for some basic data transformations such as splitting fields on a given delimiter to create new fields and pivoting data from a crosstab format into a columnar format.

A field can be split in Tableau Desktop from the **Data Source** page by selecting the dropdown for the field and clicking on either **Split** or **Custom Split**. Selecting **Split** will allow Tableau to split the field on whatever character it thinks is best. **Custom Split** opens a configuration popup, allowing the user to choose the delimiter/separator to split on, and whether to split by the first or last **N** columns or on every instance (**All**). The same options can be seen in the sheet view by right-clicking the field in the **Data** pane, selecting **Transform**, and then either **Split** or **Custom Split**.

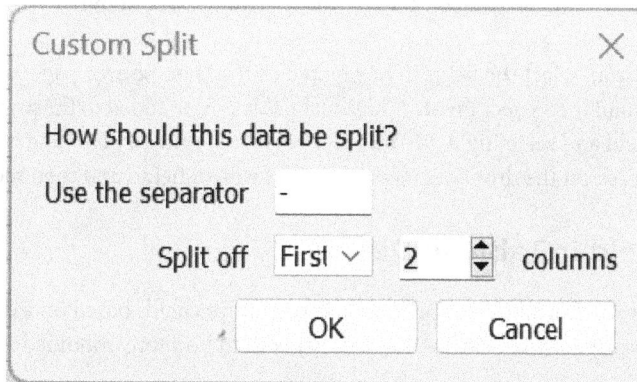

Figure 2.26: The Custom Split configuration window set up to split on - and return the first two results

Often, when data sources are in a wide format with a lot of columns, it would make more sense for Tableau to flip some of those columns to result in two new columns – the first lists the column names of those flipped and the other contains the original values. This is similar to the pivot table functionality in Excel, and it is useful when multiple columns use the same metric but refer to different categories.

An example of when it might be useful to pivot a dataset in Tableau Desktop could be a dataset containing four columns, **North**, **South**, **East**, and **West**, each containing the sales values for each region. Pivoting the four region columns in Tableau would result in a pivot field **Names** column that would contain the values **North**, **South**, **East**, and **West**, and a pivot field **Values** column that would contain the corresponding sales values. From here, the pivot field **Names** column can be renamed **Region** and the pivot field **Values** column can be renamed **Sales**.

West	South	North	East
2,152,513.35	391,721.9	501,239.89	674,982.2

Region	Sales
West	2,152,513.35
South	391,721.9
East	674,982.2
North	501,239.89

Figure 2.27: Pivoting four separate regional sales total columns to
a single region column, mapped to a sales column

To pivot data in Tableau, select the fields to be pivoted in the **Data Source** page, select the dropdown for one of the fields, and then select **Pivot**. Additional fields can be added to the pivot later by using the dropdown on the field and selecting **Add Data to Pivot**. To remove a pivot and return the dataset to its original structure, select the dropdown for either of the pivot fields, and then select **Remove Pivot**.

Splitting a Field in Tableau Desktop

In this exercise, you will learn how to split a field into multiple fields based on a delimiter in Tableau Desktop. This can be useful when a data source only needs to be lightly modified to be fit for purpose.

1. On a new sheet, duplicate the **Sample - Superstore** data source by right-clicking it in the top-left **Datapane** and selecting **Duplicate**. This will create a new data source called **Sample - Superstore (2)**.

2. Ensure **Sample - Superstore (2)** is selected (it should be highlighted gray), and then drag the **Order ID** column onto the rows shelf. A list of order IDs is now visible, and it is clear that each ID is made up of two letters (signifying the country), four digits (signifying the year), and a six-digit ID number all separated by hyphens.

3. To create **Order Country** and **Order Year** columns from the order ID, right-click the field and select **Transform**, followed by **Custom Split**.

4. Set a hyphen as the separator, and make sure to split off the first two columns, as you need a new column for both the country and the year but not the ID.

5. **Order Split - 1** and **Order Split - 2** have now been created and can be renamed **Order Country** and **Order Year**, respectively.

6. Once the fields have been correctly named, drag them onto the **Rows** shelf to confirm that the fields have been split correctly and that the two-letter country abbreviation as well as the year are now independent fields.

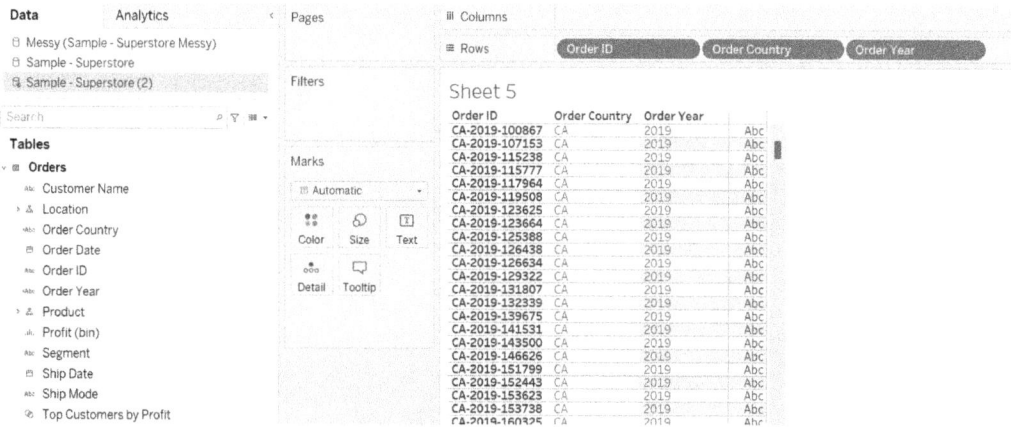

Figure 2.28: The Order ID field split by the hyphen character returning the first two columns

In this section, you have successfully split a field into two separate fields on a hyphen delimiter. The same methodology can be used to split fields on any character, and if multiple splits are required, then multiple column outputs can be specified.

Pivoting Fields in Tableau Desktop

In this exercise, you will restructure the data source in Tableau Desktop by pivoting multiple metrics. Charts in Tableau often require data to be structured in a specific way, and so the knowledge of how to restructure data sources within Tableau Desktop can be invaluable.

1. Using the same data source created via duplication in the *Splitting a Field in Tableau Desktop* exercise, navigate to the **Data Source** tab.

2. From the **Data Source** tab, scroll across the **Orders** table field listings until **Sales**, **Quantity**, **Discount**, and **Profit** are visible.

3. Select all four previously mentioned columns at once using either Shift or Ctrl + clicks.

4. Select the dropdown for any of the four fields and then select **Pivot**. The four measures have all now been pivoted and replaced with two new fields **Pivot Field Names** and **Pivot Field Values**. **Pivot Field Names** contains the name of the fields that have been pivoted; **Sales**, **Quantity**, **Discount**, and **Profit**. **Pivot Field Values** contains the corresponding metrics values.

5. Rename the **Pivot Field Names** column to **Metrics** and the **Pivot Field Values** column to **Values**. Notice how all the other columns in the dataset have four duplicate rows now, as there is a row for each of the pivoted columns.

Abc Orders **Sub-Category**	Abc Orders **Product Name**	Abc Pivot **Metrics**	# Pivot **Values**	=Abc Calculation **Order Country**	=Abc Calculation **Order Year**
Paper	Message Book, Wirebound, Fo...	Discount	0.20	US	2019
Paper	Message Book, Wirebound, Fo...	Profit	5.55	US	2019
Paper	Message Book, Wirebound, Fo...	Quantity	2.00	US	2019
Paper	Message Book, Wirebound, Fo...	Sales	16.45	US	2019
Binders	GBC Standard Plastic Binding...	Discount	0.80	US	2019
Binders	GBC Standard Plastic Binding...	Profit	-5.49	US	2019
Binders	GBC Standard Plastic Binding...	Quantity	2.00	US	2019
Binders	GBC Standard Plastic Binding...	Sales	3.54	US	2019

Figure 2.29: The metric fields have been pivoted and the pivot fields have been suitably renamed

You now know how to pivot data in Tableau Desktop. In this example, multiple distinct metrics have been pivoted into a metrics and values column, but pivoting is also often useful when the metric is common across multiple columns, for example, sales in the North, South, East, and West, with a column for each. In this case, the name and values columns represent region and sales as opposed to metric name and metric value.

Note that any calculated fields, sets, or bins that were dependent on any of the fields that have been pivoted will no longer work and will display a red exclamation mark.

Aggregating Data in Tableau Prep

Tableau Prep offers a wider range of functionalities when it comes to transforming the shape of a data source. The **Aggregate** logical step in Tableau Prep allows users to group values in a data source at a different level of detail. For example, a dataset with a row for every single order could be aggregated to return the total sales for each product. This would mean the data source has fewer records and therefore will perform faster. It also means it could be combined with another data source at the product level.

To aggregate data in Tableau Prep, add the **Aggregate** step and drag any field that should be grouped from the left-hand list of fields into the **Grouped Fields** section. Drag any field that should be aggregated into the **Aggregated Fields** section and select the method of aggregation at the top of the field. Numeric fields will usually be aggregated and all other required fields will be grouped. However, there are some cases where numeric fields could be grouped – for example, when price is intended to be used as a discrete dimension as opposed to a continuous measure in the analysis. Similarly, non-numeric fields can be aggregated to get a count of values or distinct values in that field or also a minimum or maximum value for the field. This is often useful when using dates and wanting to get a first or last date at the specified level of aggregation.

$$\Sigma$$

Aggregate 1

Figure 2.30: The Tableau Prep Aggregate step icon is a summation symbol

Aggregating Data

In this exercise, you will learn how to aggregate data in Tableau Prep. Ensuring data is at the correct level of aggregation is an important skill for anyone looking to analyze data and it is not something that can be done in Tableau Desktop.

1. To complete this exercise in Tableau Prep, connect to Tableau's readily available training data source, the **Sample - Superstore** data source, which can be located using the steps in the *Using Data Interpreter in Tableau Desktop* exercise in the *Data Interpreter* sub-section of the *Cleaning Data* section. Add this connection to Tableau Prep and drag the **Orders** and **People** datasets to the canvas.

2. Continuing the Tableau Prep flow built in the *Cleaning Data in Tableau Prep* sub-section of the *Cleaning Data* section, hover over the **Clean 1** step, click the plus icon, and then select **Aggregate** to add an **Aggregate** step.

3. Drag the **Ship Mode** and **Region** fields into the **Grouped Fields** pane.

4. Drag the **Sales** and **Ship Date** fields into the **Aggregated Fields** pane. The dataset now consists of a breakdown of total sales and a count of ship dates by **Ship Mode** and **Region**.

5. To instead return the most recent ship date by **Ship Mode** and **Region**, select **CNT** at the top of the **Ship Date** field and then select **Maximum**.

6. Add another clean step to view the new data structure. The data now consists of the four previously mentioned columns only with the values aggregated as specified. **Ship Date** can be renamed **Latest Ship Date** to make more sense.

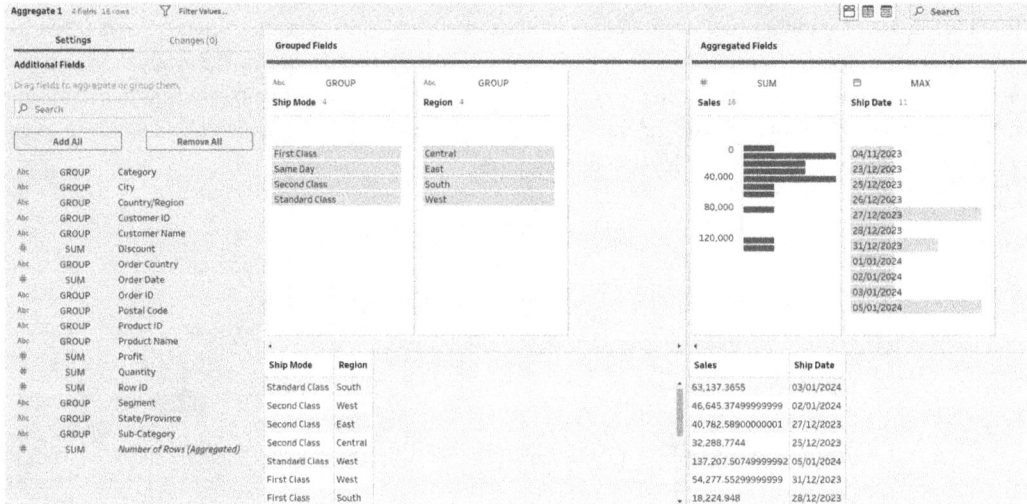

Figure 2.31: Orders data aggregated to ship mode and region level with the most recent ship date and total sales aggregated

In this section, you have aggregated data from the individual order level with thousands of rows down to the **Ship Mode** and **Region** levels, returning the total sales for all orders and the most recent (maximum) ship date. This demonstrates how the **Aggregate** tool can be used to aggregate data to make a less granular summary.

Pivoting Data in Tableau Prep

Pivoting data in Tableau Prep is similar to Tableau Desktop, but with a wider range of functionalities. The data can be pivoted from **Columns** to **Rows**, which works in the same way as in Tableau Desktop, taking a wide dataset and making it thinner and taller by pivoting the selected fields into two new columns. The first column is **Pivot Names**, which returns the names of the columns that have been pivoted, and the second created column is **Pivot Values**, which contains the corresponding original field values. Pivots can also be created with separate value columns for other metrics, all within the same pivot.

The additional functionality offered by Tableau Prep is the ability to pivot from **Rows** to **Columns**. This means taking a dataset and making it smaller and wider. A field is selected to pivot from rows to columns, meaning the distinct values of that field will now become column headers. Therefore, if a region column is selected to be pivoted and it has the values **North**, **South**, **East**, and **West**, four new columns will be created called **North**, **South**, **East**, and **West**. A value must also be selected to populate the new columns as well as a level of aggregation, so you could, for example, end up with the four region columns showing the sum of sales for each region or the average discount percentage.

To create a pivot in Tableau Prep, first add a **Pivot** step to the flow. By default, the pivot will be set to **Columns to Rows** and fields can be dragged into the **Pivoted Fields** pane. The fields will then be pivoted, and the **Pivot 1 Names** and **Pivot 1 Values** fields will be created. The pane interface will update to reflect this and more fields can be added to the **Pivot 1 Values** section if needed. To add an additional pivot value, click the plus icon in the top right of the **Pivoted Fields** pane. This will add another section called **Pivot 2 Values**, in which the additional metrics/values for pivoting (still mapped to the initial pivot names) can be added.

Before dragging and dropping fields into the **Pivoted Fields** pane, there is a hyperlink option to **Use wildcard search to pivot**. This then allows users to type into the search box a piece of text to look for anywhere in a field's name (contains) or at the start or end of a field name. This is useful if data is going to be added to a source over time as it means new files do not manually have to be added; a naming convention will just have to be followed for the filenames. An example of this would be order files added yearly to a folder, such as **Orders_2023**, **Orders_2024**, and so on. These could be brought in with a wildcard search on **Starts with Orders**.

To pivot fields in the other direction, use the dropdown in the top right of the **Pivoted Fields** pane to switch from **Columns to Rows** to **Rows to Columns**. This updates the **Pivoted Fields** pane to now show two sections. The top section, **Field that will pivot rows to columns** is the field that will have its values set as column headers. **Field to aggregate for new columns** is for the aggregated measure/metric that will populate the new columns.

Pivoting Rows to Columns in Tableau Prep

In this exercise, you will pivot data from rows to columns in Tableau Prep. You will observe that there is more functionality available for the pivot in Tableau Prep than in Tableau Desktop, and this will allow you to decide which tool is preferable when you need to restructure your data source.

To complete this exercise in Tableau Prep, connect to Tableau's readily available training data source, the **Sample - Superstore** data source, which can be located using the steps in the *Using Data Interpreter in Tableau Desktop* exercise in the *Data Interpreter* sub-section of the *Cleaning Data* section. Add this connection to Tableau Prep and drag the **Orders** and **People** datasets to the canvas.

1. Continuing the Tableau Prep flow built in the *Aggregating Data in Tableau Prep* sub-section of the *Transforming Data* section, add a **Pivot** step after the **Clean 3** step.

2. Use the dropdown to the top right of **Pivoted Fields** to switch from **Columns to Rows** to **Rows to Columns**.

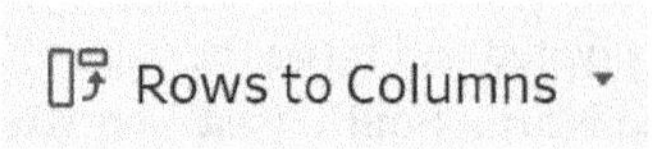

Figure 2.32: Row to Columns pivot selection

3. Drag the **Region** field into the **Field that will pivot rows to columns** section to ensure that **Region** values will be pivoted and become column headers.

4. Drag the **Sales** field into the **Field to aggregate for new columns** section so that each region will show the total sales by **Ship Mode** and **Latest Ship Date** (the remaining level of aggregation not pivoted). Add a clean step to view the new data structure. There is now a **West**, **South**, **Central**, and **East** column, each with a summed sales value by each ship mode.

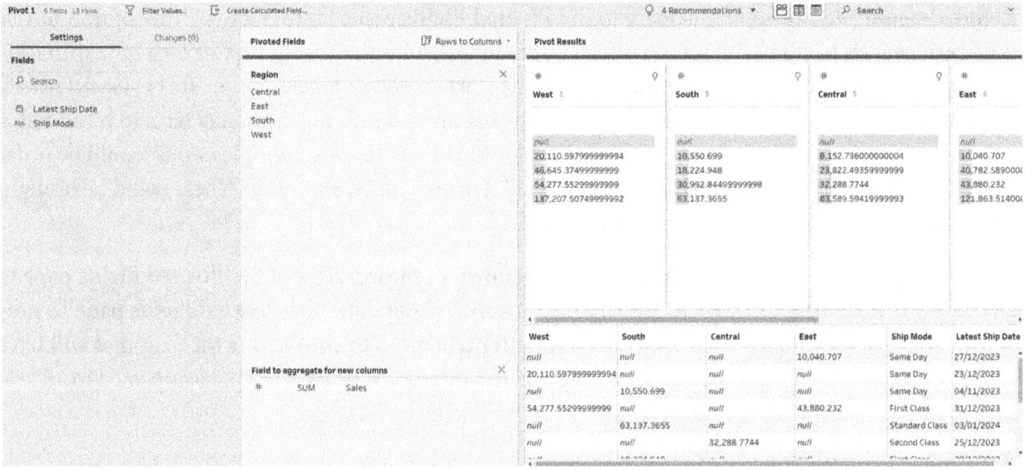

Figure 2.33: Region values have been pivoted from rows to columns with sales values summed to the ship mode level of aggregation

You have successfully pivoted the **Region** field that was previously a single column with four values from rows to columns aggregated by sales. This has resulted in four new columns (one for each value in the **Region** column) with the sales totals by ship mode. You have observed greater control over directionality and specificity when pivoting in Tableau Prep as opposed to Tableau Desktop.

Pivoting Columns to Rows in Tableau Prep

In this exercise, you will pivot data in the opposite direction, from columns to rows, and leave the **Region** column unchanged. Through this, you will have learned both methods of pivoting in Tableau Prep and will understand how to transform your data source by pivoting in whatever direction is needed.

1. Continuing from the previous exercise, add a new **Pivot** step after the **Clean 6** step just created. By default, the **Columns to Rows** setting is applied in the **Pivot** step.

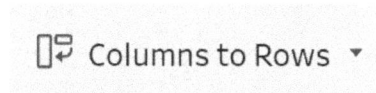

Figure 2.34: Columns to Rows pivot selection

2. Select the **Central**, **East**, **South**, and **West** fields and drag them into the **Pivoted Fields** pane. The region columns now show as **Pivot 1 Names** and **Pivot 1 Values**. These can be renamed by double-clicking.

3. Rename **Pivot 1 Names** to **Region** and **Pivot 1 Values** to **Sales**. The data has now been pivoted back to the previous state before exercise 1, with a region column containing **Central**, **East**, **South**, and **West** values and a **Sales** column aggregated at the **Region** and **Ship Mode** levels.

4. Add a clean step after the pivot to observe that the fields and data structure are now the same as the **Clean 3** step before the rows-to-columns pivot. However, there are additional rows in the data with Null sales values. This is because the initial pivot created columns for each region, but there wasn't always a corresponding sales value in the data to aggregate. The value was therefore left as null for that ship mode.

5. Pivoting the data back to the original structure does not by default get rid of the `null` values created by the rows-to-columns pivot. These can be removed in the final clean step by right-clicking the **Null** bar in **Sales** and selecting **Exclude**.

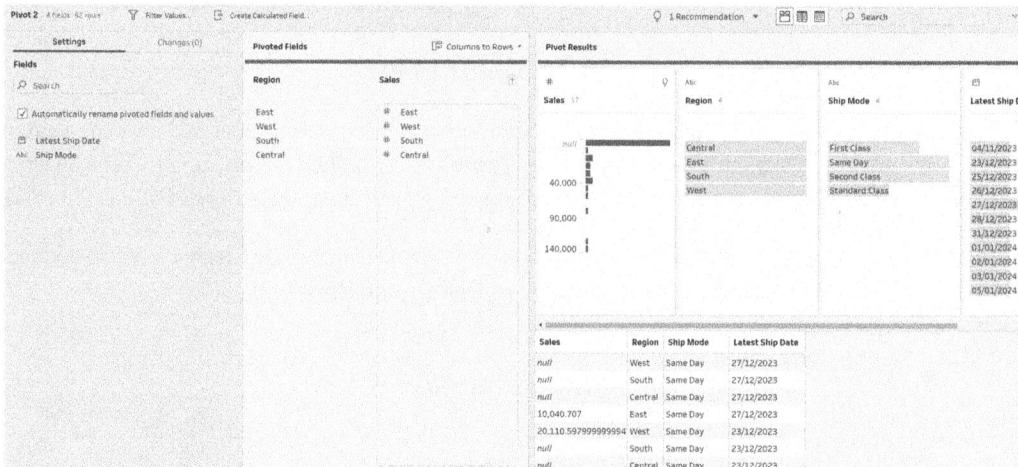

Figure 2.35: Region columns have been pivoted back to create a region and a sales column. Null values are visible in sales but can be removed in a subsequent clean step

The region data pivoted in the rows-to-columns pivot exercise has been pivoted back into a single **Region** column and a single **Sales** column using the column-to-rows direction pivot. You now know how to pivot specific columns in both directions using Tableau Prep.

Combining Data

This section covers the methodologies for combining data sources together in Tableau. How to join data sources on common fields in Tableau Desktop is covered first, followed by how to union data sources in Tableau Desktop. Relationships in Tableau Desktop and Tableau Blending will also be discussed. Finally, how to join and union data sources in Tableau Prep is looked into.

Tableau Desktop Joins

A common method for combining two data sources is to join them side by side on a common field or on multiple common fields. Tableau Desktop allows datasets from any valid data source type to be combined together (Tableau Data sources cannot be used in joins). There are multiple ways to configure a join in Tableau Desktop such as deciding which fields to match with between the two data sources and then which data to keep as a result of the join.

Choosing which fields to match on can be as simple as selecting a common field with the same name in two data sources. However, more complex joins may require multiple matched fields to ensure data is accurate. An example of this could be joining on both country and city fields across both data sources to ensure a false join is not created between records in both data sources with the same city name but different countries (for example, Birmingham is a city in both the UK and USA).

Join clauses can also be set to match records where the value from one source is not the same as the value from another. Similarly, greater than and less than (or equal to) clauses can be set up, as opposed to a simple value match. This can often be useful when it comes to joining dates that fall within a range of two date fields in another data source. An example of this could be joining a sales promotion data source to a sales order data source where the promotions data source only has a promotion start and end date fields. In this case, the sales orders can be joined to the correct promotion where the sales order date falls between the promotion's start and end date. Complex join conditions can also be set up using Tableau's calculation logic, for example, joining a full name field from one data source to a calculated concatenation of first name and last name in another.

When creating a join clause, the field types selected from each source must be of the same data type. Removing a field from a data source that is used in a join clause will cause that join clause to break.

Once the join clause has been created, it needs to be decided which data will be kept. There are four types of join possible in Tableau Desktop, and these are represented by Venn diagrams in the user interface.

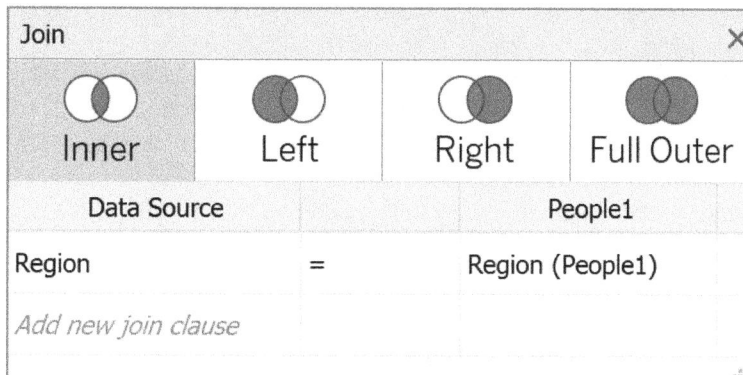

Figure 2.36: The Join configuration window showing an inner join of two
data sources where the Region field is common across both

Inner join is the default selection, and the icon shows the overlapping section of the Venn diagram colored only. Inner joins retain only data from both data sources where there is a match in the join clause.

Left joins are represented by the left circle of the Venn diagram as well as the overlapping section both being colored. Left joins retain all the data from the data source on the left-hand side of the join and bring in any records from the right-hand data source where the join clause is met.

Right joins are the opposite of left joins, retaining all data from the right-hand data source and having the right circle and overlapping sections of the Venn diagram icon colored.

The final join option is a full outer join, which is symbolized by a fully colored Venn diagram and retains all data from both data sources, combining records where the join clause is met.

In the left, right, and full outer joins, records that come from one data source only (no match on the join clause from the data source) will still have the columns from the other data source, but as there is no match in the other data source, the values for those columns will be **null**. For example, a data source with a **Region** column could be left joined to a data source with **Region** and **Sales** columns. If the first data source has a region that is not present in the second data source, then records for that region are still kept in the joined dataset as the join type is left and the first data source is on the left-hand side. The joined dataset will also have a **Sales** column, but for the region that had no match, the sales value with be **null**.

When creating a join, it is important to consider the granularity (lowest level of detail) of each data source. If data sources being joined are not at the correct level of aggregation, there can be duplication of records resulting in inaccurate measures when aggregating. An example of this could be a data source with **Country**, **City**, and **Sales** fields being joined with a data source with **City**, **Postcode**, and **Population** fields. If these data sources were joined on **City**, then the **Sales** value for the city would be duplicated for as many postcode records as there are in that city. To resolve this, the **Population** data source should be aggregated to **Total Population** and **City** fields only (removing the **Postcode** level of detail).

To create a join in Tableau Desktop, open the **Data Source** tab and make sure the desired connections (data sources) are added. Drag the main table for the join onto the canvas and then double-click it to switch to the join/union table interface. Drag the table you wish to join with onto the canvas. The table rectangles will now be connected via a Venn diagram. Select the Venn diagram to open the join configuration window, then select the join type and join clause. Multiple tables can be dragged onto the pane to create multiple joins in a single data source. However, each individual join is always between two tables.

Creating a Join in Tableau Desktop

In this exercise, you will learn how to join data sources in Tableau Desktop. This is a useful skill to learn as a single data source is often not sufficient for analysis and supplementary data is often required.

1. On the sheet from the previous exercise, create a duplicate of the **Sample - Superstore** data source by right-clicking it in the top left of the page and selecting **Duplicate**.

2. This will create a new data source called **Sample - Superstore (copy 2)**. Select this data source and then navigate to the **Data Source** tab in the bottom left.

3. The **Sample - Superstore** data source by default has a Tableau relationship set up; remove this by right-clicking and removing the **People** and **Returns** tables on the canvas.

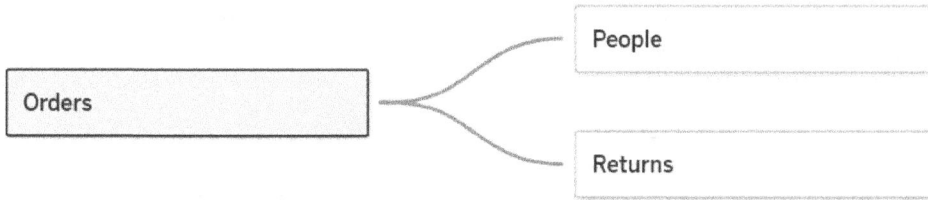

Figure 2.37: Remove the People and Returns tables from the canvas

4.   Returning to the **Sheet** view will show that the **People** and **Returns** fields are no longer present. **Regional Manager** from the **People** table, for example, is no longer present.

5.   Return to the **Data Source** tab and double-click the **Orders** table on the canvas to enter the join configuration interface.

6.   Drag the **People** table from the **Sheets** section on the left onto the canvas to create a join with the **Orders** table.

7.   Select the Venn diagram icon to configure the join.

8.   Switch the **Join** type to **Left** to ensure no **Orders** data is dropped if there is no match with the **People** data source.

9.   Set the **Join** clause to **Region**. A join has now been successfully created that combines the **Orders** and **People** data sources on matching regions.

10. On returning to the **Sheet** view, you will be able to see the **Regional Manager** field from the **People** table.

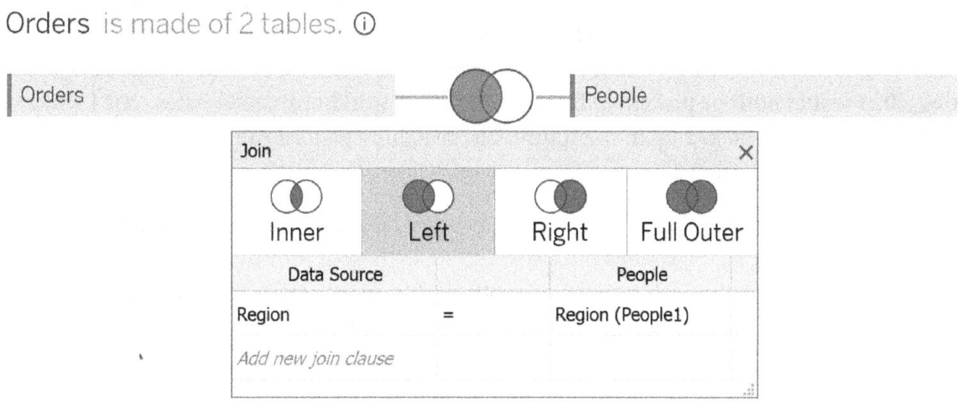

Figure 2.38: Left join created between orders and people tables where region matches region

In this section, you created a left join between the **Orders** table and the **People** table, retaining all data from the left-hand side (**Orders**). You now know how to set up a join in Tableau Desktop as well as how to configure it by join type and set the join clause.

Tableau Desktop Unions

Data sources can be combined in Tableau Desktop by placing one data source on top of the other structurally, aligning common columns, and adding `null` values in columns that are not common between data sources. This method of combining data is called unioning the data sources.

Table headers should be the same across the data sources being unioned if the data from one column should be placed on top of the same from the other data source. Data types must also be common across common columns. An example of a union might be a sales file from 2022 and a sales file from 2023. Both files have the same structure in terms of columns, each with the same name and data type. Therefore, when the union is created, the 2022 data can be placed neatly on top of the 2023 data structurally. By default, unions match columns based on the header names so the order of the fields does not matter. However, the union can be configured to generate field names automatically, in which case the union will occur based on the position of the column in the dataset.

Unions can be created by dragging the **New Union** icon on the **Data Source** page from the left-hand side onto the canvas. This will open the union configuration window and allow the user to create a manual or a wildcard (automatic) union. The manual process of creating a union works by dragging the tables from the left-hand side into the union configuration pane. These tables will be fixed in a union, but a user may require a more dynamic union to take place; for example, if a date-stamped monthly sales file was added to a folder, a fixed union would not pick it up.

To create a dynamic union that will pick up new files, select the **Wildcard (automatic)** section of the union configuration window. The wildcard search will look in the location of the data source connection that has been set up, but this can be expanded to parent folders or subfolders using the checkboxes at the bottom. The workbook names to look for can be specified as well as any sheet names. Asterisks can be used to create the dynamic search; for example, yearly sales workbooks called **Sales_2022** and **Sales_2023** would both be picked up by **Sales_***. A new workbook called **Sales_2024** would then also be automatically picked up in the future. The matching pattern can also be set to exclude any undesired data sources from a union.

Once the union has been set up, it behaves as any other table does on the canvas and can be put in relationship with other tables. Unions can also be set up in the join configuration interface by selecting a new union or by dragging a table on top of another table on the canvas. Unions in the join interface can also be joined to other tables.

Creating a Union in Tableau Desktop

In this exercise, you will learn how to vertically union data sources in Tableau Desktop. Often data sources are broken into multiple files but otherwise have the same column headers. In this case, the data needs to be unioned so as to be read as a single data source.

1. Using the **Sample - Superstore (copy 2)** data source created with the join in the *Creating a Join in Tableau Desktop* exercise, return to the **Data Source** window and delete the **Orders** join on the canvas by right-clicking and selecting **remove** (ensure you are in the default relationship configuration and not the join configuration page).

2. Now create a new union by dragging the **New Union** icon onto the canvas.

3. Drag the **Orders** table from the left-hand side onto the **Union Configuration** pane and press **OK**. A new union has been created, consisting of a single instance of the **Orders** table.

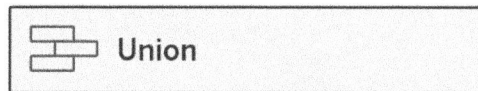

Figure 2.39: A union on the data source canvas

4. Create a new sheet and drag the **Sales** measure onto **Text** on the **Marks** card. The sheet now shows the total sales values for the dataset (2,326,534).

5. Return to the **Data Source** page using the tab in the bottom left. Edit the union by right-clicking and selecting **Edit Union**.

6. Drag the **Orders** table onto the **Union Configuration** pane. Two **Orders** tables can now be seen in the union and the **Tables in union** count has increased from one to two. This means that the data source now consists of two instances of the **Orders** table on top of each other.

7. Press **OK** and return to the newly created sheet to see that the **Sales** value has now doubled (4,653,069).

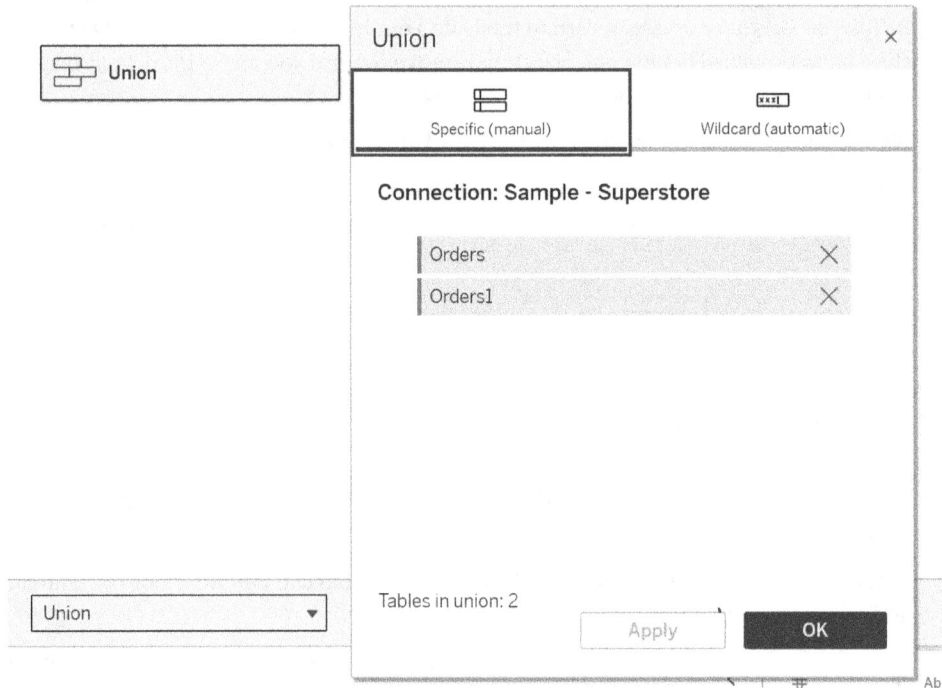

Figure 2.40: Union created that places the Orders table on top of itself

You have successfully unioned two data sources into a single larger data source and have observed how the data is impacted as a result.

Relationships

Relationships are a way to combine tables together on Tableau, side by side, on a common field(s). However, instead of creating a fixed large table that merges the two data sources, the tables are kept separate at their individual level of detail and joins are created automatically based on the fields used in the view.

Relationships are more flexible than joins as only the relationship (similar to a join clause) has to be configured. A join type does not need to be selected because what is included or excluded from the data depends on the fields used in the view. The fact that relationships keep tables distinct means no data is lost and that tables can be at different levels of aggregation because they are not physically joined, so there is no chance of data duplication. The aggregation of measures and the joins required are all calculated by Tableau; the user only needs to define the relationship that connects the tables.

To create a relationship in Tableau, drag a table to the canvas in the **Data Source** page, then drag the table it should have a relationship with to the canvas as well. The tables will be connected by a **noodle**, which can be selected to open the relationship configuration window at the bottom of the page. The configuration for relationships is similar to setting up a join clause. One or more fields from both the left and the right tables can be selected, and the join or relationship condition can be set up as equal to (match), not equal to, or less/greater than (or equal to). Relationship calculations can also be used to create fields for the matching condition, similar to joins. The fields being related must be of the same data type.

How do relationships differ from joins? Learn more

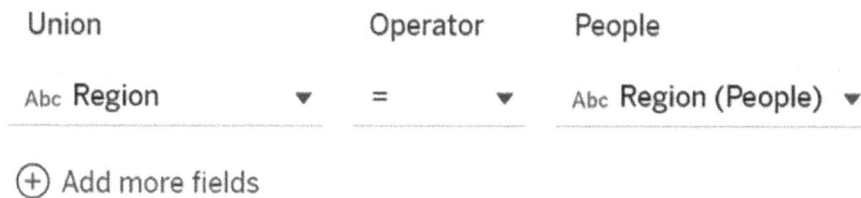

Union	Operator	People
Abc **Region** ▼	= ▼	Abc **Region (People)** ▼

⊕ Add more fields

Figure 2.41: Relationship set up between the Union and People tables on the
Region field. The Operator can be switched to change the match condition,
and Add more fields can be selected to add additional clauses

Multiple tables can be related on the canvas, but each relationship is always directly set up as table-to-table via the connecting noodle. One table is always considered the root table, and this can be switched by right-clicking a table on the canvas and selecting **Swap with root**. Tables can also be removed from relationships by right-clicking them on the canvas and selecting **Remove**.

Creating a Relationship in Tableau Desktop

You will now create a relationship between tables in a Tableau data source. Relationships allow for dynamic combinations of data sources without creating a single fixed table. This is an important skill to learn as tables are not always at the same level of aggregation.

1. Return to the **Data Source** tab in which the union was created in the *Creating a Union in Tableau Desktop* exercise from the *Tableau Desktop Unions* sub-section. There is currently a union table on the canvas that contains a union of the **Orders** table. Looking through the fields in the **Data** pane below or in the **Data** pane from a sheet view shows that the **Regional Manager** field is missing from the dataset. This is because the **People** table is not included.

2. To bring in the **People** table via a relationship, drag it from the left-hand side onto the canvas. A noodle has been created connecting the **Union** and **People** tables.

3. Selecting the noodle will allow you to configure the relationship. By default, Tableau should have picked up that the correct relationship is **Region** equals **Region** across tables. If this is not the case, add the **Region** field to both sides and set = as the operator.

4. The relationship has now been successfully created and each table can be selected separately, resulting in the **Data** pane shown below.

5. Opening a new sheet and ensuring the **Data** pane view type is **Group by Data Source Table** shows that there are two distinct tables in the data source, **People** and **Union**, and that the **Regional Manager** field is now available for use.

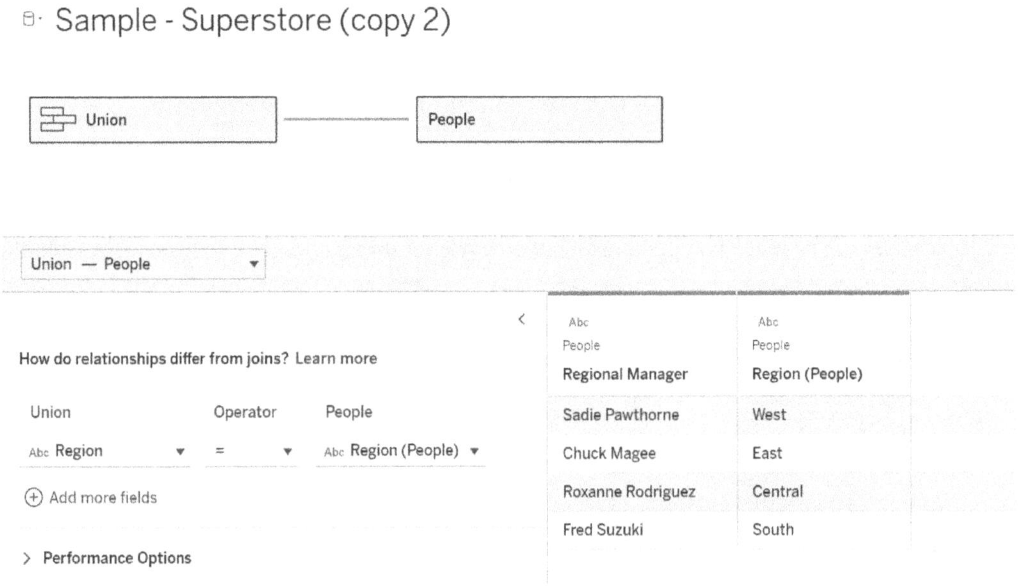

Figure 2.42: Relationship between the Union and People tables on the common Region field

A relationship has been successfully created between the unioned order tables and the **People** table on the **Region** field. Both tables are distinct (the **Data** pane shows **People** table fields only) as opposed to being merged into a single large table.

Blending

Data blending in Tableau Desktop is another method of combining data from multiple data sources in a side-by-side structure via common fields that does not create a physically joined single table. Blending functions differently to both joins and relationships in Tableau in that the data is not actually joined directly. Each source is aggregated separately and then presented together in the view.

The benefit of blending over relationships or joins is that multiple connecting fields can be established between data sources, and these can be toggled on or off with each view in a workbook. This makes them much more dynamic and therefore suitable to problems that require multiple conflicting connections at different levels of aggregation between two Tableau data sources. For example, a **Sales** data source with values by **Country** and **Product Category** could be blended with a **Sales Target** data source, also broken down by **Country** and **Product Category**. Instead of fixing a join or relationship on both **Country** and **Product Category**, the matching criteria can be defined sheet by sheet. In one sheet, the total sales by **Country** and **Product Category** could be calculated as a percentage of the total **Country Sales** target. In another, it could be calculated as a percentage of the total **Product Category** target. In another, the sales could be calculated as a percentage of the target for both **Country** and **Product Category**. Blending can also be done to combine published Tableau data sources, whereas joins and relationships cannot do the same.

Blending in Tableau Desktop means using two separate data sources on the same sheet. The first data source to be used (via dragging a field from that data source onto the view) will be considered the primary data source. Dragging a field from another data source will only be permitted if a blend relationship has been set up between the data sources. If a blend relationship has been set up, then the secondary data source field can be dragged onto the view.

Blends essentially act as a left join in the view with the primary data source being considered as the left part of the join and the secondary data source acting as the right part. This means any field from the primary data source can be used on the sheet as all data from that source is retained. Only values from the secondary data source that have matches in the primary data source can be used. This means that the data source with the more complete data source should be used as the primary data source where possible. An example of this would be a blend on a region field but one data source has no data for the **North** region. In this case, the data source with data for all regions would be optimal as the primary source, otherwise the **North** region would be excluded from the view. If a data source has a relationship in it, then it cannot act as the secondary data source in a blend.

When a blend relationship has been set up, calculations can be created across both data sources in the blend, but the secondary data source fields must be aggregated. Blended fields will have dot notation in the calculated field configuration so that their source can be identified. For example, a **Profit** field brought in from the **Sample - Superstore** data source would present as **[Sample - Superstore].[Profit]** as opposed to just **[Profit]** if **Sample Superstore** was the primary data source.

To blend two data sources, at least one linking field must be set up between them. To configure linking fields between data sources, go to **Data** in the top menu, followed by **Edit Blend Relationships**. From here, a primary data source can be selected and the secondary data source chosen from the list of other data sources. Tableau will, by default, look for common dimensions, but if the link is not found by default, the **Custom** radio button can be selected, followed by the **Add** button, and then the field from each source can be selected to link them together. Multiple links can be set up between data sources.

Once blend relationships have been set up, they can be used in a view. First, drag a field from the primary data source into the view. Then, switch to the secondary data source where a link icon will be displayed to the right of any linking fields. If the field is being used to link the data sources, then it will show as red and connected; if the field is not being used, the link will be grayed out with a line through the middle. Whether the link is active or not can be toggled by clicking the icon; doing this allows users to set the condition for combined data sources sheet by sheet.

Figure 2.43: Blended data linking field icons – the top icon shows a field that is linked for the view whereas the bottom shows a field that is not linked for the view

Primary and secondary data sources can be identified when blending by looking at the data source icons in the top left. If the data source icon has a blue circle, it is the primary data source, whereas if the circle is orange, the data source is secondary. Multiple data sources can be blended on the same sheet, but there will always be only one primary data source. Fields from secondary data sources can be identified in the view by a data source icon with an orange circle placed at the right end of the pill.

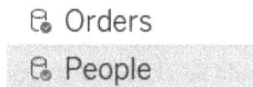

Figure 2.44: Primary versus secondary data source icons, with Orders reflecting the primary and People the secondary

Creating a Blend Relationship in Tableau

In this exercise, you will create a blend relationship between two data sources in Tableau Desktop. Knowing how to use blends in Tableau can provide a user with dynamic methods for combining data sources that can differ on a sheet-by-sheet basis.

1. Start by adding a new **Sample - Superstore** data source to the workbook and renaming it **Orders** by right-clicking the data source name and selecting **Rename** in the sheet view or by simply selecting the name at the top of the **Data Source** tab.

2. Remove both the **People** and **Returns** tables from the canvas so that only the **Orders** table remains.

3. Repeat the previous process, but instead, name the data source **People** and remove the **Orders** and **Returns** tables from the canvas, leaving only **People**.

4. On a new sheet, select **Data** from the top menu and click **Edit Blend Relationships**.

5. Set **Orders** as the **Primary** data source and select **People** in the list of choices for the **Secondary** data source. The automatically assigned relationships should now be visible in the right-hand pane.

6. By default, Tableau should have matched the **Region** fields from both data sources, but if it has not, select the **Custom** radio button followed by **Add** and then select the **Region** field for both datasets.

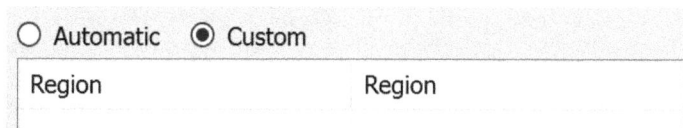

Figure 2.45: Custom blend relationship on the Region field configured

7. The blend relationship has now been confirmed as correct and can be used to create the view. Ensure that the **Orders** data source is selected, and then drag **Region** onto the **Rows** shelf.

8. Now select the **People** data source. The **Region** field in the **Data** pane has a link icon on the right-hand side. The link icon should be red, indicating that the link is active for the sheet. If it is not red, select it so that the link is active then drag the Region field onto the **Rows** shelf.

9. Dragging the **Region** field activates the **People** data source as the secondary data source in the blend, and the blue and orange data source icons are now visible in the top-left data source section. The blending has successfully worked and regions from both data sources are used in the same sheet and mapped correctly to each other.

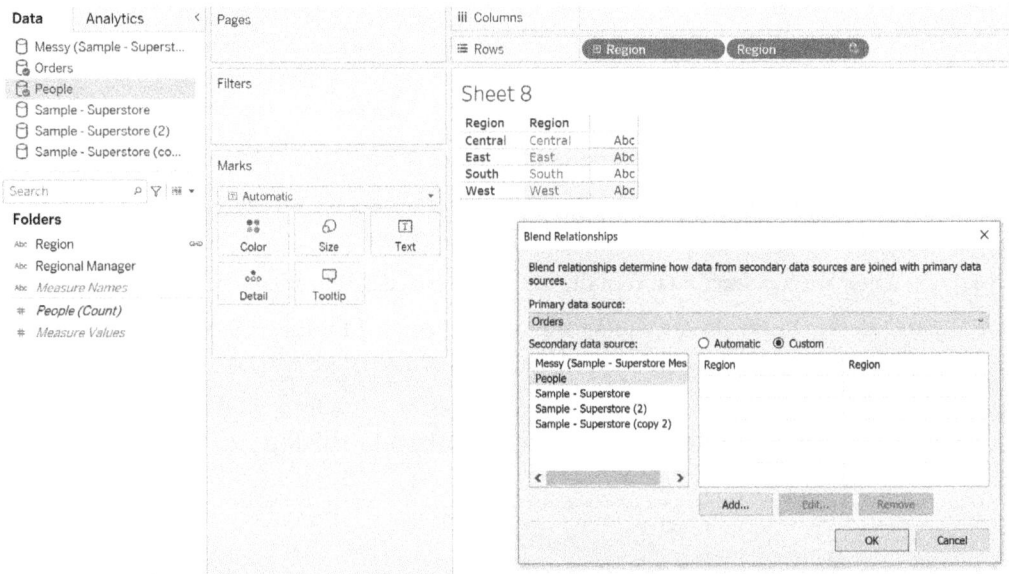

Figure 2.46: Blend relationship set up on Region, primary and secondary data sources are made clear by blue and orange coloring, and the Region linking field can be seen colored red, indicating that it is active

You now know how to create blend relationships, can tell the difference between a primary and secondary data source when blending, and know how to select and deselect linking fields.

Primary versus Secondary Data Source Choice

In this exercise, you will learn the difference between primary and secondary data sources in blending. This will help you understand the behavior of linking fields and data from secondary data sources when utilizing blending in Tableau Desktop.

1. Using the same workbook as in the *Creating a Blend Relationship in Tableau* exercise, create a new sheet.

2. Right-click the **Orders** data source, select **Edit data source filters**, and add a data source filter on the **Region** field that excludes the central region (select the **Region** field and then tick **Central** and the **Exclude** box).

3. Ensure **Orders** is the selected data source and drag the **Region** field onto the **Rows** shelf to create a list of regions. Only **East**, **South**, and **West** are shown as the **Central** region has been filtered out.

4. Now select the **People** data source and drag the **Region** field onto the **Rows** shelf. The **Central** region from the **People** data source does not show in the view as the **Orders** data source is primary while the **People** data source is secondary.

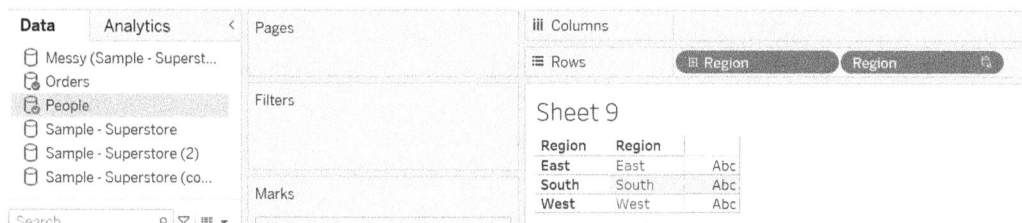

Figure 2.47: As Orders is the primary data source and Central has
been filtered for region, it does not show in the view

5. Create a new sheet and ensure the **People** data source is selected.

6. Drag **Region** onto the **Rows** shelf to create a list of all four possible regions.

7. Now select the **Orders** data source and drag the **Region** field onto the **Rows** shelf. The **East**, **South**, and **West** values are shown for the **Orders Region** field, but as **Central** has been filtered out, a `null` value is inserted. In this case, because **People** is the primary data source and **Orders** is secondary, the list of regions is taken from **People** and the **Orders** data is just the records that have a match.

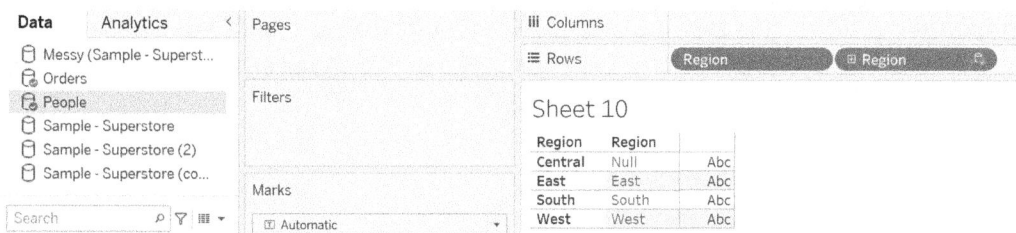

Figure 2.48: The People data source is the primary data source, therefore all regions are
shown in the view. Where the Orders data has no match, a Null record is shown

In this exercise, you have demonstrated the difference between primary and secondary data sources when blending in Tableau. It is important to use the correct primary data source to ensure data is not lost during blending due to the primary being less complete than the secondary.

Tableau Prep Joins

Tableau Prep is preferable to Tableau Desktop when creating combined data sources as it is generally better for creating data sources. If any additional data cleaning or transformation needs to occur before or after a join, then Tableau Prep is better. Tableau Prep also offers more flexibility in the join methodology/output.

In general, Tableau Prep works the same way as Tableau Desktop when it comes to a join where two tables/data sources are being combined – one from the left and one from the right. A join clause needs to be created, and the functionality is the same again, with the ability to select fields from either side and have them be compared to see whether they are equal, not equal, greater than, less than, greater than or equal to, or less than or equal to each other. Join clauses do not include Tableau calculations in Tableau Prep but a **Calculated** field can be made in a clean step prior to the join step.

When it comes to the output of the join, there are additional options in Tableau Prep. The same Venn diagram system is used but it is also possible to select just the left circle, excluding the overlapping section to create what is called a left-only join. The left-only join results in data that is from the left table and has not met the join clause with the right table only. The same can be done for the right-hand side, keeping only unmatched right table records. Left and right joins that include the inner section are still possible. Full joins where all data is selected are also still possible, and so are inner joins, in which only records that meet the join clause are retained. The final join type available in Tableau Prep is the not inner join type, which retains all data that does not meet the join clause.

To create a join in Tableau Prep, first add the join step in the first data source stream and then drag the second data source stream onto the join step in the **Add** popup. A join step can also be created directly by dragging a data step onto another data step and then dropping it on the **Join** popup. The join can then be configured in the **Settings** section of the join step by selecting the fields for the join clause from each side. The plus icon in the top right-hand corner can be used to add additional join clauses/conditions. The **Join Type** section contains a Venn diagram and the shaded sections describe how the data will be joined (inner join by default). Sections of the Venn diagram can be selected or deselected by clicking in order to create different join types.

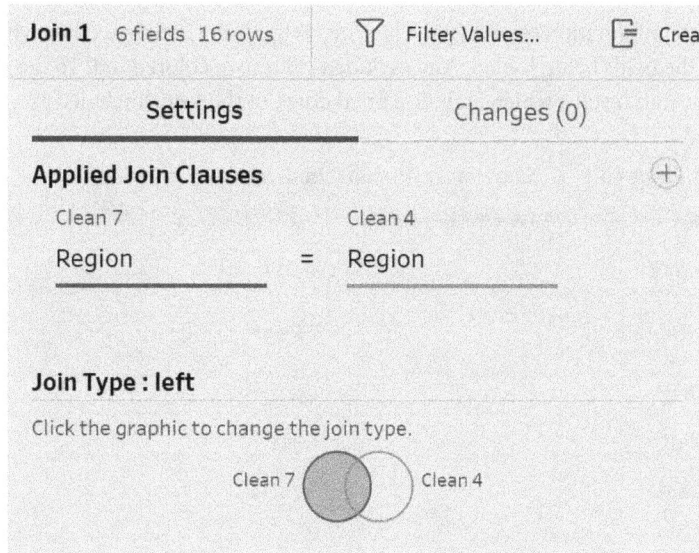

Figure 2.49: Join configuration for a left join where the Region field matches

A summary of the join results also includes a breakdown of the total number of records included and excluded from each data source as well as the total records joined.

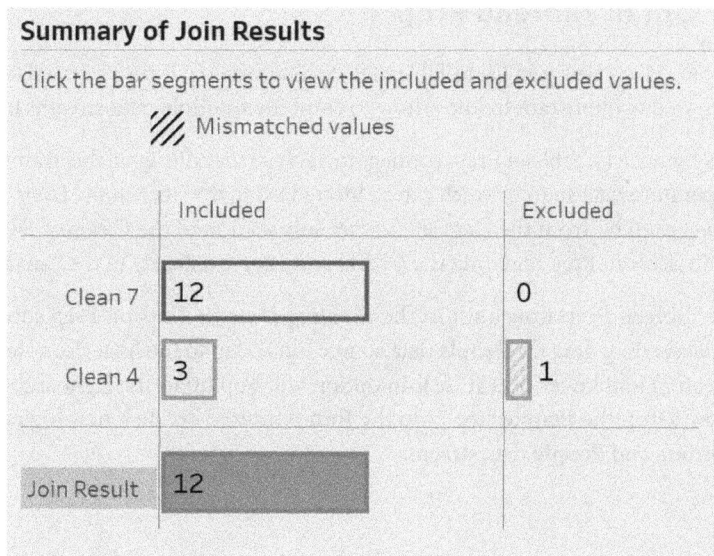

Figure 2.50: A join summary showing that 12 records were kept from the Clean 7 step and three records were kept from the Clean 4 step. One record was excluded from the Clean 4 step (left join). In total, the join therefore results in 12 records

To the right of the configuration section in the join step is the **Join Clauses** section, which includes a list of the values for the fields being joined. Any excluded values are colored red. To the right of the list of join values are the join results, which include a breakdown of the fields included as a result of the join.

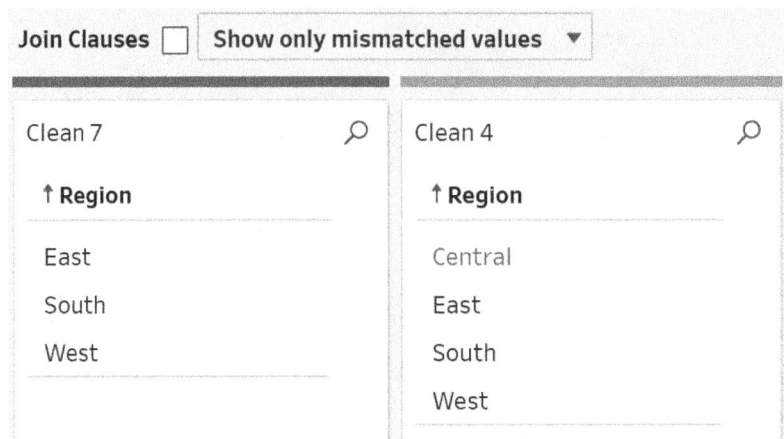

Figure 2.51: Join on the Region field where East, South, and West match across data sources but the Central region is excluded from the right data source (Clean 4 step)

Creating a Join in Tableau Prep

In this exercise, you will create a join using Tableau Prep. Tableau Prep is the preferred tool for advanced data preparation, so it is important to know how to combine multiple data streams using the tool.

To complete this exercise in Tableau Prep, connect to Tableau's readily available training data source, the **Sample - Superstore** data source, which can be located using the steps in the *Using Data Interpreter in Tableau Desktop* exercise from the *Data Interpreter* sub-section of the *Cleaning Data* section. Add this connection to Tableau Prep and drag the *Orders* and *People* datasets to the canvas.

Continuing the Tableau Prep flow built in the *Pivoting Data in Tableau Prep* sub-section of the *Transforming Data* section, drag the **People** data source input step to the final clean step in the **Orders** data stream (Clean 7) and hover over it. A **Join** option will pop up to the right and a **Union** option will pop up below. Drop the **People** step onto the **Join** popup to create a new logical join step that combines the **Orders** and **People** data streams.

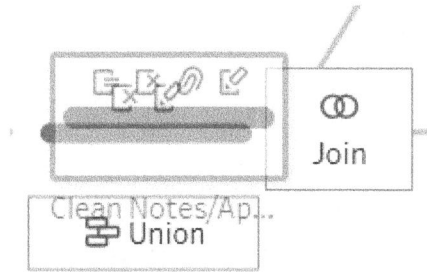

Figure 2.52: When dragging one step over another, there is an option to join
or union the data streams by dropping the step on either popup

1. Set the join clause to **Region** is equal to **Region** and make the join type a left join by selecting the left-hand circle of the Venn diagram.

2. The results of the join show that 16 records were kept from the **Orders** data stream and 4 records were taken from the **People** data stream. The **Join Clauses** section shows that all four **Region** values were included in both data sources. The **Join Results** section shows all of the fields and values included as a result of the join, and these are color-coded so the source can be identified.

Figure 2.53: Join step added to Tableau Prep to combine the Orders
and People tables on the Region field using a left join

In this exercise, you have created a join in Tableau Prep between two data sources on the common **Region** field. You now know how to join data streams in Tableau Prep and have observed the additional analytical information provided post join in Tableau Prep when compared to Tableau Desktop.

Tableau Prep Unions

Unions in Tableau Prep work the same way as in Tableau Desktop – by placing one dataset on top of the other based on field name, and where field names do not match, records are filled using null. Columns also need to be the same data type to be unioned on top of each other.

To create a union, drag one step in the flow onto another so that the **Join** and **Union** popups appear. Then, drop the step on **Union**. This will create the **Union** logical step, which has a setting pane on the left-hand side that includes an inputs legend at the top. The legend lists all of the data sources included in the union (there can be up to 10, with new ones being added by dragging and dropping onto the union step), and each distinct step is color-coded. There is then a description of the total number of fields resulting from the union and the amount of them that are mismatched, meaning not included in all data sources in the union. Mismatched fields are then listed below with a square to the right indicating both positionally and by color which data source(s) they are included in.

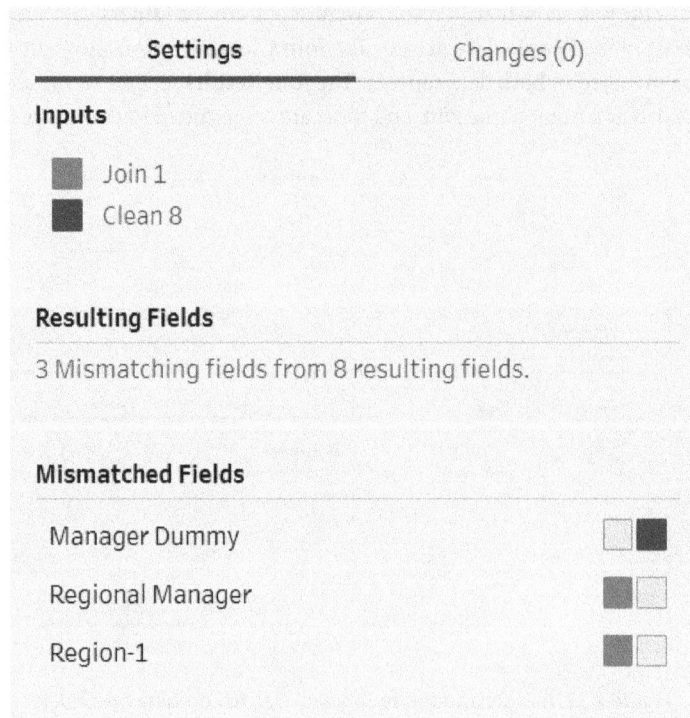

Settings	Changes (0)

Inputs

■ Join 1
■ Clean 8

Resulting Fields

3 Mismatching fields from 8 resulting fields.

Mismatched Fields

Manager Dummy   □■

Regional Manager   ■□

Region-1   ■□

Figure 2.54: Union setting configuration showing a union of two data sources resulting in 8 fields with 3 mismatched. The Manager Dummy field comes from the Clean 8 step, whereas the Regional Manager and Region-1 fields come from the Join 1 step

It may be the case that fields that should have actually been unioned were not as a result of differing column names. For example, **Sales Manager** and **Sales Mgr** may represent the same field but as the names do not match exactly, they are not automatically unioned. This can be rectified by selecting the field with the correct name in the **Mismatched Fields** section and then hovering over the field that should match it and clicking the plus icon that appears. They can also be selected and right-clicked, after which you can then select **Merge Fields**.

To the right of the **Settings** pane in the **Union** step is **Union Results**. **Union Results** contains the full list of fields that result from the union, and these are color-coded in accordance with the inputs legend. Mismatched fields will have blank coloring indicating which data sources do not include that field. A **Table Names** field is also added, which generates a data source name from each input step and maps it to the relevant records.

Creating a Union in Tableau Prep

In this exercise, you will create a union in Tableau Prep. The functionality is similar to Tableau Desktop but as Tableau Prep is the preferred tool for data preparation, it is important to know how to create a union in Tableau Prep in addition to Tableau Desktop.

To complete this exercise in Tableau Prep, connect to Tableau's readily available training data source, the **Sample - Superstore** data source, which can be located using the steps in the *Using Data Interpreter in Tableau Desktop* exercise in the *Data Interpreter* sub-section of the *Cleaning Data* section. Add this connection to Tableau Prep and drag the **Orders** and **People** datasets to the canvas.

1. Continuing the Tableau Prep flow from the previous exercise in the *Tableau Prep Joins* section, add a clean step to the **Orders** data stream after step **Clean 7** and before, but separate to, the previously created **Join 1** step. Do this by hovering over **Clean 7** and selecting the plus symbol followed by **Clean Step**.

2. In the new clean step (**Clean 8**), click **Create Calculated Field** and set the field name to **Manager Dummy**. In the main configuration canvas, write **Dummy**. This creates a dummy regional manager column that has a mismatched name.

3. To create the union, drag the **Clean 8** step just created to the **Join 1** step created in the previous exercise and hover over it. A **Join** option will pop up to the right and a **Union** option will pop up below. Drop the step onto the **Union** popup to create a new logical union step that combines the orders with a dummy regional manager field and the previously joined **Orders** and **People** data streams.

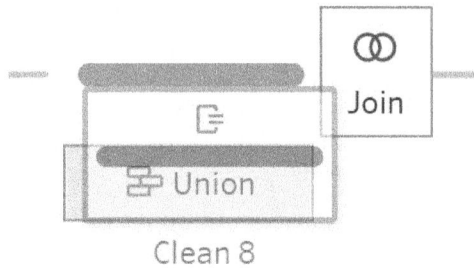

Figure 2.55: When dragging one step over the other, there is an option to join
or union the data streams by dropping the step on either popup

4. The **Inputs** section contains the two unioned steps, **Join 1** and **Clean 8**, and color codes them. There are a total of eight fields as a result of the union, and three of them are mismatched.

5. Looking at the mismatched fields below, and also in **Union Results**, it would make sense to merge the **Manager Dummy** and **Regional Manager** fields. Select the **Regional Manager** field first, as this is the correct name for the field, then hover over the **Manager Dummy** field and click the plus icon to merge them together.

6. The **Region-1** field is a mismatched field that was a duplication of the **Region** field during the join step (both sources had a **Region** field, so the **People Region** field was renamed **Region-1**). We can remove the **Region-1** field by right-clicking it in **Union Results** and selecting **Remove**. The union is now complete with no mismatched fields.

7. We can confirm the union was successful by comparing value counts across steps. Each region has a count of four records in the **Join 1** step but eight records in the **Union 1** step. This is because the data has essentially been put on top of each other.

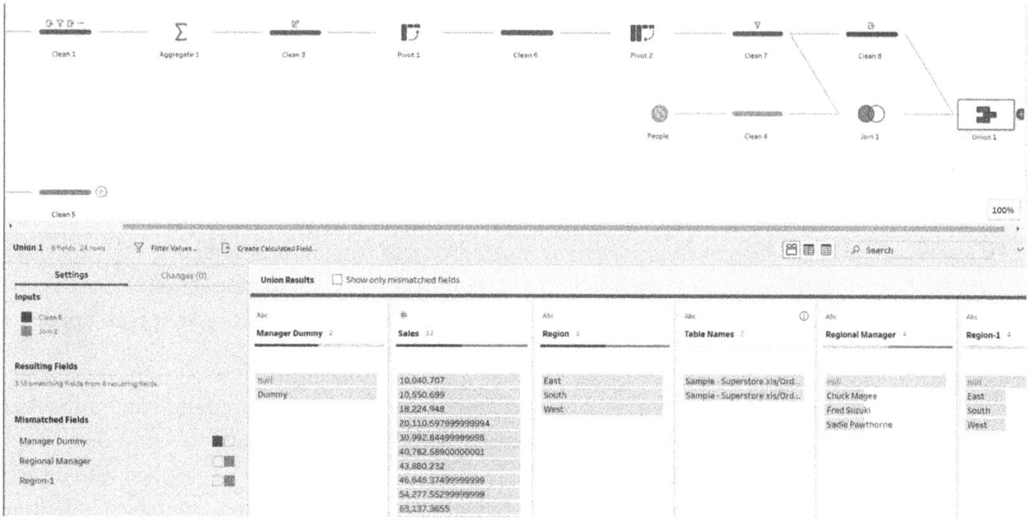

Figure 2.56: Two data streams have been unioned and mismatched fields have been merged or removed

In this exercise, you have created a blend in Tableau Prep. You have observed the advanced functionality that Tableau offers, such as merging fields.

Summary

It is important to consider data quality before conducting analysis in Tableau. If data has not been assessed for consistency, accuracy, and completeness, then the insights derived from it cannot be trusted.

There are multiple data types for fields in Tableau, including string, numeric, Boolean, geographic, date, and date and time. Fields behave differently depending on data type but also depending on color. Blue fields are discrete, which means that they essentially consist of a finite number of grouping values. Green fields are continuous, which means there is theoretically an infinite number of values that are possible. Fields can also be either dimensions or measures. Dimensions are used to break up views, whereas measures are used to aggregate metrics.

Tableau Desktop offers a wide range of functionality for cleaning, transforming, and combining data to ensure that it is ready for analysis. Fields can be cleaned by renaming, filtering out unneeded values, setting default properties, and utilizing aliases where needed. Fields can also be grouped into folders for organizational purposes. In addition to these manual data cleansing operations, Tableau Desktop provides **Data Interpreter** to automatically clean Excel, CSV, PDF, and Google Sheets files.

Data sources can also be transformed in Tableau Desktop by splitting fields on specific delimiters to create new fields, and by pivoting the data to make it taller and thinner as opposed to wide.

Multiple sources of data can also be combined in Tableau Desktop via joins and unions to create new, larger data sources, or by blends or relationships to create dynamic data sources that allow tables to retain their original level of detail.

If more advanced data cleaning, transformation, or combination is required, then Tableau Prep is preferable. Tableau Prep includes multiple data cleaning functionalities not available in Tableau Desktop, such as grouping values within a field, updating all values in a field, changing the case of characters, and removing characters for all values within a field. Tableau Prep offers the ability to aggregate data to a higher level of detail and enables more advanced pivoting that can go both ways. Tableau Prep also has enhanced functionality when combining data, such as joins where the join clause is not met and merging mismatched fields when unioning.

Exam Readiness Drill – Chapter Review Questions

Apart from a solid understanding of key concepts, being able to think quickly under time pressure is a skill that will help you ace your certification exam. That is why working on these skills early on in your learning journey is key.

Chapter review questions are designed to improve your test-taking skills progressively with each chapter you learn and review your understanding of key concepts in the chapter at the same time. You'll find these at the end of each chapter.

> **How to Access these Resources**
>
> To learn how to access these resources, head over to the chapter titled *Chapter 9, Accessing the Online Practice Resources*.

To open the Chapter Review Questions for this chapter, perform the following steps:

1. Click the link – `https://packt.link/TDA_CH02`.

    Alternatively, you can scan the following **QR code** (*Figure 2.57*):

Figure 2.57: QR code that opens Chapter Review Questions for logged-in users

2. Once you log in, you'll see a page similar to the one shown in *Figure 2.58*:

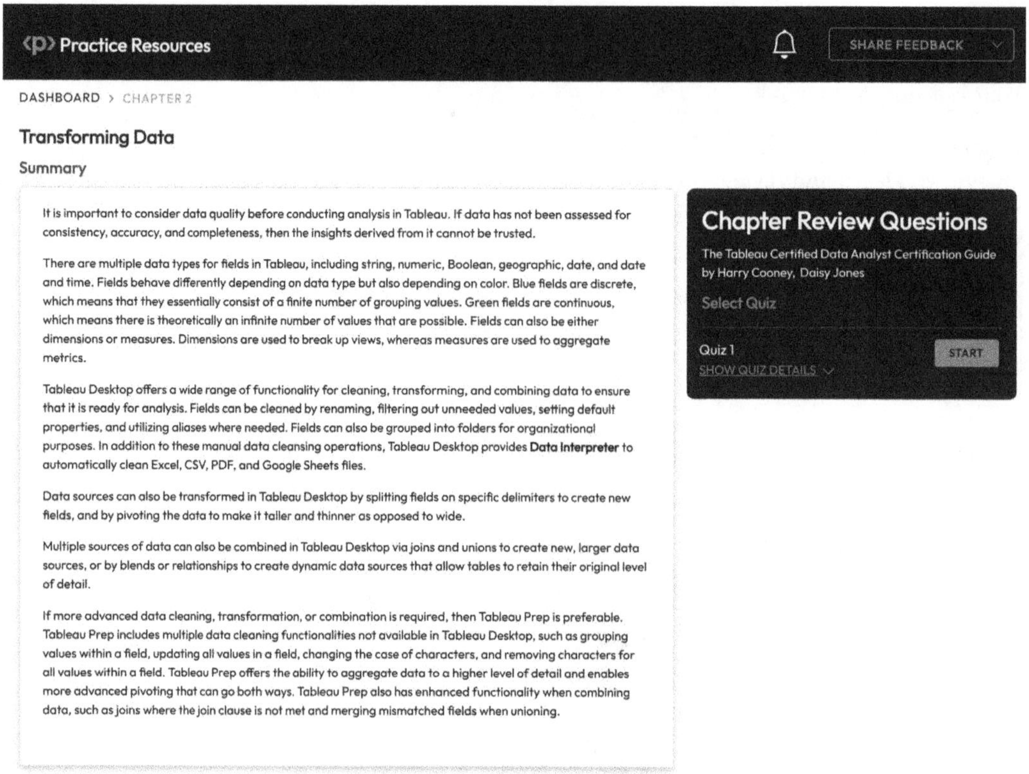

Figure 2.58: Chapter Review Questions for Chapter 2

3. Once ready, start the following practice drills, re-attempting the quiz multiple times.

Exam Readiness Drill

For the first three attempts, don't worry about the time limit.

ATTEMPT 1

The first time, aim for at least **40%**. Look at the answers you got wrong and read the relevant sections in the chapter again to fix your learning gaps.

ATTEMPT 2

The second time, aim for at least **60%**. Look at the answers you got wrong and read the relevant sections in the chapter again to fix any remaining learning gaps.

ATTEMPT 3

The third time, aim for at least **75%**. Once you score 75% or more, you start working on your timing.

> **Tip**
>
> You may take more than **three** attempts to reach 75%. That's okay. Just review the relevant sections in the chapter till you get there.

Working On Timing

Target: Your aim is to keep the score the same while trying to answer these questions as quickly as possible. Here's an example of how your next attempts should look like:

Attempt	Score	Time Taken
Attempt 5	77%	21 mins 30 seconds
Attempt 6	78%	18 mins 34 seconds
Attempt 7	76%	14 mins 44 seconds

Table 2.2: Sample timing practice drills on the online platform

> **Note**
>
> The time limits shown in the above table are just examples. Set your own time limits with each attempt based on the time limit of the quiz on the website.

With each new attempt, your score should stay above **75%** while your "time taken" to complete should "decrease". Repeat as many attempts as you want till you feel confident dealing with the time pressure.

3
Calculations

Introduction

This chapter will take you through the different types of calculated fields that can be created in **Tableau Desktop**. This includes aggregations, fixed levels of detail calculations, numeric calculations, string calculations, date logic, Boolean logic, and table calculations. We can use calculations to analyze, filter, and even format our data. For example, we can create a calculated field that allows the user to highlight specific points in the data where color can be added to emphasize those points. Calculated fields are part of *Domain 2: Explore and Analyse Data*, which contributes to 46% of the exam, so it is imperative for the candidate to understand this section to the fullest extent.

Defining Calculations

Before delving into the wide expanse of calculations and how to build certain types of them, it is important to understand the definitions of the types of calculations and how they can be used across all Tableau platforms.

Data sources in Tableau can be supplemented within Tableau Desktop by creating new fields using Tableau's calculated fields. **Calculated fields** involve writing out logic in Tableau's own calculated field language, resulting in a new column/field that functions in the same way as any data source field with Tableau Desktop. Calculated fields can be used in views, as filters, within other fields, and so on. An example of a calculated field could be a data source breaking down orders with the cost price and the quantity purchased for each order. A total sales field is not included in the data source but can be created in Tableau using a calculated field that calculates order quantity multiplied by cost price.

Charts can be made dynamic using Tableau's calculated field logic via **table calculations**, meaning that how data is configured within a chart depends upon how the chart has been interacted with. Table calculations can be configured via a menu and a chart can be created that reacts to data in the view in a specific way. Table calculations can also be stored as calculated fields that are then configured to be calculated in a specific direction, across or down the table.

Calculated fields offer a variety of methods for aggregating/summarizing numeric data, starting with the basic measure aggregations, such as sum and average. This is useful as it means logic is not restricted to a row-by-row calculation but can instead be calculated based on the level of detail in a view. Using our earlier example of an orders data source, if the profit ratio needs to be calculated by dividing profit by sales, doing so row by row will mean the metric will only ever be useful at the order level. Aggregation within calculations allows the user to get the sum of profit divided by the sum of sales across the whole data source, meaning an accurate calculation at any level of detail.

Calculated fields also offer advanced analytical options such as the standard deviation of a field or the correlation between two fields. Fixed level of detail calculations allows users to aggregate a specific measure at a fixed dimension level within a dataset. These functionalities enable users to produce more advanced visuals in terms of analytics and fixing values at specific levels of detail.

When it comes to creating new calculated fields in Tableau, there are also many numeric options available. These can include general mathematical functions such as finding the square root of a number. This can also be used to format numeric fields such as rounding to a specific number of decimal places or setting null values to zero. Circle-related logic can also be implemented in Tableau using the `pi` function, as well as `sin`, `cos`, `tan`, and others.

Tableau's calculated fields also allow for logic beyond just numeric fields. There are many functions available for the formatting of text/string fields, such as parsing out pieces of text positionally, searching for specific strings, updating the case of lettering, and using regex for advanced textual functions.

Date fields can also be created using calculated fields, and similarly, date logic can be applied to existing date fields. A common example of date logic is taking date parts from a date field, be it returning the year, the month, or the day from that date field. Dates can be created in Tableau from strings, numbers, and the current date and time. Date logic can also be applied to round dates to a specific period, as well as to add or take away a specific unit of time from date fields.

There are a number of calculated field methods that allow users to convert values from one data type to another. Date and spatial fields can also be created from scratch.

Boolean logic is available in Tableau for the creation of boolean fields based on whether nulls are present, whether a value in a list is present, and more. `IF` and `CASE` statements also allow for more advanced conditional logic with multiple conditions and multiple possible outputs based on those conditions.

Calculated Fields

Calculated fields in Tableau can be created by selecting the caret at the top right of the **Data pane** followed by **Create Calculated Field**. This will open up the calculated field configuration window in which the logic can be entered. There is also a help section that documents every calculation/function available, as well as an example of how to implement it.

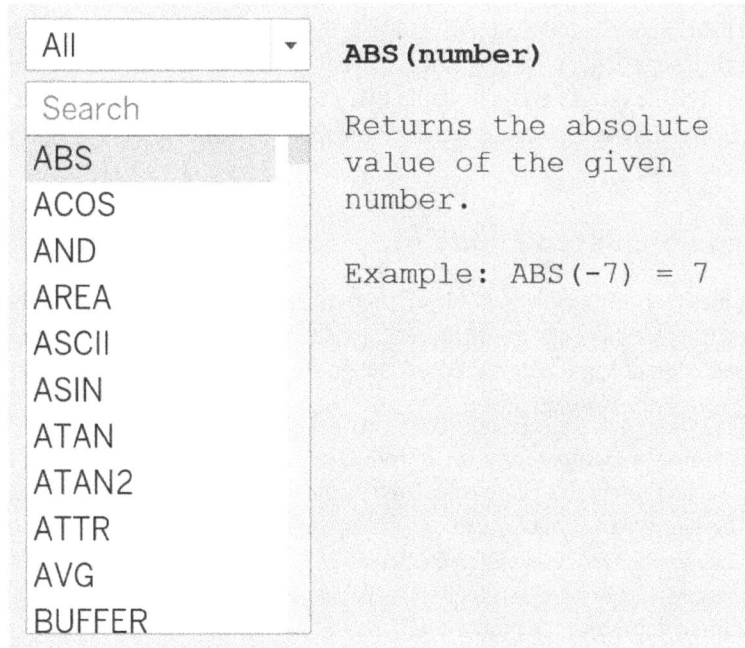

Figure 3.1: Calculated field help section – use the dropdown to select the calculation type category

Calculated fields allow users to combine existing fields and apply transformational functions to the values referenced. Values can be hardcoded into the calculation and the outcome will depend on the data type used. For example, typing 2 + 2 into a calculated field will cause the calculated field to return 4. However, if the 2s are strings, **2 + 2**, then the output of the calculated field would be 22. It is not possible to combine different data types; for example, 2 + 2. It is therefore important to consider the data types being used in the calculation as well as what the desired output should be.

In addition to hardcoded values, fields from the data source can be referenced in calculated fields. For example, **[Example field] + 2** is a calculated field that would add two to every value row by row from the **Example field** column.

It is important also to consider aggregation in Tableau's calculated fields. In the previous example, the **Example field** column is not aggregated, therefore the calculated field adds 2 to each value row by row. If the calculated field is formatted as **SUM([Example field]) + 2**, then the output would be different. Instead of adding 2 to each value row by row, the values are first summed to a total, then 2 is added to that.

If the **Example field** column had the values 1, 2, and 3, then the first example would create a calculated field returning the values 3, 4, and 5. If this was placed on a view, the total value returned would be 12, as 3 + 4 + 5 = 12. The second example would first sum up 1, 2, and 3 and then add 2 to that total, returning 8. It is therefore important to consider whether fields should be aggregated or not when creating calculated fields.

Aggregated Calculated Fields

When placing measures on a view in Tableau, they must be aggregated. This can be done within calculated fields themselves. Measures that are aggregated in calculated fields will not be given the option of the selection of aggregation methods when placed on the view because they have been pre-aggregated within the calculation.

In the orders data source example from the introduction, aggregations in calculated fields could be used for things such as total sales being calculated as the result of the aggregated sum of cost prices multiplied by the aggregated sum of quantity sold. Aggregations could also be used to find the first order date per customer. Fixed level of detail calculations could be used to get a fixed first order date per customer that can then be used as a reference against every order date for that customer, allowing for the calculation of days since the first order.

All standard measure aggregation options and advanced aggregation methods such as finding the median or variance of values in a field are available. Fixed level of detail calculations are advanced Tableau calculated field types that allow users to fix a field at a specific level of aggregation.

If any field in a calculated field is aggregated, then all other fields included in the calculated field must also be aggregated.

Measure Aggregation

Standard aggregation methods when placing a measure on a view in Tableau are also available within calculated fields themselves. This section will cover the aggregation methods available in Tableau, including both their functionality and how to write them. Any mathematical calculation that multiplies or divides fields in Tableau is done using aggregation to avoid creating values fixed at the lowest level of detail. Aggregations are also incredibly versatile, allowing users to get the first and last dates in a given dataset, for example.

The SUM function takes a single numeric input and returns the summed total of values in the input (null values are ignored). It is formatted as SUM(num) and an example would be a field with the values 1, null, and 3, formatted as SUM([Field]), returning 4.

The AVG function also takes a numeric input and returns the average of the values, ignoring null values. It is formatted as AVG(num) and the same example with the values 1, null, and 3 would be formatted as AVG([Field]), returning 2.

The MIN and MAX functions can both be given either one or two inputs, which can be of any data type. If given two inputs, MIN will return the minimum of the two and MAX will return the maximum. If a single input is given, then MIN will return the minimum of the values within the input and MAX will return the maximum. The minimum refers to either the smallest number, the first date, or the letter with the earliest location in the alphabet depending on the data type input. Both functions are formatted the same, MIN(value1, [value2]) or MAX(value1, [value2]), with the second input value being optional. An example of two inputs is MIN("Angry", "Zesty"), returning Angry, while MAX("Angry", "Zesty") would return Zesty. An example with a single value using the example field with the values 1, null, and 3 is MIN([Field]), returning 1, and MAX([Field]), returning 3.

The COUNT function takes a single input of any data type and counts the number of items input (excluding null values). It is formatted as COUNT(value) and the example field with the values 1, null, and 3 would be formatted as COUNT([Field]) and would return 2.

The COUNTD function is the same as the COUNT function but counts unique values only. It is formatted as COUNTD(value). An example field containing the values A, B, A, A, and C would be formatted as COUNTD([Field]) and would return 3 as there are three unique/distinct values in the field. The COUNT function would return 5 as there are 5 total values in the field.

The ATTR function returns the same value that is used as the input if that is the only value within (for all rows). Null values are ignored. If there is more than one value within the input, then an asterisk is returned. It is formatted as ATTR(value). If an example field with the values 1, null, and 3 was used as the input, then * would be returned, as there are two values in the input – 1 and 3. If another field, **Field 2**, had values 1, 1, and 1 as the input, then the output would be 1. This function is useful when a field only ever has one value.

The COLLECT function takes a spatial field as the input and combines all the values within. It is formatted as COLLECT(spatial) and ignores null values. A field containing geospatial points could be aggregated using COLLECT.

Using the SUM and MIN Functions

In this exercise, you will create calculated fields using aggregations. Specifically, you will be using the SUM and MIN aggregations. This exercise will provide you with enough experience in the configuration of aggregation calculations to apply and create them yourself with other aggregation functions. It is important for Tableau users to understand aggregations and how they work as how the data is aggregated will impact the end values that display when you are creating a chart. For example, a MIN aggregation will display the lowest value in the field on a chart whereas the sum will display the summed total for the field.

On a new sheet, drag the **Customer Name** field from the **Data pane** onto the **Rows** shelf to create a list of customers.

1. Drag the **Sales** measure onto **Text** on the **Marks card** to list the total sales value for each customer.

2. Drag the **Quantity** value on top of the sales numbers on the table to create a table with a list of customers showing total sales and total quantity.

3. To show the sales per unit, create a calculated field by selecting the caret at the top right of the **Data pane**, followed by **Create Calculated Field.**

4. In the calculated field configuration popup, give the field the name SUM, then type SUM([Sales]) / SUM([Quantity]) into the main configuration pane. Then, press **OK.**

5. Drag the newly created **SUM** field on top of the **Sales** and **Quantity** part of the table on the view to add it to the table. The sales per customer are now visible in the table.

6. To find the first order date for each customer, create a new calculated field and call it MIN.

7. In the configuration pane, type MIN([Order Date], then press **OK** and drag the newly created **MIN** field onto the **Rows** shelf. The first order date is now visible for each customer.

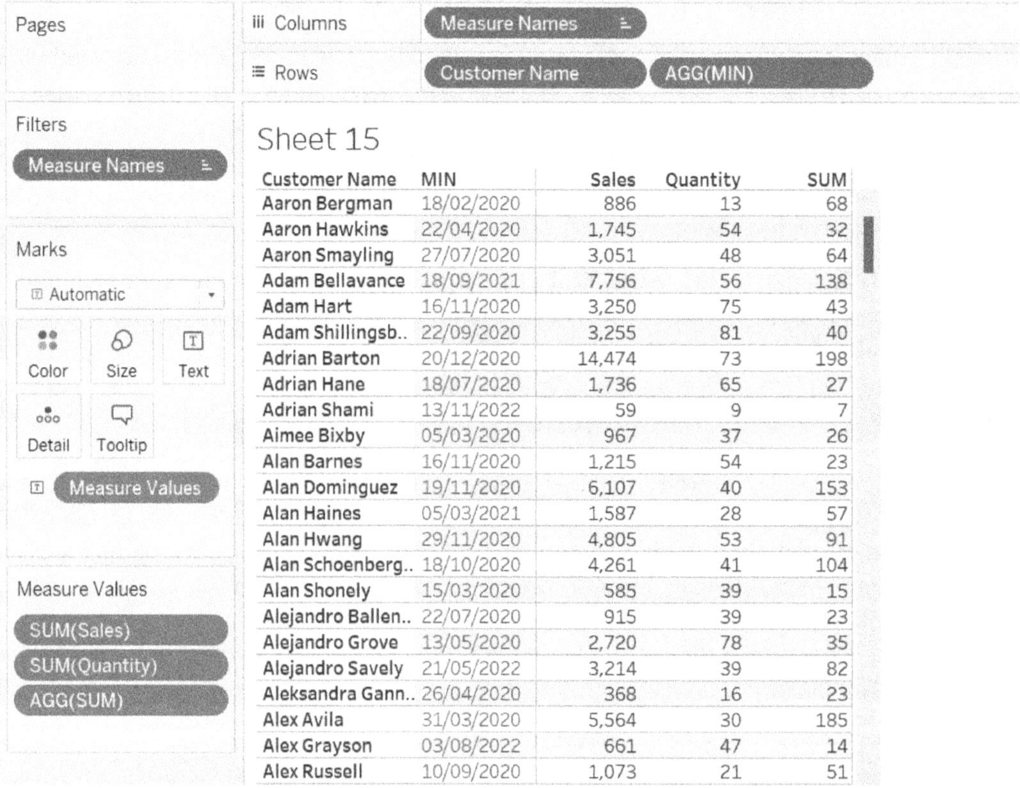

| Pages | | | | | iii Columns | | Measure Names | |
| | | | | | ≡ Rows | | Customer Name | AGG(MIN) |

Filters

Measure Names

Sheet 15

Customer Name	MIN	Sales	Quantity	SUM
Aaron Bergman	18/02/2020	886	13	68
Aaron Hawkins	22/04/2020	1,745	54	32
Aaron Smayling	27/07/2020	3,051	48	64
Adam Bellavance	18/09/2021	7,756	56	138
Adam Hart	16/11/2020	3,250	75	43
Adam Shillingsb..	22/09/2020	3,255	81	40
Adrian Barton	20/12/2020	14,474	73	198
Adrian Hane	18/07/2020	1,736	65	27
Adrian Shami	13/11/2022	59	9	7
Aimee Bixby	05/03/2020	967	37	26
Alan Barnes	16/11/2020	1,215	54	23
Alan Dominguez	19/11/2020	6,107	40	153
Alan Haines	05/03/2021	1,587	28	57
Alan Hwang	29/11/2020	4,805	53	91
Alan Schoenberg..	18/10/2020	4,261	41	104
Alan Shonely	15/03/2020	585	39	15
Alejandro Ballen..	22/07/2020	915	39	23
Alejandro Grove	13/05/2020	2,720	78	35
Alejandro Savely	21/05/2022	3,214	39	82
Aleksandra Gann..	26/04/2020	368	16	23
Alex Avila	31/03/2020	5,564	30	185
Alex Grayson	03/08/2022	661	47	14
Alex Russell	10/09/2020	1,073	21	51

Marks

Automatic

Color Size Text

Detail Tooltip

Measure Values

Measure Values

SUM(Sales)

SUM(Quantity)

AGG(SUM)

Figure 3.2: The SUM function is used to get the total sales, which is then divided by the total quantity, and the MIN function is used to get the first order date for each customer

In this exercise, you used the SUM function in a calculated field to get total sales by total quantity. You also used the MIN function to get the first order date in the data source. As the view has been created at the customer level, the sales by quantity and the first order date are specific to each customer.

Math Functions

In addition to the common field aggregation option available on Tableau, there are also advanced aggregation fields that perform mathematical functions on the values used as inputs during aggregation.

The MEDIAN function takes a numeric field input and returns the median across all the values in the field (ignoring null records). It is formatted as MEDIAN(num) and using the example field with the values 1, null, and 3, the function would be formatted as MEDIAN([Field]) and would return 2.

The CORR function takes two input fields and calculates the **Pearson correlation coefficient** between the values in both fields, resulting in an output between -1 and 1. It is formatted as CORR(field1, field2).

The PERCENTILE function takes a field input and a numeric input that can be any value between 0 and 1 (inclusive). It returns the percentile value from the field relating to the specified numeric input.

The STDEV function takes a single field input and returns the standard deviation for the values within based on a sample of the population. It is formatted as STDEV(field). The STDEVP function returns the standard deviation for all values based on a biased population and is formatted the same, STDEVP(field).

The VAR function takes a single field input and returns the statistical variance for all values within based on a sample of the population. To return the statistical variance based on the entire population, VARP can be used. Both are formatted the same way, VAR(field) and VARP(field).

The COVAR function takes two fields as inputs and returns the sample covariance of the fields. If both field inputs are the same, then the value returned will show value distribution instead (equivalent to using the VAR function). The COVARP function also takes two field inputs but returns the population covariance of the fields as opposed to the sample covariance. If both fields are the same when using the COVARP function, then the value distribution will also be shown (equivalent to the VARP function). The functions are written in the format COVAR(field1, field2) and COVARP(field1, field2).

Using the MEDIAN and STDEV Functions

In this exercise, you will use the MEDIAN and STDEV mathematical functions. Learning how to use these functions will enable you to use all mathematical functions in Tableau Desktop, facilitating correlational and variation analysis:

1. 1. On a new sheet, drag the **Sub-Category** field onto the **Rows** shelf to create a list of **sub-categories**.

2. 2. Drag the **Sales** column on to **Text on the Marks card** to show the total sales for each **sub-category**.

3. 3. Now right-click and hold on **Sales**, then drag and drop it on top of the sales numbers on the table.

4. 4. The aggregation selection popup will appear. Select **AVG** to show the average sales per sub-category as well as the total.

5. 5. To show the median sales as well, create a calculated field called **MEDIAN.** Add the MEDIAN([Sales]) **logic,** click **OK**, then drop the **MEDIAN** field onto the table. The total, average, and median sales are now all visible.

6. To add the standard deviation of sales within each **sub-category,** create a new calculated field called **STDEV**.

7. Add the STDEV([Sales]) logic, click **OK**, and then drag the field onto the view. The standard deviation of the sales for each **sub-category** is now visible as well.

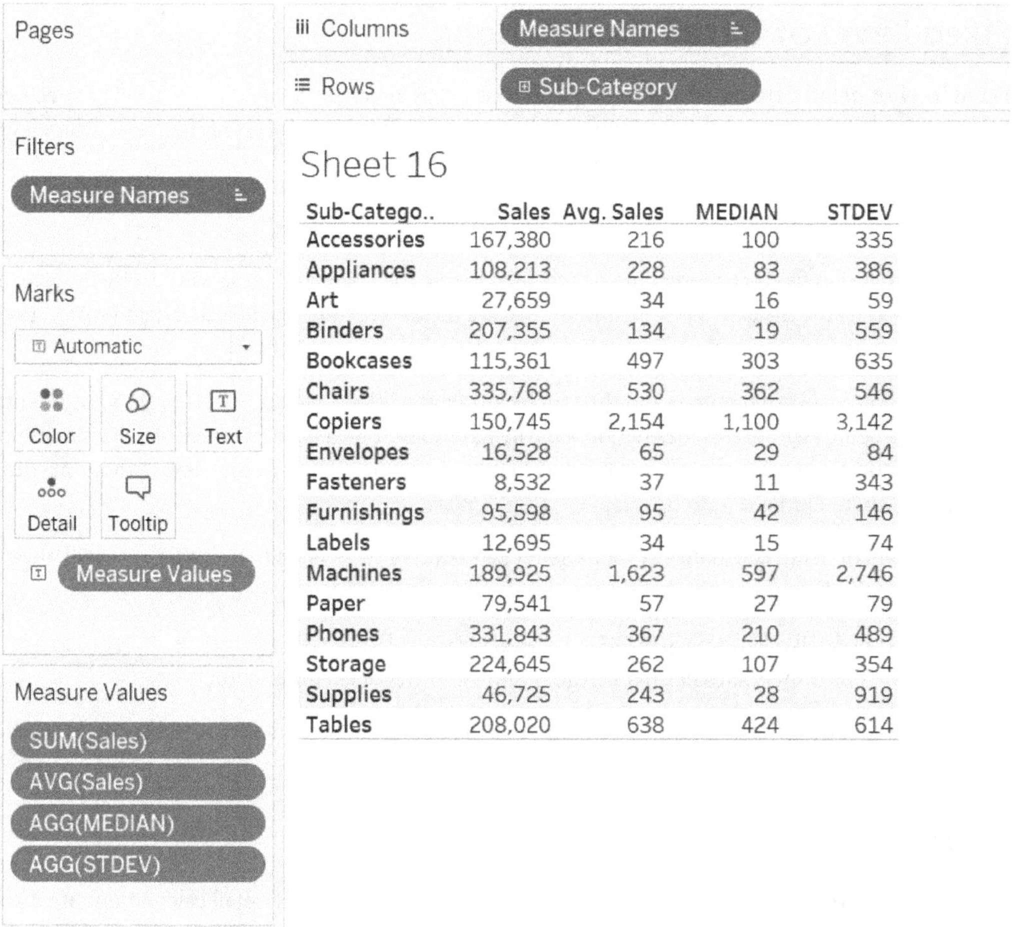

Pages		iii Columns	Measure Names
		≡ Rows	⊞ Sub-Category

Filters

Measure Names

Sheet 16

Sub-Catego..	Sales	Avg. Sales	MEDIAN	STDEV
Accessories	167,380	216	100	335
Appliances	108,213	228	83	386
Art	27,659	34	16	59
Binders	207,355	134	19	559
Bookcases	115,361	497	303	635
Chairs	335,768	530	362	546
Copiers	150,745	2,154	1,100	3,142
Envelopes	16,528	65	29	84
Fasteners	8,532	37	11	343
Furnishings	95,598	95	42	146
Labels	12,695	34	15	74
Machines	189,925	1,623	597	2,746
Paper	79,541	57	27	79
Phones	331,843	367	210	489
Storage	224,645	262	107	354
Supplies	46,725	243	28	919
Tables	208,020	638	424	614

Marks

Automatic

Color Size Text

Detail Tooltip

Measure Values

Measure Values

SUM(Sales)
AVG(Sales)
AGG(MEDIAN)
AGG(STDEV)

Figure 3.3: Sub-Category table created showing the total and average sales as well as the median and standard deviation created using the MEDIAN and STDEV functions

You have successfully utilized two mathematical functions in Tableau and now know how to find the median for a field as well as the standard deviation. You can apply the same techniques with other mathematical functions to calculate variations and correlations.

Fixed Level of Detail Calculations

Fixed level of detail calculated fields are fields that fix an aggregated value to the level of specified dimensions as opposed to the measure being aggregated to the dimensions on the view. Fixed level of detail calculations are written in the following format:

```
{ FIXED [Dimension1], [Dimension_n] : AGG([field]) }
```

As many dimension fields as needed can be included before the colon and these will specify the level of aggregation to fix the measure at. If no dimensions are included in the fixed level of detail calculation, then the aggregation will be fixed across the whole dataset.

An aggregation is always required and this can be any aggregation (SUM, AVG, MIN, MAX, and so on) and the field being aggregated does not necessarily have to be a measure. For example, the maximum date per customer could be found and fixed at that level and then used in the view without needing to bring the customer field onto the view.

The fixed level of detail calculation `{ FIXED [Category] : SUM([Sales]) }` will always return the total sales per category.

The fixed level of detail calculation `{ FIXED [Category], [Sub-Category] : SUM([Sales]) }` will always return the total sales per category and sub-category combination.

The fixed level of detail calculation `{ FIXED : SUM([Sales]) }` will return the total sales for the whole data set.

Creating a Fixed Level of Detail Calculation

In this exercise, you will create a fixed level of detail calculation. Level of detail calculations are a key feature in Tableau that allows you to fix values that can be used as reference points or specific totals.

1. On a new sheet, drag the **Category** dimension onto the **Rows** shelf to create a list of categories, then drag the **Sales** measure onto **Text** on the **Marks card** to show total sales by category.

2. Create a fixed level of detail calculation by creating a new calculated field, calling it **FIXED**, and adding the `{ FIXED [Category] : SUM([Sales]) }` **logic.**

3. Click **OK**, then drag and drop the **FIXED** calculated field on top of the sales values in the view to create a table showing the total sales per category and the fixed calculation.

4. Both the summed sales value and the fixed calculation values are the same, as the fixed calculation returns the total summed sales per category and the view reflects the same thing.

5. Add a **Sub-Category** field to the **Rows** shelf to the right of **Category**. This lower level of detail breakdown of the view illustrates the functionality of fixed calculations. The summed sales values have updated to correspond to each sub-category, while the fixed level of detail calculations values have remained the same. This is because the fixed calculation has fixed the summed sales value at the category level and this will remain consistent regardless of changes to the view.

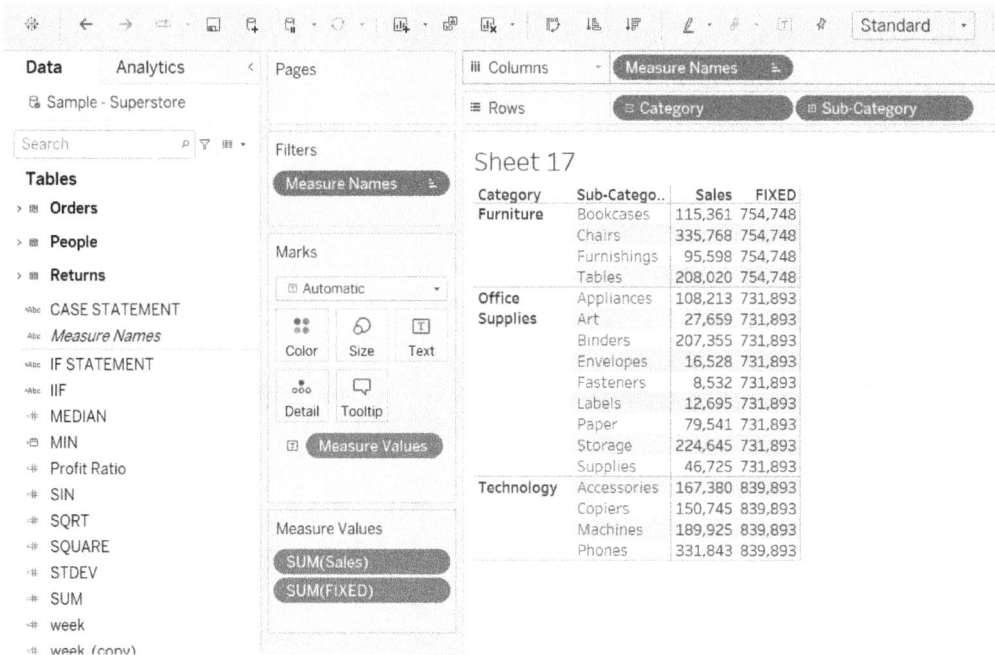

Figure 3.4: The FIXED sales values can be observed to be fixed at the
category total level regardless of the level of detail in the view

You have successfully created a fixed level of detail calculation and have observed the functionality by placing the calculation on the view and changing the level of detail. You now know how to create calculations that fix the total values at whatever level of detail is required.

Numeric Calculated Fields

There is a wide variety of number functions available in Tableau's calculated fields. Users can take any numeric value, be it typed in the calculated field or taken from another field, and apply a mathematical function to it. These range from functions for numerical transformation to number formatting options and logic relating to circles.

Referring back to the orders data source example from the introduction, there are many numeric types of calculation logic that could be applied to supplement the data source. For example, the DIV function could be used to divide total profit by total sales to calculate a profit ratio. Numeric functions can also be used to format the data correctly, for example, by using the ROUND function to specify a correct number of decimal places for metrics in the data source.

Transformational Mathematical Functions

There are a variety of mathematical functions available for use in calculated fields in Tableau. Mathematical functions take a numeric value and apply a transformation to it, resulting in a new numeric value.

The DIV function takes two integer values and divides the first by the second. The DIV function is formatted as DIV(num1, num2). An example of the DIV function is DIV(25, 5), returning 5.

The POWER function takes two numeric values and raises the first to the power of the second. The function takes the format POWER(num1, num2). An example of the POWER function is POWER(10,2), returning 100.

There is also a SQUARE function, which is similar to the POWER function but only takes one value and returns the square of it. SQUARE is formatted as SQUARE(num) and an example is SQUARE(3), which returns the value 9.

The opposite of the SQUARE function is the SQRT function, which returns the square root of the number passed in. It follows the same format as the SQUARE function, SQRT(num). An example of the SQRT function is SQRT(9), returning 3.

The LOG function returns the logarithm of the number within either to a specified base or to the default of base 10 if not specified. The LOG function is formatted as LOG(num1, [num2]) with num2 being optional. An example of the LOG function is LOG(25, 5), which returns 2.

The LN function takes a single numeric input and returns the natural logarithm of that input. It is formatted as LN(num) and an example is LN(7.5), returning 2.10149.

The EXP function takes a single numeric value and returns e raised to the power of that value. The EXP function is formatted as EXP(num) and an example is EXP(2.10149), returning 7.5.

Using the SQUARE and SQRT Functions

In this section, you will utilize both the SQUARE and SQRT transformational mathematical functions. It is useful to learn about these basic math functions to prepare you for any use cases where mathematical transformations are required within Tableau Desktop. The syntax used is similar to that of other mathematical transformation functions and so can be applied generally.

1. On a new sheet, create a new calculated field by selecting the caret at the top right of the **Data pane** followed by **Create Calculated Field**.

2. In the calculated field configuration pane, name the calculated field SQUARE and then type SQUARE(2) into the calculated field pane.

3. Click **OK**, then drag the newly created **SQUARE** calculated field from the **Data pane** and drop it on **Text** on the **Marks card**. The number 4 now shows as text on the view as the calculated field returns the square of 2.

4. Create a new calculated field called SQRT and type SQRT([SQUARE]). This new calculated field returns the square root of the **SQUARE** calculated field, therefore dropping it on **Text** will result in the number 2 showing.

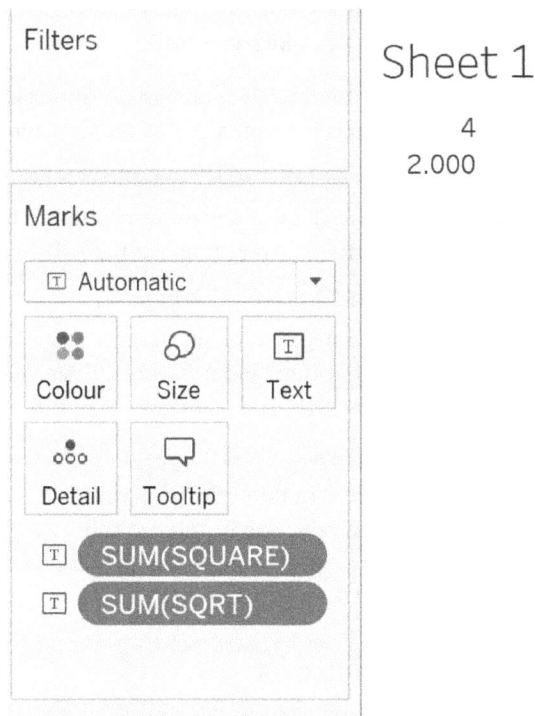

Figure 3.5: SQUARE and SQRT functions used in separate calculated fields
with the SQRT column canceling out the SQUARE function

You have now demonstrated the functionality of the SQUARE function and then canceled out that same functionality with the SQRT function. You will be able to utilize similar syntax to apply other mathematical transformations such as divisions and logarithms.

Field Format Functions

There are also mathematical functions available in Tableau's calculated fields that transform a numeric value to update how the value is formatted.

The ROUND function takes two numeric values. The first value is a decimal number and the second is an integer that represents the number of digits the first value should be rounded to. The ROUND function is formatted as ROUND(num, num) and an example is ROUND(2.123, 2), which would return 2.12.

The FLOOR function takes a decimal and rounds it down to the nearest integer. It is formatted as FLOOR(num) and an example is FLOOR(2.9), which returns 2.

The CEILING function does the opposite of the FLOOR function, rounding the given decimal up to the nearest integer. The CEILING function is formatted as CEILING(num) and an example is CEILING(2.1), which returns 3.

The ABS function returns the absolute/positive value of any number put into the function. It takes any numeric value and is formatted as ABS(num). An example of the ABS function is ABS(-3), which returns 3.

The SIGN function returns the sign, positive or negative, of any numeric value put into the function. The values returned by the function are either -1 if the input number is negative, 0 if the input is 0, or 1 if the value is positive. It is formatted as SIGN(num) and an example is SIGN(-87), which would return -1.

The ZN function replaces null or empty values in numeric fields with 0. If a non-null numeric value is input into the function, then that same value will be returned. The ZN function is formatted as ZN(num). A field called **Dummy** containing the values 1, 2, 3, and null would be formatted ZN([Dummy]) and would return 1, 2, 3, and 0.

Using the ROUND and ABS Functions

In this section, you will use the ROUND function to round numeric data to two decimal places and the ABS function to convert all numeric values to a positive (greater than 0) value. These are useful functions to learn about in and of themselves but also have a similar syntax to other numeric field formatting functions, so the takeaways from this exercise can also be applied more generally.

1. On a new sheet, right-click on the **Profit** field, then drag and drop it onto the **Rows** shelf.

2. The aggregation popup will appear. Choose the top option, **Do not aggregate profit**.

3. By default, a Gantt chart is created, but right-clicking the field and then selecting **Discrete** from the dropdown will make the field discrete, resulting in a list of all profit values from the data set.

4. Scrolling up and down shows that there are more than two decimal places for most values and there are both positive and negative values.

5. Create a calculated field called ROUND and type ROUND([Profit],2) into it, then click **OK**.

6. Right-click and drag the new field onto the **Rows** shelf to the right of **Profit** and choose not to aggregate in the same way.

7. Update the field to be discrete in the same way as for the **Profit** field. Comparing the **Profit** numbers to the updated **Profit** numbers in the **ROUND** field shows the function has successfully rounded the **Profit** values to two decimal places.

8. Create a new calculated field called ABS and type in ABS([Profit]), then click **OK**.

9. Repeat the steps to place the **ABS** field onto the **Rows** shelf with no aggregation and convert it to discrete to show that the ABS function has converted negative **Profit** values to positive ones (already positive values are unchanged).

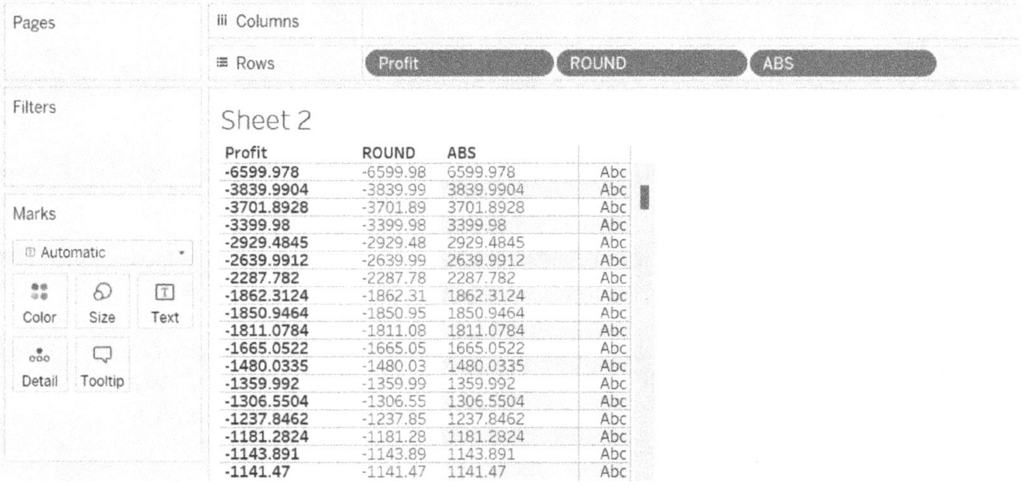

Figure 3.6: ROUND and ABS functions applied to the Profit field and the effects observed

In this exercise, you utilized the ROUND and ABS functions to update the formatting of a value returned from another field. There are many similar functions that you will now be able to use in the same way, such as the ZN function, which returns 0 in the place of any null values.

Trigonometry

There are a variety of trigonometric functions in Tableau that allow users to calculate angles, ratios, and lengths based on the values provided.

The PI function returns the value of the pi constant (3.14159) and is formatted as PI(). An example would be using the PI function to calculate the circumference of a circle with a diameter of 100. The calculation would be formatted as 100 * PI() and would return 314.159.

The RADIANS function takes a single input value, formatted as RADIANS(num), converts it from an angle (degrees), and returns the radians. An example would be RADIANS(90), which returns 1.571.

The DEGREES function does the opposite of RADIANS, taking an angle in radians and converting it to degrees. It takes a single input formatted as DEGREES(num), an example being DEGREES(1.571), which returns 90.

The SIN function returns the sine of an angle value provided in the format SIN(num). An example is SIN(RADIANS(30)), which returns 0.5.

The ASIN function does the opposite, returning the arcsine or angle of the input value. The input value is considered the sine value and the angle in radians is returned. To get the original 30 value from the previous example, the ASIN function could be used in the following calculation: DEGREES(ASIN(0.5)).

The COS function takes a single value, which is considered an angle, formatted as COS(num), and returns the cosine of that angle. An example is COS(RADIANS(30)), which returns 0.866. The opposite ACOS function takes the same number of inputs, ACOS(num), and the input is considered the cosine. The cosine input value is returned as the angle in radians.

The TAN function returns the tangent of an angle, which is input as a single value into the function in the format TAN(num). An example is TAN(RADIANS(30)), which returns 0.577.

The ATAN function is the inverse of the TAN function, returning the arctangent or angle of the input value. It is formatted as ATAN(num), of which an example is DEGREES(ATAN(0.57735)), which returns 30.

The ATAN2 function also returns the arctangent but takes two inputs, calculating the angle between them in radians. It is formatted as ATAN2(num1, num2), of which an example is ATAN2(5, 4), which returns 0.896.

The COT function takes a single input as an angle in radians and returns the cotangent. It is written in the format COT(num) and an example is COT(RADIANS(30)), which returns 1.732.

Using the SIN and RADIANS Functions

In this exercise, you will use the SIN and RADIANS trigonometric functions. Trigonometric functions allow for creative circular visuals for any data in Tableau and so have uses beyond data sources in which trigonometry is directly relevant.

1. On a new sheet, create a new calculated field called SIN, type in SIN(RADIANS(30)), and then click **OK**.

2. The creation of the calculated field and use of the function shows that multiple functions can be used in the same calculated fields, even nested within each other. Right-click the **SIN** calculated field in the **Data** pane and convert it to discrete before dragging it onto the **Rows** shelf.

3. The value **0.5** displays for the field as the `RADIANS` function converts the value 30 to the equivalent value in radians. The `SIN` function then takes the value created from the `RADIANS` function and returns the sine of the angle, which is 0.5.

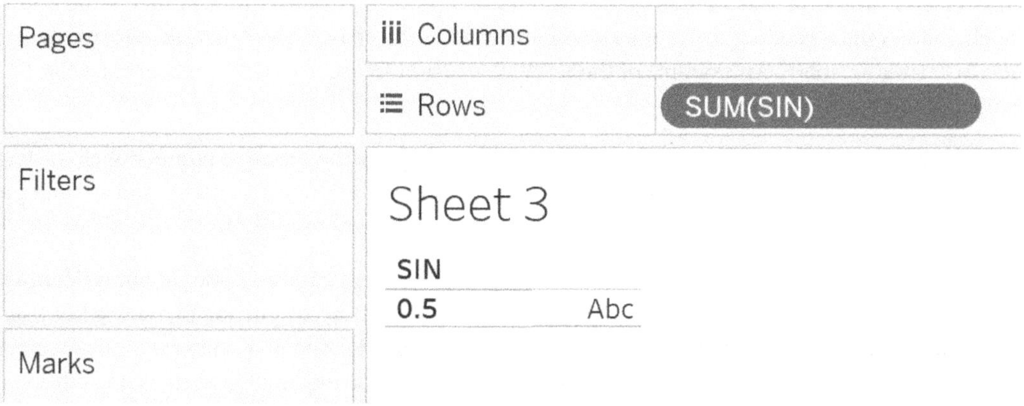

Figure 3.7: Nested functions used to find the sine of a 30-degree
angle using the SIN and RADIANS functions

In this section, you used two trigonometric functions in the same calculated fields, showing that two functions can be used at once and also demonstrating the trigonometric abilities of Tableau Desktop's calculated fields.

String Calculated Fields

When it comes to string-based logic in Tableau Desktop, users have many options available. If sections of a string need to be parsed out, this can be done positionally. Similarly, string fields can be cleaned positionally or by searching for specific sub-strings. There are a variety of functions available for searching through strings using different methods. The case of lettering can also be modified to uppercase, lowercase, or title case using string functions. For advanced string manipulation, there are also regex functions available.

An example use of string functions in calculated fields that would be useful in the orders data source mentioned in the introduction is the `SPLIT` function, which could be used to split customer names in the data source on the space delimiter, resulting in two fields, one for first names and one for last names. There are also string functions that look for specific substrings. For example, `CONTAINS` could be used to check IDs for specific substrings and group them accordingly.

Positional Functions

Many string functions in Tableau use the index position of a string to enact some kind of transformation.

The `RIGHT` function takes a string and a number input. The number input reflects the count of characters to be retained from the right-hand side of the string. The function is formatted as `RIGHT(str, num)`. An example of the `RIGHT` function in use is `RIGHT("Example", 5)`, which would return `ample`.

The `LEFT` function is similar to the `RIGHT` function but retains the specified number of characters from the left-hand side instead. It is written in the same format, `LEFT(str, num)`, and an example is `LEFT("Example", 4)`, returning `Exam`.

The `MID` function takes a string input, a numeric input, and also another optional numeric input. Similar to the `LEFT` and `RIGHT` functions, the `MID` function extracts a portion of a string but from a specified midpoint in the string as opposed to from either side. The first numeric input specifies the point from which to start taking the string extract from the string input. If there is no additional numeric input, then the rest of the string from the specified midpoint will be retained. If an additional numeric input is provided, then it will be used to count the number of characters from the midpoint to retain, inclusive of the midpoint (essentially providing the length of the new string extract). The `MID` function is formatted as `MID(str, num, [num2])`. An example of the `MID` function is `MID("Example", 3)`, which would return `ample`, whereas `MID("Example", 3, 3)` returns `amp`.

The `TRIM` function removes leading and trailing spaces from a string provided as the input. The function is formatted as `TRIM(str)` and an example is `TRIM("    Example    ")`, returning `Example`.

`LTRIM` and `RTRIM` act in the same way as the `TRIM` function but only remove spaces from one direction. `LTRIM` removes leading spaces whereas `RTRIM` removes any trailing spaces. Both are written in the same format as `TRIM`, taking a single input. Using the same example, `LTRIM("    Example    ")` returns `"Example    "` and `RTRIM("    Example    ")` returns `"    Example"`.

Using the RIGHT, LEFT, and MID Functions

In this exercise, you will use the `RIGHT`, `LEFT`, and `MID` string functions. These functions are commonly used to parse specific substrings from string values and are invaluable for basic data transformations within Tableau Desktop.

1. On a new sheet, drag the **Order ID** column onto the **Rows** shelf to create a list of order IDs.

2. To parse out the ID number part of **Order ID**, create a calculated field called **RIGHT** and type in `RIGHT([Order ID], 6)`, then click **OK**.

3. Drag the **RIGHT** field onto the **Rows** shelf to display the six rightmost characters of the **Order ID** field.

4. To extract the country code from the order ID, create a calculated field called LEFT, type in LEFT([Order ID], 2), and then click **OK**.

5. Drag the **LEFT** field onto columns to display the two leftmost characters from the **Order ID** field.

6. To extract the year from **Order ID**, create a new calculated field called **MID**, type in MID([Order ID], 4, 4), and then click **OK**.

7. Place the **MID** column on the **Rows** shelf to show the extracted 4-character string, starting from the fourth character in the **ORDER ID** field.

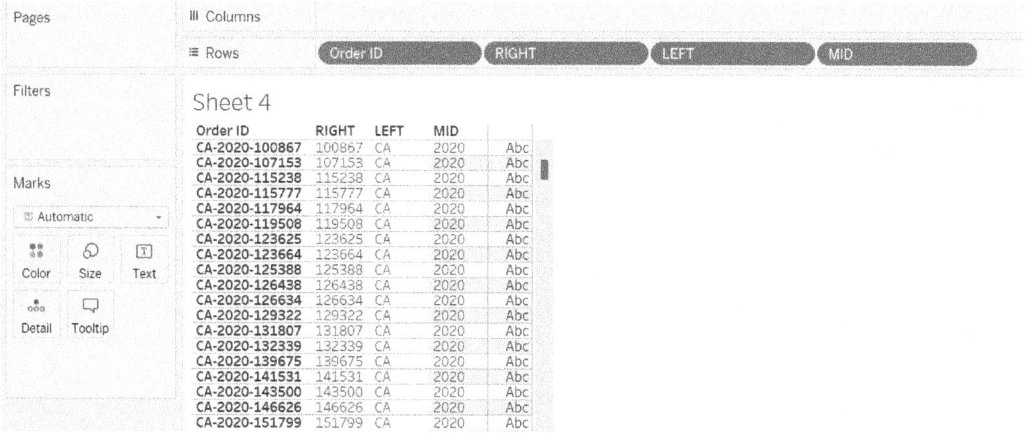

Figure 3.8: Order ID subsections have been parsed out using RIGHT, LEFT, and MID functions

In this exercise, you used the positional-based string functions to extract a specific substring from a larger string value. These functions have many use cases and can be used in conjunction with each other to get evermore specific intersections of strings positionally.

Searching

There are additional string functions in Tableau's calculated fields that allow for tests or transformations to be run, based on a search for a specific substring within a string.

The CONTAINS function allows users to search for a specific substring within a string and returns TRUE or FALSE based on whether the substring is present. The CONTAINS function takes two inputs, the string to search and the substring to look for. It is formatted as CONTAINS(str, substr) with an example being CONTAINS("Example", "amp"), returning TRUE, whereas CONTAINS("Example", "dample") returns FALSE.

The REPLACE function is similar to the CONTAINS function but contains an additional string input that replaces the substring in the original string if found (the match must be exact, including the letter casing). The result of the calculation is the updated original string. It is formatted as REPLACE(str, substr, replacementstr) and an example would be REPLACE("Example", "ple", "ination"), returning Examination.

The FIND function also looks for a substring within the string, but if found, returns the index position the string can be found at. If the substring is not found then 0 is returned, otherwise the index position (the first character equaling one) of the first instance of the substring is returned. The string and substring inputs are required but there is also an optional numeric input that specifies at what point in the string to start the search for the substring. If a midpoint to start searching from is specified, the index of each character will not be impacted in the output. The function is formatted as FIND(str, substr, [num]). An example without the starting index input is FIND("Example Exam", "am"), which would return 3. An example with the starting index is FIND("Example Exam", "am", 4), which would return 11.

The FINDNTH function is similar to the FIND function in that it searches for a substring within a string, but the numeric input is required as opposed to being optional. The numeric input in the FINDNTH function refers to the nth occurrence of the substring, allowing users to look for a second, third, and fourth occurrence, and so on, of a substring within a string. The function returns the index position of this nth occurring substring. It is formatted as FINDNTH(str, substr, num) and an example is FINDNTH("Example Exam", "am", 2), which returns 11, whereas FINDNTH("Example Exam", "am", 1) would return 3.

The STARTSWITH and ENDSWITH functions look for substrings at the start or end of a string respectively and return TRUE if the substring is present and FALSE if not. They are written in the same format, STARTSWITH(str, substr) and ENDSWITH(str, substr). Leading and trailing whitespace is ignored in the check. An example of each is STARTSWITH("Example", "Exam") and ENDSWITH("Example", "ample"), both returning TRUE.

Using the CONTAINS, STARTSWITH, and REPLACE Functions

In this exercise, you will use functions that search for specific substrings within a string as opposed to taking a substring from a specific position within a string. These functions offer a lot of versatility when dealing with strings and any user that wants to work with strings that are highly customizable and dynamic will need to understand how they work.

1. On a new sheet, drag the **Sub-Category** field onto the **Rows** shelf to display a list of sub-categories.

2. To find sub-categories that contain the er substring, create a new calculated field called **CONTAINS**, type CONTAINS([Sub-Category], "er"), and click **OK**.

3. Drag the new **CONTAINS** field onto **Rows** to show **TRUE** values where er is contained in the sub-category name.

4. To run a similar check, looking for sub-categories that begin with the letter A, create a new calculation called STARTSWITH, type in STARTSWITH([Sub-Category], "A"), and then click **OK**.

5. Drag the new **STARTSWITH** column onto the **Rows** shelf to show where the condition is met and where it is not.

6. Finally, to replace a substring within **Sub-Category**, create a calculated field called REPLACE with the text REPLACE([Sub-Category], "in", "REPLACEMENT_TEXT"), and click **OK**.

7. Putting the new **REPLACE** field onto the **Columns** shelf shows the sub-categories but with any instance of the in substring replaced with the text **REPLACEMENT_TEXT**.

Sub-Catego..	CONTAINS	STARTSWITH	REPLACE	
Accessories	False	True	Accessories	Abc
Appliances	False	True	Appliances	Abc
Art	False	True	Art	Abc
Binders	True	False	BREPLACEMENT_TEXTders	Abc
Bookcases	False	False	Bookcases	Abc
Chairs	False	False	Chairs	Abc
Copiers	True	False	Copiers	Abc
Envelopes	False	False	Envelopes	Abc
Fasteners	True	False	Fasteners	Abc
Furnishings	False	False	FurnishREPLACEMENT_TEXTgs	Abc
Labels	False	False	Labels	Abc
Machines	False	False	MachREPLACEMENT_TEXTes	Abc
Paper	True	False	Paper	Abc
Phones	False	False	Phones	Abc
Storage	False	False	Storage	Abc
Supplies	False	False	Supplies	Abc
Tables	False	False	Tables	Abc

Figure 3.9: Substrings found within, at the start of, and replaced, using
the CONTAINS, STARTSWITH, and REPLACE functions

In this exercise, you used the CONTAINS function to return TRUE where a specific substring is found, used the STARTSWITH function to return TRUE where a specific substring is found at the start of a function, and used REPLACE to replace specific substrings with another string. You now have a greater understanding of the syntax and functionality of string search type functions in Tableau Desktop.

Case

The case of strings in Tableau can also be updated using the UPPER, LOWER, and PROPER functions. Each takes a single string as the input and outputs the same string with the case transformed.

The UPPER function takes a string input and outputs it with all characters in uppercase. It's formatted as UPPER(str) and an example is UPPER("Example"), returning EXAMPLE.

The LOWER function is the same as the UPPER function but outputs the string all in lowercase as opposed to uppercase. It is formatted as LOWER(str) and the same example, LOWER("Example"), would return example.

The PROPER function capitalizes the first letter of each word in the string provided. It treats spaces and any non-alphanumeric characters as breaks between words, meaning the next letter will be capitalized. It is formatted as PROPER(str) and is often useful to standardize the capitalization of first and last names. An example is PROPER("jim o'connell"), which would return Jim O'Connell.

Using the UPPER and LOWER Functions

In this section, you will modify the case of strings, converting all letters to either uppercase or lowercase. Case transformation functions are useful for standardizing string data in Tableau Desktop.

1. On a new sheet, place the **Customer Name** field onto the **Rows** shelf to display a list of customers' first and last names.

2. Create a new calculated field called UPPER and type in UPPER([Customer Name]), then click **OK**.

3. Repeat the previous steps but name the calculated field LOWER and replace the UPPER function with LOWER.

4. Place both new fields onto the **Rows** shelf to see the customer names displayed in all uppercase and lowercase.

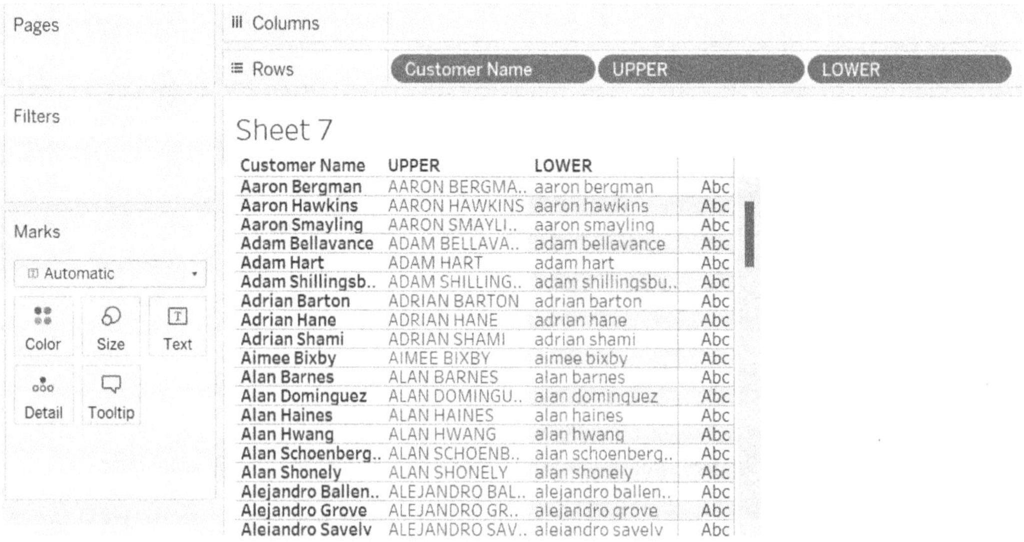

Figure 3.10: Customer names recreated in uppercase and lowercase
using the UPPER and LOWER functions

In this exercise, you modified the case of customer names to both uppercase and lowercase and you now know how to standardize the case of a string field.

Other

Additional string-related Tableau calculated field logic allows for fields to be split and the length of strings to be returned and includes spaces in input strings. It also permits **American Standard Code for Information Interchange (ASCII)** codes, which codify all string characters, to be utilized.

The SPLIT function allows users to parse out a section of a string split by a specified delimiter. The SPLIT function requires a string input, the delimiter input, and a number input to specify which subsection to take the substring from. If the value is 1, then the subsection of the string before the first occurrence of the delimiter will be taken, whereas if it is 2, the subsection after the first occurrence of the delimiter will be taken. The number can also be negative, in which case it will count backward from the last delimiter as opposed to forward from the first. The SPLIT function is formatted as SPLIT(str, delimstr, num) and an example would be SPLIT("Ex-am-ple", "-", 1), returning Ex, while SPLIT("Ex-am-ple", "-", -2) would return am and SPLIT("Ex-am-ple", "-", 3) would return ple.

The LEN function returns the length of a string as an integer. The only input is the string and the function is formatted as LEN(str). An example would be LEN(Example), which returns 7.

The SPACE function takes a number input only and returns a string with that number of spaces. It is formatted as SPACE(num) and an example would be SPACE(4), which returns four space characters.

The CHAR function takes a single numeric input, which refers to a character's ASCII code. The function returns the character encoded by the given ASCII code's numeric value. It is formatted as CHAR(num) and an example of its use is CHAR(72), returning H.

ASCII does the opposite of the CHAR function, taking a string input and returning the ASCII code for the first character in that string. It is formatted as ASCII(str) and an example would be ASCII("H"), returning the number 72.

Using the SPLIT and LEN Functions

In this exercise, you will use the SPLIT function to return sections of a string that has been split by a specified delimiter. You will also use the LEN function to return the length of a string. Both of these functions are useful during data preparation before analysis, with the SPLIT function being useful for data transformation and LEN being useful for data validation in terms of confirming string lengths.

1. On a new sheet, add the **Order ID** field to the view by dropping it on the **Rows** shelf. This creates a list of order IDs that can be split using the hyphen delimiter.

2. To get the year from the order ID code, create a calculated field called SPLIT with the SPLIT([Order ID], ""-"", 2) logic, then click **OK**.

3. Place the **SPLIT** field on the view to see the year split off from the **Order ID** column.

4. All year values should be four characters long. To ensure this is the case, create a new calculated field called LEN and add the LEN([SPLIT]) logic, then click **OK**.

5. Right-click the **LEN** field and convert it to **Dimension** so that it will not be aggregated when being placed on the view (there may be multiple rows with the same order ID, and therefore year, and because of this, aggregation would result in the length value being summed per row).

6. Place the **LEN** field on the **Rows** shelf to show the length of the year values extracted via the **SPLIT** method from each order ID.

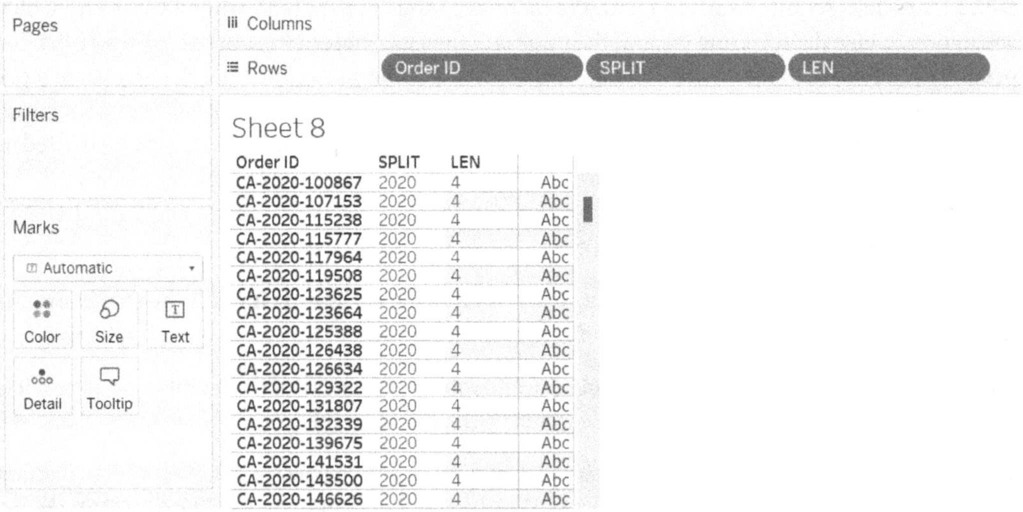

Figure 3.11: Year values parsed out from the Order ID field using the SPLIT function and then the length of each resulting year value checked using the LEN function

In this section, you split the **Order ID** column on the hyphen delimiter, returning the second section resulting from the split. You also used the LEN function to count the length of the result of the split. You now know how to use both functions and can apply what you have learned to similar functions, such as the SPACE function, which returns a given number of empty space values.

Date Calculated Fields

Date logic in Tableau Desktop allows for the conversion of date fields into specific date parts. The current date and time can be returned by creating a calculated field. There is also logic for the manipulation of existing dates, such as adding or subtracting periods of time from a given date.

Returning to the order example data source from the introduction, date functions could be used to parse out the order year as a specific column from the order date and the number of days between when an order is placed and shipped could be calculated.

Date Parts

Many date functions in Tableau allow users to take date fields and return a specific date part from them.

The DATEPART function takes a string input, used to specify the relevant date part, and a date input from which to calculate the date part. The string input must be a date part name, all in lowercase (e.g., year, quarter, month, week, or day). The function returns the specified date part for the given date as an integer. There is an optional string input value to specify when a week should start, either sunday or monday. Specifying when a week should start will impact the integer returned when using the week date part. The function is formatted as DATEPART(datepartstr, date, [str]). An example is DATEPART("year", #2024-01-13#), which returns 2024, whereas DATEPART("day", #2024-01-13#) returns 13. Note that dates can be written manually in calculated fields in the format shown (#YYYY-MM-DD#).

The DATENAME function is essentially the same as the DATEPART function, but instead of returning the integer for the relevant date part, the name is returned as a string. It is formatted as DATEPART(datepartstr, date, [str]) with the optional start-of-week string parameter again present. Using the same example as earlier, DATENAME("year", #2024-01-13#) returns 2024 as a string as opposed to an integer, and DATEPART("month", #2024-01-13#) returns January.

The YEAR, QUARTER, MONTH, WEEK, and DAY functions are all shortened functions for returning the relevant date parts. Each function takes a single date input value and returns the corresponding date part as an integer. It is formatted as FUNCTION(date), with some examples being YEAR(#2024-01-13#), MONTH(#2024-01-13#), and DAY(#2024-01-13#), returning 2024, 1, and 13 respectively.

To return integer date parts using the ISO8601 week-based date part, ISOYEAR, ISOQUARTER, ISOWEEKDAY, and ISOWEEK can all be used, taking a single date input in the format FUNCTION(date). For example, ISOWEEK(#2024-01-13#) returns 2.

Using the YEAR, MONTH, DAY, and DATEPART Functions

In this exercise, you will return specific date parts from a date field using a range of different functions. Dates are incredibly common in data sets, and it is important for analysts to understand how to extract the specific parts of a date they want to utilize from a date field.

1. On a new sheet, right-click and drag **Order Date** from the **Data pane** onto the **Rows** shelf.

2. The aggregation configuration will pop up. Select Order Date (Discrete) to display a list of each order date in the data set.

3. To parse out the year date part from each date value, create a calculated field, call it YEAR, and type in YEAR([Order Date]).

4. Click **OK** to create the calculated field, then right-click it and convert it to a dimension to prevent aggregation of the integer value returned.

5. Repeat the previous steps for **MONTH** and **DAY**, replacing the name and function used in each calculation.

6. Drag each of the three newly created calculated fields onto the **Rows** shelf to display each of the date parts alongside the order dates they are parsed from.

7. Create a new calculated field called DATEPART, add the DATEPART("year", [Order Date]) logic, then click **OK**.

8. Make the **DATEPART** calculated field a dimension and drag it onto the **Rows** shelf. The integer value reflects the year of the order date and matches the output of the **YEAR** field.

9. Edit the **DATEPART** calculation and update the date part from **year** to **month** and then **day** to observe the impact of different date part specifications.

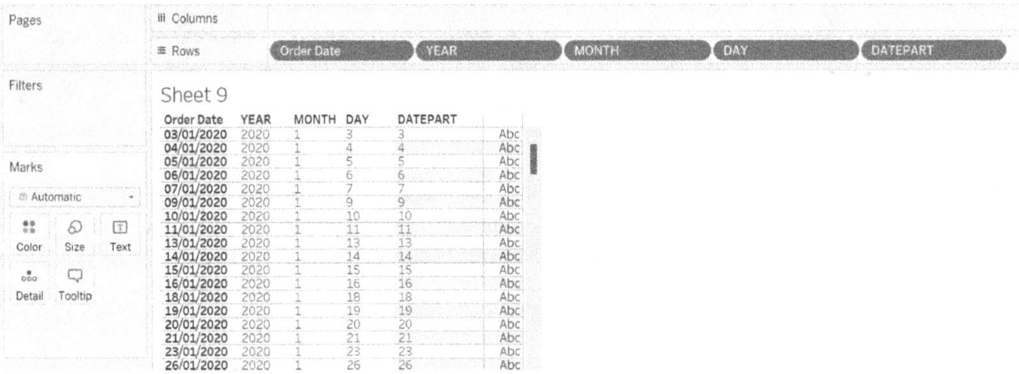

Figure 3.12: Date parts parsed from the Order Date field using the
YEAR, MONTH, DAY, and DATEPART functions

In this section, you used two methodologies to extract the year, month, and day date parts from the **Order Date** field. You will be able to use the same logic to extract other date parts if needed, such as weeks and quarters.

Date Logic

In addition to creating date parts from dates, there is a variety of other date-specific logic that can be applied using functions in Tableau's calculated fields.

The DATETRUNC function essentially aggregates or truncates a date to a different level that is specified as part of the calculation. It takes a date part input that specifies which level to truncate the date to, the date to be truncated, and there is an optional start-of-week string input for specifying whether weeks should start on a Sunday or Monday. It is formatted as DATETRUNC(datepartstr, date, [str]). An example would be DATETRUNC("month", #2024-01-13#), returning the truncated date #2024-01-01#, where the day level of detail has been removed and the date has aggregated to the relevant month.

The DATEADD function allows users to add a specified number of the given date part to a date. It takes the date part input as a string, a numeric input for the amount to add, and the date to be added to. It is formatted as DATEADD(datepartstr, num, date) and an example would be DATEADD("month", 5, #2024-01-13#), which returns #2024-06-13#, where 5 months have been added to the provided date. The number can also be negative – for example, DATEADD("year", -1, #2024-01-13#) returns #2023-01-13#.

The DATEDIFF function returns the difference between two provided dates using the specified date part. There is an optional start-of-week string input and the function is formatted as DATEDIFF(datepartstr, date1, date2, [str]) where date1 is subtracted from date2. An example is DATEDIFF("year", #2024-01-13#, #2025-01-13#), which returns 1. Switching the date part string input from year to month would return 12.

The TODAY and NOW functions return the current date and datetimes respectively and are formatted as TODAY() and NOW(). These functions are often useful as dynamic reference points in other date calculations.

Using the DATETRUNC and DATEADD Functions

In this exercise, you will apply date logic to truncate dates to a specified level and will also add a specified interval to date values. These functions are useful for KPI logic pertaining to dates, such as calculating year-to-date or year-on-year values. Date functions also generally utilize similar syntax to these functions, so these are good exemplars that will aid you in understanding date logic functions more generally.

1. On a new sheet, place **Order Date** on discrete and not aggregate rows to create a list of order dates.

2. Create a new calculated field called DATETRUNC with the DATETRUNC(""month"", [Order Date]) logic, and then click **OK**.

3. If the field is a datetime type, convert it to a date type.

4. Place the field on discrete and not aggregated rows in the same way as **Order Date**. The dates are comparable with the **DATETRUNC** dates, reflecting the same date but truncated to the month level, and therefore losing the day level of detail.

5. Create a new calculated field called DATEADD with the DATEADD("day", 5, [Order Date]) logic, then click **OK**. Convert the field to a date type if it is automatically set as datetime.

6. Place the field on discrete and not aggregated rows and observe that it reflects the **Order Date** values but with five days added.

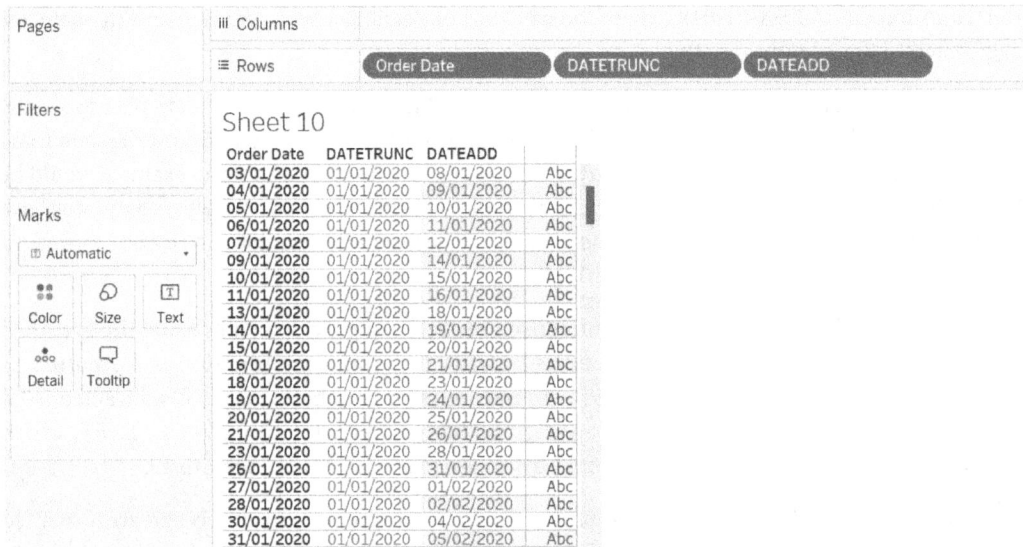

Figure 3.13: Order Date truncated to month level and 5 days added
using the DATETRUNC and DATEADD functions

In this exercise, you truncated order dates down to the year level and added five days to each **Order Date** value using the DATEADD function. The functions used in this exercise are often used in conjunction with fixed level of detail calculations to calculate year-on-year comparisons.

Using the DATEDIFF and TODAY Functions

In this exercise, you will use the DATEDIFF and TODAY functions. These functions are commonly used together to find the difference between a date in the data set and the current date.

1. On the same sheet created for the *Using the SUM and the MIN functions* exercise, create a new calculated field called DATEDIFF.

2. Populate the calculated field with the DATEDIFF("year", [Order Date], TODAY()) logic and click **OK**. This creates a calculated field that calculates the difference in years between each order date and the current day's date.

3. Convert the field to a dimension so that it is not aggregated, then place it on the **Rows** shelf. The difference between the order date and today's date in years can be observed.

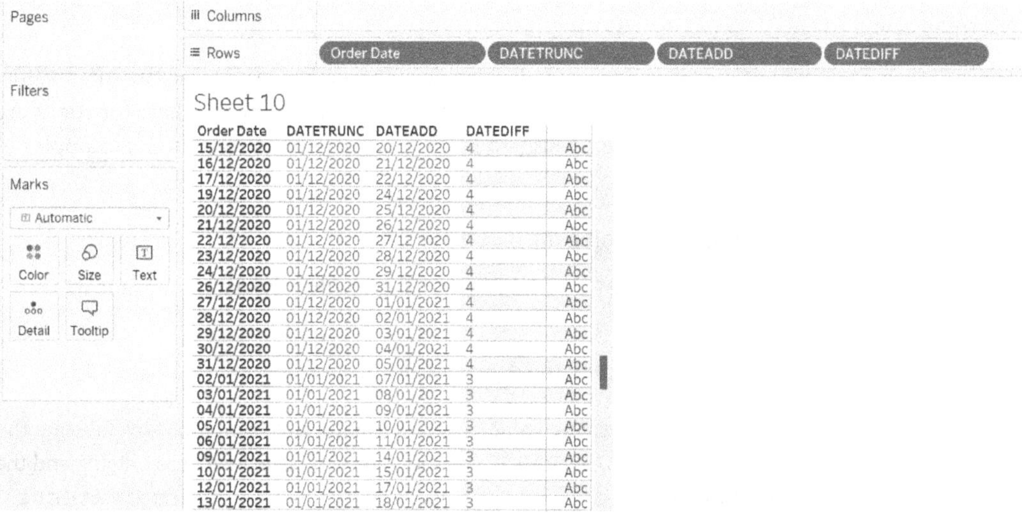

Figure 3.14: The difference in years between the Order Date values in the data
and today's date is calculated using the DATEDIFF and TODAY functions

In this exercise, you used the DATEDIFF and TODAY functions to calculate the difference in years between order dates and today's date. Note that your numbers may be different than those shown in the screenshot as the date returned by the TODAY function is dynamic.

Type Conversion

There are functions in Tableau to convert any field or value to another data type as long as it can be configured sufficiently. Dates and spatial fields can also be created from scratch.

In the order example data source, a type conversion function used could be the STR function, which would convert a numeric ID field into a string, as technically, IDs should not be treated as numeric values.

Data Types

For every data type in Tableau, there is a calculated field function to convert from one type to another.

The STR function converts a value to a string data type. It is formatted as STR (value) and outputs the input value as a string. This is often useful for ID fields, which may be read in by Tableau as a numeric data type. An example would be an ID field with numeric values such as 10001 being formatted as STR ([ID]), resulting in the ID being cast as strings, 10001.

The INT and FLOAT functions are similar to the STR function, but instead convert values to numeric outputs. FLOAT converts values to floating point numbers (decimals) and INT creates integers. They are formatted as INT(value) and FLOAT(value) with an example being INT("5.2"), returning 5, and FLOAT("5.2"), returning 5.2.

The DATE and DATETIME functions return a date based on a number, string, or date input value. The input value must be in a predefined date pattern that Tableau can recognize. These functions are formatted as DATE(value) and DATETIME(value), respectively. Here are some examples of values that would be accepted:

- DATE("13/01/2024") returning the date value #2024-01-13#
- DATETIME(January 13, 2024 12:00:00) returning the datetime value #2024-01-13 12:00:00
- DATE(#2024-01-13#) returning the date value #2024-01-13#

If a date field needs to be created from a string value but the format is not recognized by Tableau, the DATEPARSE function can be used. The DATEPARSE function takes the date format as a string and the string input to be converted to the date in the format DATEPARSE(dateformatstr, datestr). An example is DATEPARSE("yyyyMMdd", "20240113"), which returns #2024-01-13#.

Similar to the DATEPARSE function, the MAKEDATE function creates a date field but instead of requiring a date format, it takes numeric inputs for year, month, and day and outputs the subsequent date. It is formatted as MAKEDATE(num1, num2, num3) where num1 refers to the year, num2 the month, and num3 the day. An example would be MAKEDATE(2024, 1, 13), which returns #2024-01-13#.

Similar to the MAKEDATE function, the MAKETIME function takes three numbers for hour, minute, and second to create a datetime field (with the default date being #1899-01-01#). It is formatted as MAKETIME(num1, num2, num3) with num1 referring to hour, num2 to minute, and num3 to second. An example is MAKETIME(16, 52, 34), returning #1899-01-01 16:52:34#.

The MAKEDATETIME function takes a date input, which can be a date, datetime, or string, and a time input, which must be a datetime. It takes the date from the date input and the time from the time input and creates a datetime field. It is formatted as MAKEDATETIME(date, time). An example of its use is MAKEDATETIME(#2024-01-13#, #12:00:00#), returning #2024-01-13 12:00:00#.

Using the STR and MAKEDATE Functions

In this exercise, you will use the STR and MAKEDATE type conversion functions to convert a field to a string and a date respectively. Type conversion functions allow you to convert field types if that is required for your data and because they are specific to calculated fields, they can be used to convert an existing field within the calculation only. An example of this would be converting a numeric field to a string in a calculated field so that a string function can be applied to the values of the numeric field.

1. On a new sheet, drop the **Order ID** field onto the **Rows** shelf to create a list of order IDs, and then drop the **Discount** field onto **Text** on the **Marks card** to show the discount for each order.

2. To create a discount percentage string field that can be used as a descriptive dimension, create a calculated field and call it STR.

3. Populate the calculated field with the following logic: STR([Discount] * 100) + "%". The logic multiplies the discount value by 100 to get a value between 1 and 100, as opposed to a decimal, and then converts the resulting number to a string. A percentage sign is then added at the end of the string.

4. Click **OK**, then place the **STR** field on the **Rows** shelf to display the percentage discount categories for each order.

5. To create a date field from the **Order ID** field, first create a calculated field called MAKEDATE.

6. Type the MAKEDATE(INT(SPLIT([Order ID], "-", 2)), 1, 1) logic. This logic splits the order ID on the – delimiter and retains the second subsection, which is the year part of the order ID. This returns a string, so it is wrapped in the INT function to convert the year to an integer. The resulting year integer is used as the first argument in the MAKEDATE function, followed by a 1 to set the month to January and another 1 to set the date to the 1st. The resulting output is a date column that reflects the year of the order date, with the month and day always set to January 1st.

7. Click **OK** and place the **MAKEDATE** field on the **Rows** shelf as a discrete field with no aggregation to display the year of each order.

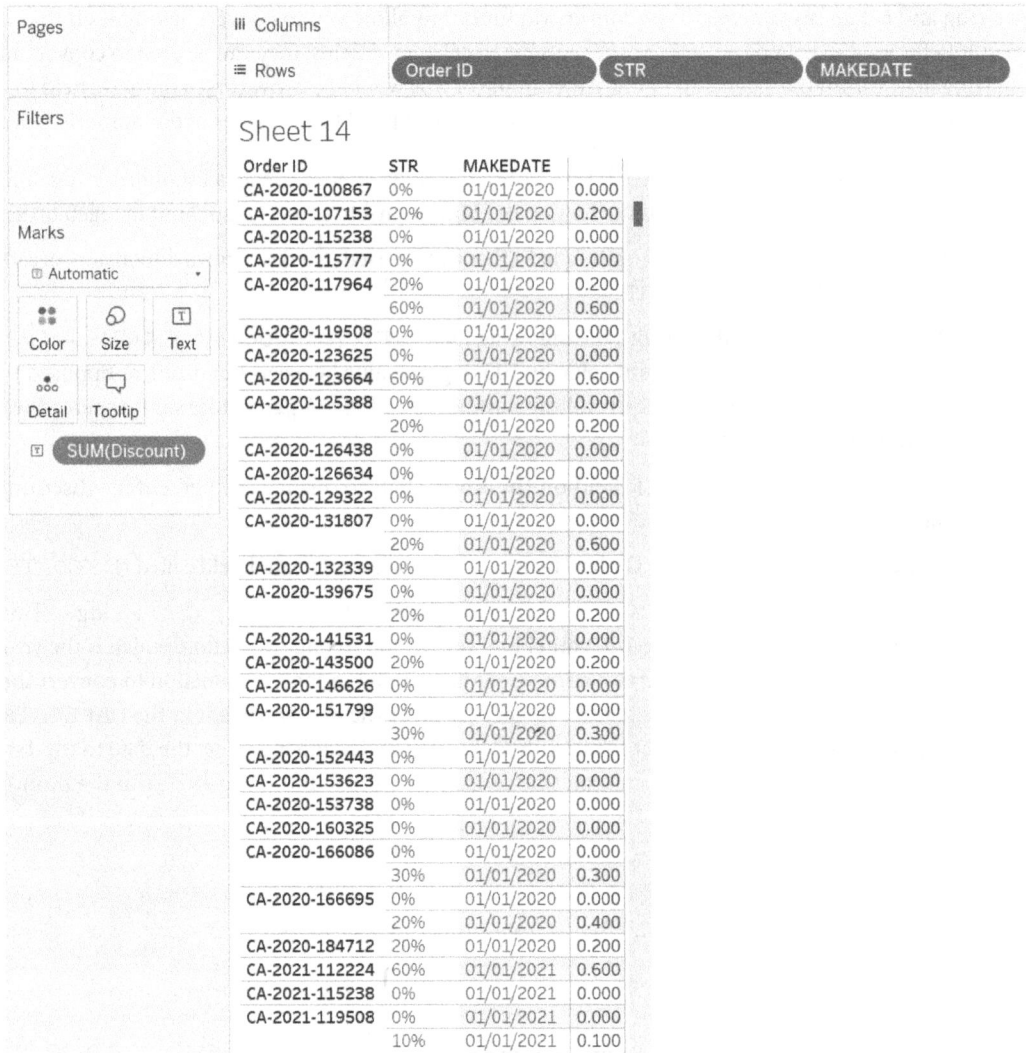

Pages				iii Columns						
				≡ Rows		Order ID		STR		MAKEDATE

Filters			Sheet 14			
			Order ID	STR	MAKEDATE	
			CA-2020-100867	0%	01/01/2020	0.000
			CA-2020-107153	20%	01/01/2020	0.200
Marks			CA-2020-115238	0%	01/01/2020	0.000
			CA-2020-115777	0%	01/01/2020	0.000
⊡ Automatic ▾			CA-2020-117964	20%	01/01/2020	0.200
				60%	01/01/2020	0.600
::	⌬	⊤	CA-2020-119508	0%	01/01/2020	0.000
Color	Size	Text	CA-2020-123625	0%	01/01/2020	0.000
			CA-2020-123664	60%	01/01/2020	0.600
₀₀₀	⌕		CA-2020-125388	0%	01/01/2020	0.000
Detail	Tooltip			20%	01/01/2020	0.200
⊤ SUM(Discount)			CA-2020-126438	0%	01/01/2020	0.000
			CA-2020-126634	0%	01/01/2020	0.000
			CA-2020-129322	0%	01/01/2020	0.000
			CA-2020-131807	0%	01/01/2020	0.000
				20%	01/01/2020	0.600
			CA-2020-132339	0%	01/01/2020	0.000
			CA-2020-139675	0%	01/01/2020	0.000
				20%	01/01/2020	0.200
			CA-2020-141531	0%	01/01/2020	0.000
			CA-2020-143500	20%	01/01/2020	0.200
			CA-2020-146626	0%	01/01/2020	0.000
			CA-2020-151799	0%	01/01/2020	0.000
				30%	01/01/2020	0.300
			CA-2020-152443	0%	01/01/2020	0.000
			CA-2020-153623	0%	01/01/2020	0.000
			CA-2020-153738	0%	01/01/2020	0.000
			CA-2020-160325	0%	01/01/2020	0.000
			CA-2020-166086	0%	01/01/2020	0.000
				30%	01/01/2020	0.300
			CA-2020-166695	0%	01/01/2020	0.000
				20%	01/01/2020	0.400
			CA-2020-184712	20%	01/01/2020	0.200
			CA-2021-112224	60%	01/01/2021	0.600
			CA-2021-115238	0%	01/01/2021	0.000
			CA-2021-119508	0%	01/01/2021	0.000
				10%	01/01/2021	0.100

Figure 3.15: STR function used to create a string field for discounts and
MAKEDATE used to create a date field for the order year

In this exercise, you used type conversions in calculations so that logic can be applied that would not normally be applicable given the data type of the field being referenced. STR was used to convert the result of a numeric calculation to a string before adding % to the end of the string. The year was also parsed out of the ID column using the SPLIT function, followed by being converted to a numeric integer using INT, and then finally referenced as the year in a MAKEDATE function. This exercise demonstrated how logic can be combined and how type conversions can be utilized to ensure functions are applicable.

Boolean Calculated Fields

Tableau's calculated fields can be configured to give a true or false result using a variety of boolean functions, such as *is a given value a date?*, *is a given value a null?*, and so on. The boolean-type logic can also be expanded into longer-form statements with multiple conditions and more possible outputs than true or false. These logical statements can take the form of either IF or CASE statements.

Boolean or logical calculated field functions could be used in the orders example data source from the introduction to implement logic that flags orders over a specific total sales value. The ISNULL function could also be used to filter out any null orders that have entered the data source by accident.

Boolean Functions

There are a variety of Boolean functions available in Tableau's calculated fields that return TRUE or FALSE based on the results of a specific test.

The ISNULL function returns TRUE if the value input is null and FALSE if an actual value is present. It is formatted as ISNULL(value) and an example of a field with the values 1, null, and 3 would be formatted as ISNULL([Field]) and would return FALSE, TRUE, and FALSE respectively.

The ISDATE function works similarly to ISNULL but returns TRUE if the string input is in a valid date format and otherwise FALSE. It is formatted as ISDATE(str) with an example being ISDATE("2023-01-13"), returning TRUE. Note that whether a date is considered valid depends on the locale setting in Tableau: for example, ISDATE("2023-01-13") would return TRUE in the UK but FALSE in the USA, whereas ISDATE("2023-13-01") would be the other way round.

The IN function checks whether the values in one expression match a value in the second expression. The IN function is formatted in an unusual way, expr1 IN expr2, where the expressions can be a set, a manually typed list of values, or a combined field. An example of this in use is [Category] IN ("Furniture", "Appliances", "Decorations"), which would return TRUE for any categories in the **Category** field with the values Furniture, Appliances, or Decorations, and FALSE for any other value.

Using the IN Function

In this exercise, you will use the IN function to create a Boolean field that returns TRUE for customers that are members of a set. The IN function is useful for checking whether values can be found in an expression and results in a Boolean that can be used as a filter.

1. On a new sheet, place the **Customer Name** field on the **Rows** shelf to create a list of customers.

2. Create a calculated field called IN with the [Customer Name] IN [Top Customers by Profit] logic to return a TRUE or FALSE value based on whether the given customer name is in the set.

3. Place the **IN** field on the **Rows** shelf, as well as the **Top Customer by Profit** set to confirm that customers that are in the set have returned a TRUE value.

Figure 3.16: The IN function used to look for customer names in the
Top Customer by Profit set, then confirmed to match up

In this exercise, you used the IN function to find customers that are members of a set. The IN function can also be used to find values in fields and always results in a Boolean. ISNULL and ISDATE are fields that similarly result in a Boolean and can be used in the same way to check for null values and date values respectively.

IF Logic

IF logic in Tableau's calculated fields allows for more advanced conditional logic that goes beyond a single condition returning TRUE or FALSE. With the IF logic, outputs can be specified as a result of different conditions being met and multiple conditions can be set up.

The IFNULL function takes a value to check for nulls and an output value for when a null is found. If no null is found then the original value is output. Otherwise, if a null is found in the first value, then the second value is output. It is formatted as IFNULL(value1, value2) and an example of its use is a field with the values 1, null, and 3 being formatted as IFNULL([Field], 0) and returning 1, 0, and 3.

The IIF function allows the user to specify a logical test (boolean): an output for when the test returns TRUE, and another output for FALSE. There is also an optional input to return when the test returns null; if this is not included in the input, then a null test results in a null output. It is formatted as IIF(test, output1, output2, [nulloutput]) where output1 determines the output of a TRUE test result and output2 determines the output of a FALSE test result. An example is a field with the values 1, null, and 3 being put in the IIF([Field]>2, [Field] * -1, [Field] * 1) calculated field and returning 1, null, and -3. In the previous example, the test returns FALSE for the input value 1, which is therefore multiplied by 1, resulting in an output of 1. The second value is null so the test result and output are null. The third value, 3, returns TRUE in the test, and therefore the value 3 is multiplied by -1, resulting in -3.

The IF statement in Tableau works similarly to the IIF function in that there is a test and the resulting outputs. It offers more flexibility than IIF in that both multiple tests and multiple outcomes can be specified. It uses the following Tableau calculated field keywords in its syntax: THEN, ELSEIF, ELSE, and END. It is formatted as follows:

```
IF test1 THEN output1
[ELSEIF test THEN outputn]
[ELSE defaultoutput]
END
```

The required elements of an IF statement are the initial IF clause, followed by the test, followed by THEN, and then the output for when the test returns TRUE. The final required element is the END keyword to signify the end of the IF statement. An IF statement with these minimal elements will return the output if the test is TRUE and null if the test is FALSE.

The basic IF statement outlined above can be expanded upon with the ELSE keyword, which allows the user to specify a default output for when the test returns FALSE, as opposed to the null value that is returned when the ELSE keyword is not included.

Additional tests can be added to the IF statement using the ELSEIF keyword followed by the new test, then the THEN keyword, and then the output for when that test is TRUE. Several ELSEIF additional tests can be added to the IF statement to meet any specific requirements. If multiple tests are passed (i.e., return TRUE) then the output for the field will be the first test that was passed in the statement.

The AND and OR keywords can be used to apply multiple conditions within a single test. For example, test1 AND test2 returns TRUE only if both tests are TRUE. By contrast, test1 OR test2 returns TRUE if either test is TRUE.

The following is an example IF statement for the field with the values 1, null, and 3:

```
IF ISNULL([Field]) THEN 0
ELSEIF [Field] > 2 THEN [Field] * -1
ELSE [Field] * 1
END
```

This IF statement will return 1, 0, and -3 as the value 1 fails both the first and second tests and so is multiplied by 1, resulting in an output of 1. The null value passes the first test and so the output is 0. The value 3 fails the first test but passes the second test and so is multiplied by -1, resulting in -3.

Using the IIF Function and Creating an IF Statement

In this exercise, you will use both the IIF function and an IF statement and will be able to observe the difference in functionality and syntax. Conditional logic is useful to know in Tableau as it allows users to customize visual outcomes based on logical rules.

1. On a new sheet, drag the **Sub-Category** field onto the **Rows** shelf and the **Sales** field onto **Text** on the **Marks card** to create a table showing the total sales by sub-category.

2. Create a new calculated field called IIF populated with the IIF(SUM([Sales]) > 100000, "High Sales", "Low Sales") logic and click **OK**.

3. Place the **IIF** field onto the **Rows** shelf and observe that the sub-categories are now labeled **High Sales** or **Low Sales** based on whether there is a total of over 100k sales or not.

4. To create an IF statement with an additional test, create a new calculated field and call it IF STATEMENT. Populate the calculated field with the following logic:

```
IF SUM([Sales]) >= 150000 THEN "High Sales"
ELSEIF  SUM([Sales]) < 150000 AND SUM([Sales]) >= 100000 THEN
"Medium Sales"
ELSE "Low Sales"
END
```

5. Click **OK** to create the calculated field and then place it on the **Rows** shelf. The `IF` statement now categorizes sub-categories into high, medium, and low sales.

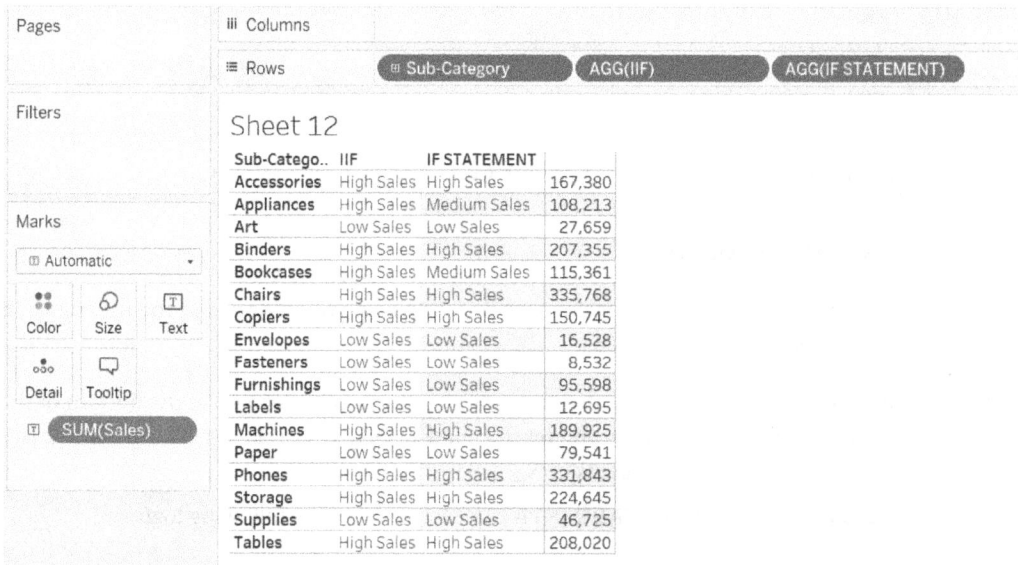

Sub-Catego..	IIF	IF STATEMENT	
Accessories	High Sales	High Sales	167,380
Appliances	High Sales	Medium Sales	108,213
Art	Low Sales	Low Sales	27,659
Binders	High Sales	High Sales	207,355
Bookcases	High Sales	Medium Sales	115,361
Chairs	High Sales	High Sales	335,768
Copiers	High Sales	High Sales	150,745
Envelopes	Low Sales	Low Sales	16,528
Fasteners	Low Sales	Low Sales	8,532
Furnishings	Low Sales	Low Sales	95,598
Labels	Low Sales	Low Sales	12,695
Machines	High Sales	High Sales	189,925
Paper	Low Sales	Low Sales	79,541
Phones	High Sales	High Sales	331,843
Storage	High Sales	High Sales	224,645
Supplies	Low Sales	Low Sales	46,725
Tables	High Sales	High Sales	208,020

Figure 3.17: IIF function and IF statement created to categorize sub-categories by total sales amounts

You have now successfully created both an `IIF` and an `IF` statement and have observed the differences in syntax. You also demonstrated that `IF` statements, while less efficient to type, have the possibility for a wider range of conditions through the `ELSEIF` clause.

CASE Statements

`CASE` statements are similar to `IF` statements, but instead of creating outputs based on tests, the values within an expression are evaluated, and if they match with a given value, then the statement outputs the corresponding result. The WHEN, THEN, ELSE, and END keywords are required and `CASE` statements are formatted as follows:

```
CASE expr
WHEN value1 THEN output1
[WHEN value THEN output]
[ELSE defaultoutput]
END
```

An example using the field with the values 1, null, and 3 is given here:

```
CASE [Field]
WHEN 1 THEN 1
WHEN 3 THEN -3
ELSE 0
END
```

This would return 1 for the value 1, 0 for the null value, and -3 for the value 3.

Writing a CASE Statement

In this section, you will create a CASE statement. This will help you understand the difference in syntax when compared to an IF statement.

1. On a new sheet, show the **Top Customer** parameter by right-clicking it at the bottom of the **Data pane** and selecting **Show Parameter**. It is now displayed as a slider between 5 and 20 on the view and increases in increments of 5.

2. Create a calculated field called CASE STATEMENT and add the following logic:

```
CASE [Top Customers]
WHEN 5 THEN "5 Customers - Lowest value"
WHEN 20 THEN "10 Customers - Highest value"
ELSE STR([Top Customers]) + " Customers"
END
```

3. Place the **CASE STATEMENT** field on **Text** on the **Marks card** to view the result of the CASE statement.

4. Change the **Top Customers** value by using the slider to switch between 5, 10, 15, and 20 to see the impact of the CASE statement.

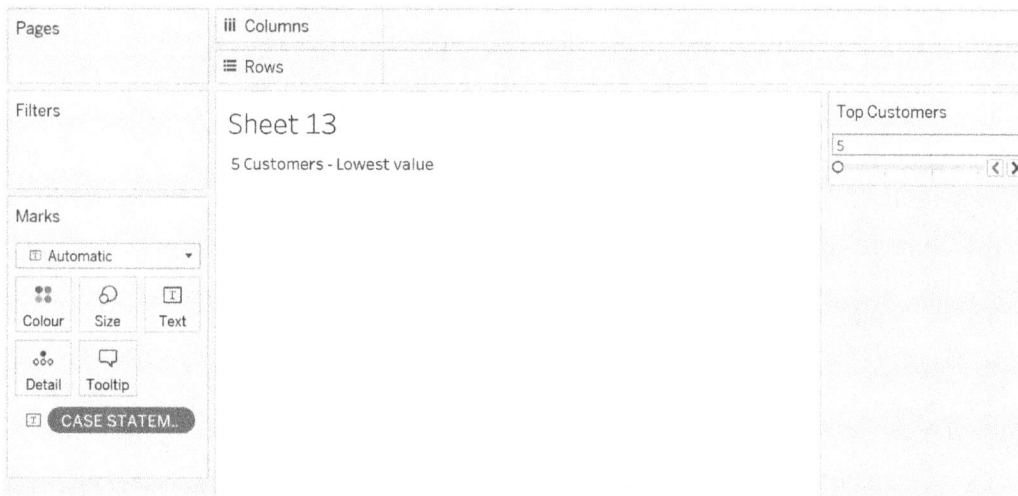

Figure 3.18: CASE statement that returns the number of customers selected in the
parameter as well as the lowest or highest value when 5 or 20 is selected

In this exercise, you created a CASE statement that returns differing values based on the selection of
a parameter. CASE statements allow for dynamic visuals based on which condition is satisfied.

Table Calculations

There will be situations where a calculation needs to be applied to a visual, but the calculation needs
to compute on fields that are currently available in the view, disregarding any measures of dimensions
that are filtered out of the visual.

Table calculations help achieve this goal. These calculations can assist a developer when trying to
calculate ranks and while running totals and percentages of totals. This section will go through each
table calculation, discussing what they achieve and how to create them.

Building a Table Calculation

Table calculations are available only for measures, as their purpose is to calculate the values present
in the visual. Therefore, it is important to make sure that all values and categorical fields are present
in the view, either on the **Columns** or **Rows** shelves or in the details of the **Marks card**.

Once the fields are present, right-click on the measure desired to create a calculation and then select **Add Table Calculation**. Once this is done, you can choose how the table calculation is run and on what level. Consider *Figure 3.19* and read through what each section does:

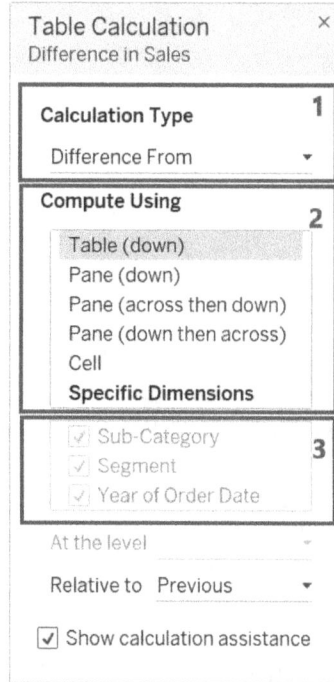

Figure 3.19: Table Calculation configuration window

There are three sections to this window. The first one is where the user specifies the calculation that needs to be performed. The options are as follows. Their effects and some additional functions will be described later in this section.

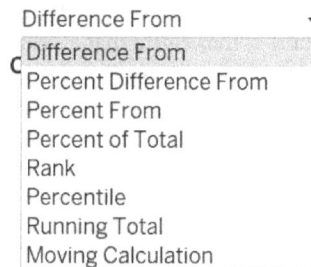

Figure 3.20: Select the calculation type

You are then prompted to select how your calculation will be computed. This tells Tableau how to read the data and carry out the specified calculation. The following screengrabs show the way these calculations are applied.

Month of Order Date	Segment	Order Date			
		2018	2019	2020	2021
January	Consumer		7,347	-8,248	17,185
	Corporate		1,681	6,102	2,736
	Home Office		-5,091	2,514	5,508
February	Consumer		6,101	6,531	-10,211
	Corporate		1,382	302	5,092
	Home Office		-51	4,195	2,441
March	Consumer		5,550	7,426	10,197
	Corporate		5,225	-2,106	2,348
	Home Office		-27,740	7,669	-5,388
April	Consumer		13,518	-8,551	-5,040
	Corporate		-10,011	12,056	-1,374
	Home Office		2,393	1,050	4,186
May	Consumer		3,060	11,295	-11,115
	Corporate		-3,687	12,997	-4,212
	Home Office		2,110	2,563	2,601
June	Consumer		-13,099	11,064	-3,616
	Corporate		5,433	-1,616	11,907
	Home Office		-2,132	6,099	4,346

Figure 3.21: Table down

Month of O..	Segment	Order Date 2018	2019	2020	2021
January	Consumer		7,347	-8,248	17,185
	Corporate		1,681	6,103	2,736
	Home Office		-5,091	2,514	5,508
February	Consumer		6,101	6,531	-10,211
	Corporate		1,382	302	5,092
	Home Office		-51	4,195	2,441
March	Consumer		5,550	7,426	10,197
	Corporate		5,225	-2,106	2,348
	Home Office		-27,740	7,669	-5,388
April	Consumer		13,518	-8,551	-5,040
	Corporate		-10,011	12,056	-1,374
	Home Office		2,393	1,050	4,186
May	Consumer		8,060	11,295	-11,115
	Corporate		-3,687	12,997	-4,212
	Home Office		2,110	2,563	2,501
June	Consumer		-13,099	11,064	-3,616
	Corporate		5,433	-1,616	11,907
	Home Office		-2,132	6,099	4,346

Figure 3.22: Table across

Month of O..	Segment	Order Date 2018	2019	2020	2021
January	Consumer		7,347	8,248	17,185
	Corporate	-21,510	1,681	6,103	2,736
	Home Office	6,614	5,091	2,514	5,508
February	Consumer	5,370	6,101	6,531	10,211
	Corporate	4,405	1,382	302	5,092
	Home Office	-7,791	-51	4,195	2,441

Figure 3.23: Table across then down

Month of O..	Segment	Order Date			
		2018	2019	2020	2021
January	Consumer		2,661	-13,025	11,079
	Corporate	-5,226	-10,893	3,458	-10,991
	Home Office	3,906	-2,866	-6,454	-3,633
February	Consumer	-2,440	8,752	12,770	-2,949
	Corporate	-1,984	-6,704	12,933	2,370
	Home Office	-1,015	-2,448	1,445	-1,206
March	Consumer	11,701	17,302	20,533	28,239
	Corporate	-763	-1,087	-10,619	-18,467
	Home Office	21,608	-11,357	-1,582	-9,318
April	Consumer	-23,606	17,661	1,431	1,779
	Corporate	5,023	-18,505	2,102	5,768
	Home Office	-9,077	3,327	-7,680	-2,119
May	Consumer	7,764	13,431	23,677	8,376
	Corporate	-3,677	-15,424	-13,712	-6,820
	Home Office	-7,455	-1,657	-12,091	-5,278

Figure 3.24: Table down then across

Month of O..	Segment	2018	2019	2020	2021
January	Consumer				
	Corporate	-5,226	-10,893	3,458	-10,991
	Home Office	3,906	-2,866	-6,454	-3,683
February	Consumer				
	Corporate	-1,984	-6,704	-12,933	2,370
	Home Office	-1,015	-2,448	1,445	-1,206
March	Consumer				
	Corporate	-763	-1,087	-10,619	-18,467
	Home Office	21,608	-11,357	-1,582	-9,318
April	Consumer				
	Corporate	5,023	-18,505	2,102	5,768
	Home Office	-9,077	3,327	-7,680	-2,119

Figure 3.25: Pane down

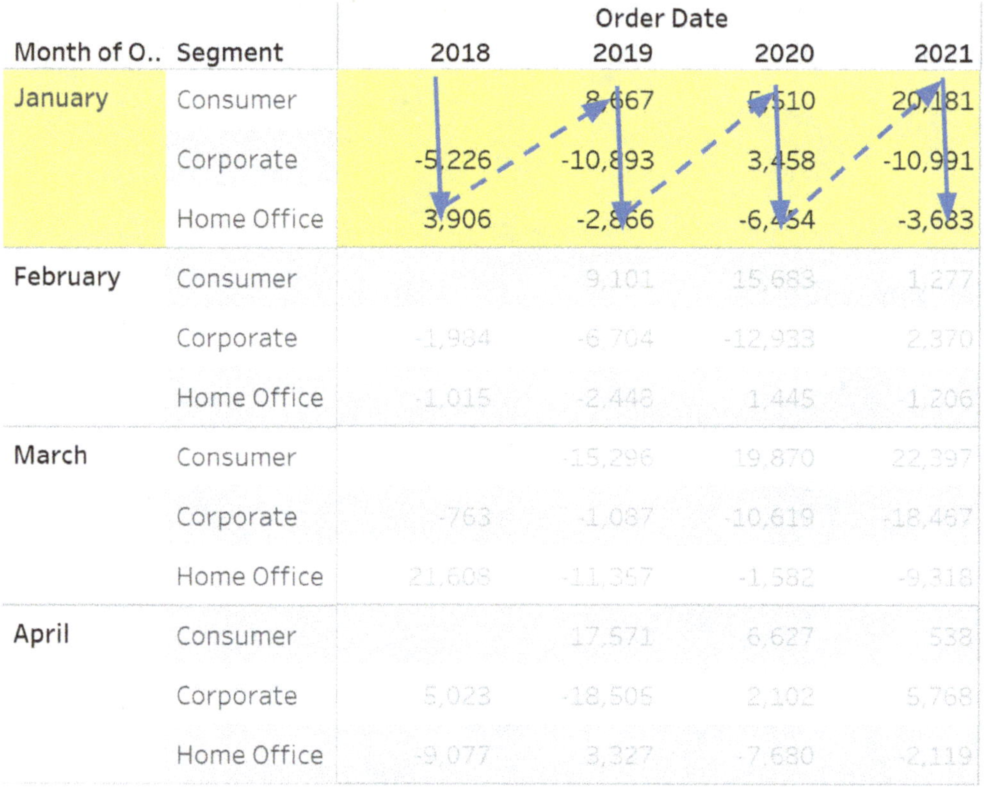

Month of O..	Segment	Order Date			
		2018	2019	2020	2021
January	Consumer		8,667	5,510	20,181
	Corporate	-5,226	-10,893	3,458	-10,991
	Home Office	3,906	-2,866	-6,454	-3,683
February	Consumer		9,101	15,683	1,277
	Corporate	-1,984	-6,704	-12,933	2,370
	Home Office	-1,015	-2,448	1,445	1,206
March	Consumer		15,296	19,870	22,397
	Corporate	-763	-1,087	-10,619	-18,467
	Home Office	21,608	-11,357	-1,582	-9,318
April	Consumer		17,571	6,627	538
	Corporate	5,023	-18,505	2,102	5,768
	Home Office	-9,077	3,327	-7,680	-2,119

Figure 3.26: Pane down then across

Month of O..	Segment	Order Date			
		2018	2019	2020	2021
January	Consumer		7,347	-8,248	17,185
	Corporate	11,510	1,681	6,103	2,706
	Home Office	6,614	5,091	2,514	5,598
February	Consumer		6,101	6,531	-10,211
	Corporate	-4,405	1,382	302	5,092
	Home Office	7,791	-51	4,195	2,441
March	Consumer		5,550	7,426	10,197
	Corporate	-23,935	5,225	-2,106	2,348
	Home Office	16,140	-27,740	7,669	-5,388
April	Consumer		13,518	-8,551	-5,040
	Corporate	5,097	-10,011	12,056	-1,374
	Home Office	-9,748	2,393	1,050	4,186

Figure 3.27: Pane across then down

When you select **Specific Dimensions**, you are next prompted to select what specific dimensions are required for the visualization, giving you more control over the calculation. This is different from selecting the directional options that we just saw, as it uses the selected fields to determine how the calculation is performed.

You can then choose at which level the calculation should be computed:

- **Deepest**: This option will produce a calculation to the finest granularity of the data available in the view. This is the default option.
- **Other fields**: Though **Deepest** is the default option, the fields available in the view are offered as other options. This gives you the control to select the level at which the table calculation operates.

Note that a developer can select multiple fields and even reorder the sequence of how the calculation is run.

The next section will look into each table calculation function that may come up in the exam.

Moving Calculations

A **moving calculation** runs by creating an aggregate across a table based on how a user configures the table calculation. This aggregation can be a sum, average, minimum, or maximum.

This kind of calculation is used to smooth the data, meaning that instead of seeing the finer points within the data, the plot line is smoothed into a progressive line to show a trend over time. The following shows an example of using a running average.

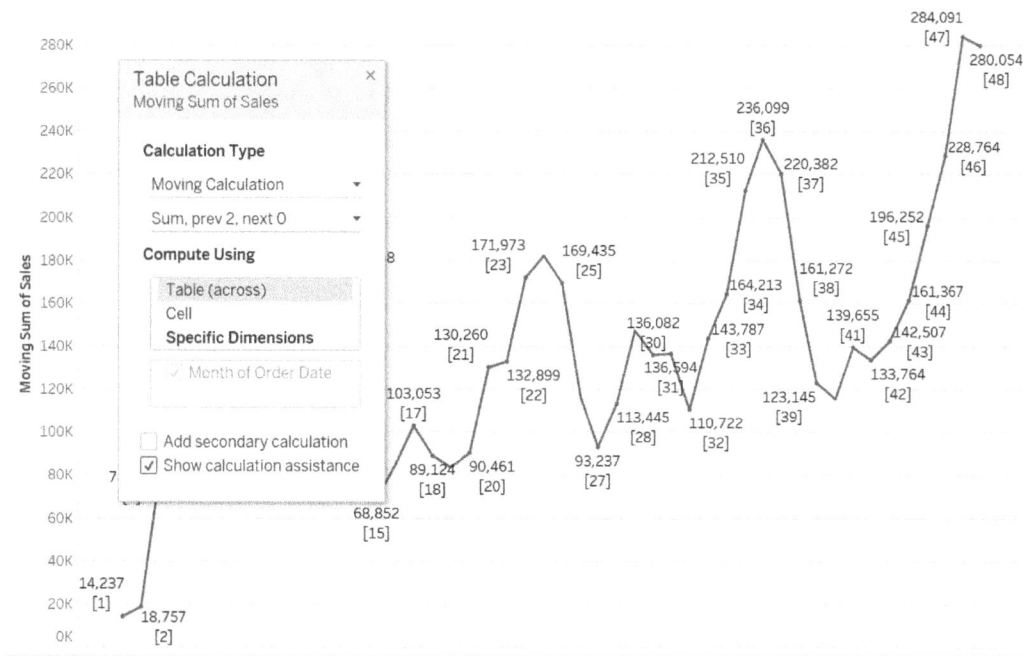

Figure 3.28: Table calculation of moving sum

Now you can zoom in on the table calculation, as shown in *Figure 3.29*, to see how this is calculated.

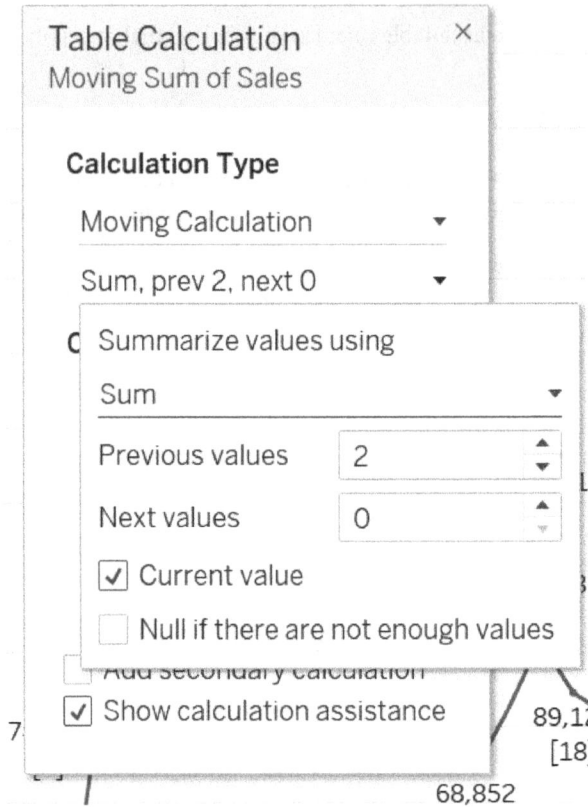

Table Calculation ×
Moving Sum of Sales

Calculation Type

Moving Calculation ▾

Sum, prev 2, next 0 ▾

Summarize values using

Sum ▾

Previous values   2   ▴▾

Next values   0   ▴▾

☑ Current value

☐ Null if there are not enough values

☑ Show calculation assistance

89,1:
[18
68,852

Figure 3.29: Table Calculation pane

The way this has been calculated is by summing the previous two values, including the current value, as that option has been checked in the configuration pane. You can update the values as you wish by editing the table calculation.

Percent of Total

One popular use of table calculations is to help identify values as a percentage of the total. This can be useful when trying to calculate what a categorical item contributes to a whole in terms of sales or any other kind of value.

Take *Figure 3.30*, which shows the sum of sales each year, broken down by category.

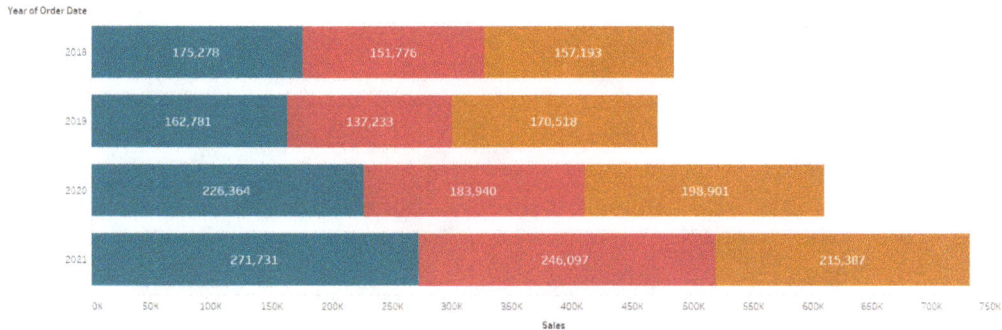

Figure 3.30: Example of a sum of sales table

Though the values are present, it can be challenging to see what these values contribute to a whole. By right-clicking the **Sales** field, adding a **Percent of Total** table calculation, and specifying the direction of the calculation (e.g., across rows or columns), the results will be as shown in *Figure 3.31*.

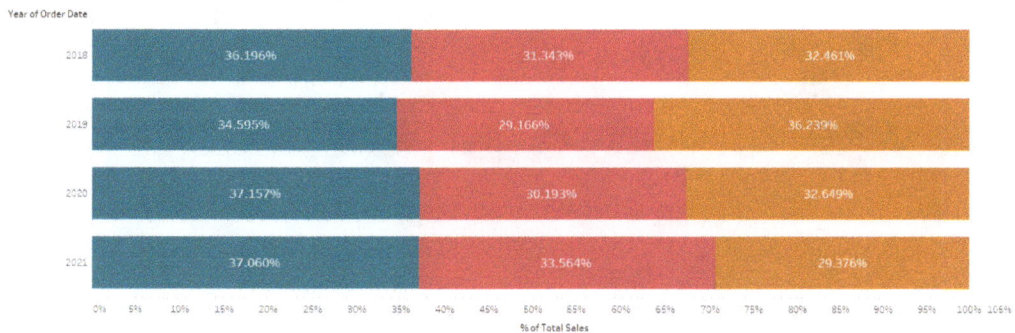

Figure 3.31: Example of a Percent of Total calculation on categories by sales

From this view, it is visually clearer which category contributes to the highest percentage of sales.

Running Sum

A running sum calculation creates a sum of the previous value to the value in the partition, thus accumulating the data points that can show trends or insights in the data.

Year of Order Date	Segment	Running Sum of Sales along Table (Down)
2018	Consumer	266,097
	Corporate	394,532
	Home Office	484,247
2019	Consumer	750,783
	Corporate	879,541
	Home Office	954,780
2020	Consumer	1,251,644
	Corporate	1,458,750
	Home Office	1,563,986
2021	Consumer	1,895,890
	Corporate	2,137,738
	Home Office	2,297,201

Figure 3.32: Example of Running Sum

Using a sum is the default option. You can also choose to use an average, minimum, or maximum:

- **Average**: This option will average the current and previous values

- **Minimum** and **Maximum**: These two options will replace the value with either the lowest or highest value in the partition

By setting the **Restarting every** option, you can select the dimension in the view by which the data is grouped, causing the calculation to restart for each partition defined by that dimension. This is shown in *Figure 3.33*:

Year of Order Date	Segment	Running Sum of Sales restarting every Year of Order Date
2018	Consumer	266,097
	Corporate	394,532
	Home Office	484,247
2019	Consumer	266,536
	Corporate	395,293
	Home Office	470,533
2020	Consumer	296,864
	Corporate	503,970
	Home Office	609,206
2021	Consumer	331,905
	Corporate	573,753
	Home Office	783,215

Figure 3.33: Running sum restarting every year

Difference and Percent of Difference

These functions allow for comparison between data points, highlighting changes in the values and showing either a value or percent difference. They are especially useful for tracking trends over time.

Year of Order Date	Segment	Difference in Sales from the Previous along Year of Order Date, Segment
2018	Consumer	
	Corporate	-137,662
	Home Office	-38,719
2019	Consumer	176,820
	Corporate	-137,779
	Home Office	-53,518
2020	Consumer	221,625
	Corporate	-89,758
	Home Office	-101,871
2021	Consumer	226,669
	Corporate	-90,057
	Home Office	-82,385

Figure 3.34: Difference from the previous value

Percentile

A percentile calculation calculates the rank of each value as a percentile within the view. Take note that this will create a rank with a starting point of 0% to a maximum of 100%. *Figure 3.35* shows how this will look:

Year of Order Date	Segment	Percentile of Sales along Table (Down)
2018	Consumer	72.73%
	Corporate	27.27%
	Home Office	9.09%
2019	Consumer	81.82%
	Corporate	36.36%
	Home Office	0.00%
2020	Consumer	90.91%
	Corporate	54.55%
	Home Office	18.18%
2021	Consumer	100.00%
	Corporate	63.64%
	Home Office	45.45%

Figure 3.35: Percent ranking

Custom Table Calculations

The following functions are ones that, instead of being calculated in a measure pill (the green pill-shaped icon used in a Tableau worksheet), utilize a calculated field.

INDEX

An INDEX function allows the user to calculate a row-level count for a table. This index starts at 1 and continues the count down the table. There will not be a recount per section unless you specify how the table calculation is calculated.

Year of Order Date	Month of Order Date	Index	Sales	Profit	Profit Ratio
2018	January	1	29	-9	-31.54%
	April	2	7	2	33.75%
	May	3	16	8	49.00%
	July	4	5	-12	-250.00%
	August	5	14	7	48.00%
	September	6	24	5	21.11%
	November	7	216	37	17.19%
	December	8	153	-2	-1.39%
2019	February	9	44	11	25.82%
	March	10	74	19	26.03%
	April	11	48	9	17.74%
	May	12	71	28	39.93%
	June	13	43	21	49.00%
	July	14	52	-1	-1.46%
	August	15	37	12	32.13%
	September	16	23	-7	-30.40%
	October	17	32	-22	-70.00%
	November	18	29	7	25.37%
	December	19	100	21	20.42%
2020	February	20	11	5	47.00%
	March	21	65	26	40.27%

Figure 3.36: Example table of rankings

The function of this calculation is INDEX ().

You can create this calculation to find a specific row. For example, if you needed to filter the first 25 rows of data, you could use a boolean calculation via `INDEX() <= 25`. The following screenshot shows the result:

Year of Order Date	Month of Order Date	Index	Sales	Profit	Profit Ratio
2018	January	1	14,237	2,450	17.21%
	February	2	4,520	862	19.08%
	March	3	55,691	499	0.90%
	April	4	28,295	3,489	12.33%
	May	5	23,648	2,739	11.58%
	June	6	34,595	4,977	14.39%
	July	7	33,946	-841	-2.48%
	August	8	27,909	5,318	19.05%
	September	9	81,777	8,328	10.18%
	October	10	31,453	3,448	10.96%
	November	11	78,629	9,292	11.82%
	December	12	69,546	8,984	12.92%
2019	January	13	18,174	-3,281	-18.05%
	February	14	11,951	2,814	23.54%
	March	15	38,726	9,732	25.13%
	April	16	34,195	4,187	12.25%
	May	17	30,132	4,668	15.49%
	June	18	24,797	3,336	13.45%
	July	19	28,765	3,289	11.43%
	August	20	36,898	5,356	14.52%
	September	21	64,596	8,209	12.71%
	October	22	31,405	2,817	8.97%
	November	23	75,973	12,475	16.42%
	December	24	74,920	8,017	10.70%
2020	January	25	18,542	2,825	15.23%
	February	26	22,979	5,005	21.78%
	March	27	51,716	3,612	6.98%
	April	28	38,750	2,978	7.68%
	May	29	56,988	8,662	15.20%
	June	30	40,345	4,750	11.77%
	July	31	39,262	4,433	11.29%
	August	32	31,115	2,062	6.63%
	September	33	73,410	9,329	12.71%
	October	34	59,688	16,243	27.21%
	November	35	79,412	4,011	5.05%
	December	36	96,999	17,885	18.44%
2021	January	37	43,971	7,140	16.24%
	February	38	20,301	1,614	7.95%

Figure 3.37: Table highlighting the first 25 rows

Rank

There are multiple ways in Tableau to sort the data in a field into ascending or descending order, and one method is by using the RANK function. RANK assigns each row a ranking number based on the expression specified. For example, the following calculation, referenced in *Figure 3.38*, will result in a field that assigns ranks based on sales:

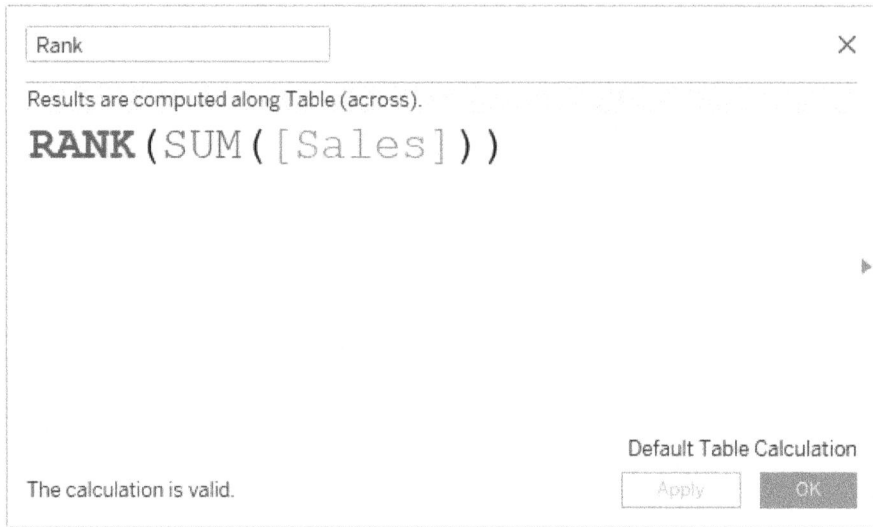

Figure 3.38: Tableau calculation for RANK

There are other levels for ranks, depending on the requirement. Take the following set of numbers and see how they would look under the different RANK functions:

(7,4,6,6,1)

- RANK_DENSE: The default ranking for this function is descending so the result will be (4,3,2,2,1), where identical values are assigned the same value

- RANK_MODIFIED: The default ranking is also descending, as previously, so the result will look like this (5,4,3,3,1), in which identical ranks will be assigned the same value

- RANK_PERCENTILE: The default ranking for this function is ascending, which then ranks the partitions by a percent value (0,0.5,0.5,0.75,1)

- RANK_UNIQUE: Unlike DENSE and MODIFIED, RANK_UNIQUE assigns a different value for each partition (5,4,3,2,1) and ranks the values in descending order by default

It is important to note that `null` values are ignored when using ranking functions. They will not be numbered or count toward the total number of records.

First-and Last

The `First` and `Last` functions return the number of rows in the partition. The `First` function starts at 0 and counts in descending order. Meanwhile, the `Last` function returns an index of the number of rows from the last number, descending to 0.

Year of Order Date	Segment	Sales	First along Table (Down)	Last along Table (Down)
2018	Consumer	266,097	0	11
	Corporate	128,435	-1	10
	Home Office	89,716	-2	9
2019	Consumer	266,536	-3	8
	Corporate	128,757	-4	7
	Home Office	75,239	-5	6
2020	Consumer	296,864	-6	5
	Corporate	207,106	-7	4
	Home Office	105,235	-8	3
2021	Consumer	331,905	-9	2
	Corporate	241,848	-10	1
	Home Office	159,463	-11	0

Figure 3.39: Example of the First and Last functions

These functions commonly appear in the exam, so it is imperative to explore them and see the results they produce when editing table calculations and how the calculations are computed.

Summary

Tableau allows for the creation of additional fields to supplement existing data sets using its calculated fields feature. Calculated fields can range from using simple pieces of logic to manipulate or transform existing fields to complex conditional statements and calculations fixing values at specific levels of aggregation.

Data types in calculated fields must be consistent and so must the aggregation of fields used. When the fields referenced in calculations are not aggregated, then the logic is applied row by row. When the fields are aggregated, then the aggregation takes place first, followed by the logic applied. Calculated fields can also be created without any reference to existing fields; for example, if a single string was to be referenced as a field, that string could be typed into and saved as a calculated field.

Aggregation in calculated fields allows you to pre-aggregate measures as well as to conduct logic on aggregated measures such as dividing one total value by another. Advanced mathematical aggregations can also be applied for statistical analysis. If an aggregation needs to be fixed regardless of the level of detail on the view, then Tableau's fixed level of detail calculations can be used.

The numerical functions available in Tableau's calculated fields are for transforming existing numeric inputs as well as formatting or updating the values of existing numerical inputs. There is also a multitude of trigonometry functions available in Tableau.

String functions in Tableau allow users to extract parts of a string based on indexes and positions. Whitespace can also be trimmed from either side of a string input. Substrings within strings can be identified as present and replaced if need be and the exact position of a substring can also be identified. There are also functions for splitting out parts of a string and returning the length as well as the ASCII code of a character.

Date fields can be created in Tableau from fields already in a date format, from integers representing days, months, and years, and also from strings if the format Tableau requires to read the string is supplied. Date parts can also be extracted from strings, both as names and integers. The difference between two dates can easily be calculated and so can new dates, given a starting date and the desired interval.

Tableau does not always read field types in the desired format, so there are a variety of functions available to convert fields from one format to another. This can also be useful for any temporary conversions required for logic within calculations that may not be desired at the data source level.

Conditional logic in Tableau can be as simple as identifying null values, dates, or items in a list. However, this can be expanded out to IF statements and CASE statements that have multiple outputs given multiple possible conditions.

Exam Readiness Drill – Chapter Review Questions

Apart from a solid understanding of key concepts, being able to think quickly under time pressure is a skill that will help you ace your certification exam. That is why working on these skills early on in your learning journey is key.

Chapter review questions are designed to improve your test-taking skills progressively with each chapter you learn and review your understanding of key concepts in the chapter at the same time. You'll find these at the end of each chapter.

> **How to Access these Resources**
>
> To learn how to access these resources, head over to the chapter titled *Chapter 9, Accessing the Online Practice Resources.*

To open the Chapter Review Questions for this chapter, perform the following steps:

1. Click the link – `https://packt.link/TDA_CH03`.

    Alternatively, you can scan the following **QR code** (*Figure 3.40*):

Figure 3.40: QR code that opens Chapter Review Questions for logged-in users

2. Once you log in, you'll see a page similar to the one shown in *Figure 3.41*:

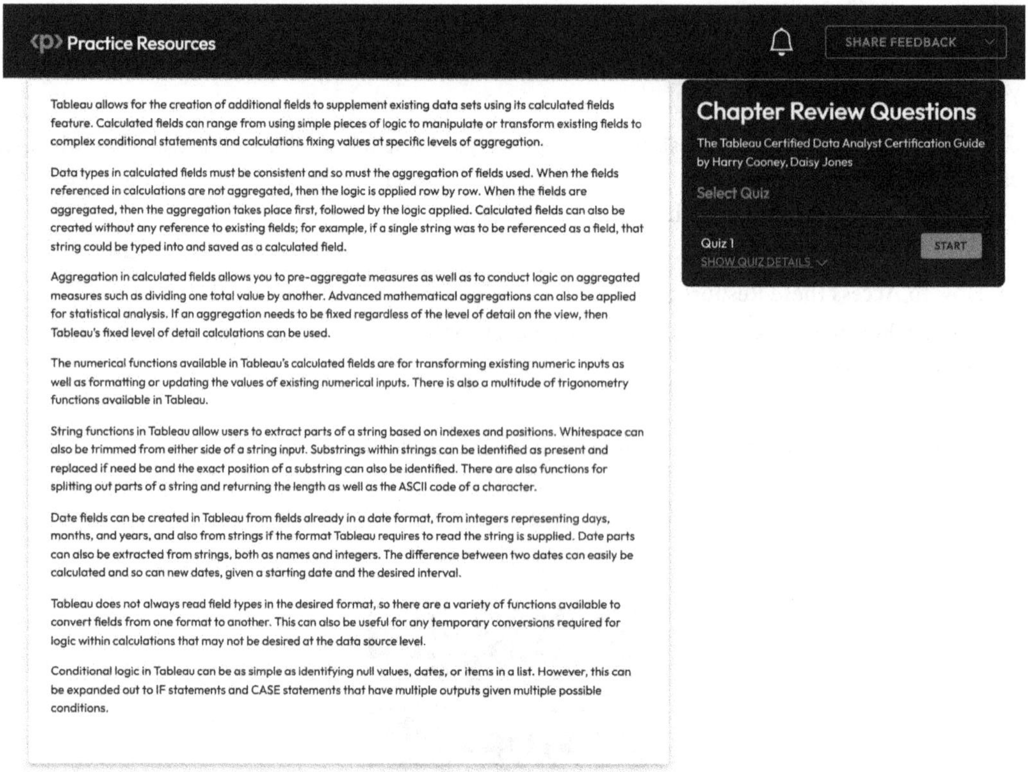

‹p› Practice Resources   🔔   SHARE FEEDBACK ⌄

Tableau allows for the creation of additional fields to supplement existing data sets using its calculated fields feature. Calculated fields can range from using simple pieces of logic to manipulate or transform existing fields to complex conditional statements and calculations fixing values at specific levels of aggregation.

Data types in calculated fields must be consistent and so must the aggregation of fields used. When the fields referenced in calculations are not aggregated, then the logic is applied row by row. When the fields are aggregated, then the aggregation takes place first, followed by the logic applied. Calculated fields can also be created without any reference to existing fields; for example, if a single string was to be referenced as a field, that string could be typed into and saved as a calculated field.

Aggregation in calculated fields allows you to pre-aggregate measures as well as to conduct logic on aggregated measures such as dividing one total value by another. Advanced mathematical aggregations can also be applied for statistical analysis. If an aggregation needs to be fixed regardless of the level of detail on the view, then Tableau's fixed level of detail calculations can be used.

The numerical functions available in Tableau's calculated fields are for transforming existing numeric inputs as well as formatting or updating the values of existing numerical inputs. There is also a multitude of trigonometry functions available in Tableau.

String functions in Tableau allow users to extract parts of a string based on indexes and positions. Whitespace can also be trimmed from either side of a string input. Substrings within strings can be identified as present and replaced if need be and the exact position of a substring can also be identified. There are also functions for splitting out parts of a string and returning the length as well as the ASCII code of a character.

Date fields can be created in Tableau from fields already in a date format, from integers representing days, months, and years, and also from strings if the format Tableau requires to read the string is supplied. Date parts can also be extracted from strings, both as names and integers. The difference between two dates can easily be calculated and so can new dates, given a starting date and the desired interval.

Tableau does not always read field types in the desired format, so there are a variety of functions available to convert fields from one format to another. This can also be useful for any temporary conversions required for logic within calculations that may not be desired at the data source level.

Conditional logic in Tableau can be as simple as identifying null values, dates, or items in a list. However, this can be expanded out to IF statements and CASE statements that have multiple outputs given multiple possible conditions.

Chapter Review Questions

The Tableau Certified Data Analyst Certification Guide by Harry Cooney, Daisy Jones

Select Quiz

Quiz 1   START
SHOW QUIZ DETAILS ⌄

Figure 3.41: Chapter Review Questions for Chapter 3

3. Once ready, start the following practice drills, re-attempting the quiz multiple times.

Exam Readiness Drill

For the first three attempts, don't worry about the time limit.

ATTEMPT 1

The first time, aim for at least **40%**. Look at the answers you got wrong and read the relevant sections in the chapter again to fix your learning gaps.

ATTEMPT 2

The second time, aim for at least **60%**. Look at the answers you got wrong and read the relevant sections in the chapter again to fix any remaining learning gaps.

ATTEMPT 3

The third time, aim for at least **75%**. Once you score 75% or more, you start working on your timing.

> Tip
>
> You may take more than **three** attempts to reach 75%. That's okay. Just review the relevant sections in the chapter till you get there.

Working On Timing

Target: Your aim is to keep the score the same while trying to answer these questions as quickly as possible. Here's an example of how your next attempts should look like:

Attempt	Score	Time Taken
Attempt 5	77%	21 mins 30 seconds
Attempt 6	78%	18 mins 34 seconds
Attempt 7	76%	14 mins 44 seconds

Table 3.1: Sample timing practice drills on the online platform

> Note
>
> The time limits shown in the above table are just examples. Set your own time limits with each attempt based on the time limit of the quiz on the website.

With each new attempt, your score should stay above **75%** while your "time taken" to complete should "decrease". Repeat as many attempts as you want till you feel confident dealing with the time pressure.

4
Grouping and Filtering

Introduction

This chapter will cover the various methods available in Tableau Desktop for grouping values together both across and within fields. This will be followed by a description of the different filtering functionalities in Tableau, including how to configure different types, how to apply filters to multiple worksheets, and the order by which filters operate. The chapter will finish with an explanation of Tableau's parameters and the common use cases for them.

Tableau Desktop's great strength is the ease with which users can manipulate their data and visualize it in different ways. Tableau Desktop allows users to quickly create custom fields based on groupings of values in existing fields, all within the user interface. Filtering is as simple as dragging and dropping, but great flexibility is afforded by different types of filtering and configuring the order in which this occurs. Parameters can easily be added to any workbook as user-selected single-value variables that provide even greater user interactivity.

When it comes to grouping data in Tableau Desktop, there are multiple options available to users. The creation of a Boolean (`true` or `false`) field based on conditions met in another field can be accomplished using sets. Grouping the values of one dimension into new categories (at a higher level of aggregation) in another field can be achieved using Tableau's groups. If two dimensions are related, with one subsuming the other, then the dimensions can be grouped together into a hierarchy. The grouping of numeric fields can also be accomplished by grouping ranges of values into bins.

When it comes to filtering the data shown in a view within Tableau, there are many options available. The type of field influences the filtering method. Numeric filter types such as **more than, less than,** or **between** a range can be implemented using measure filters. Dimensions can be set to include or exclude specific values but can also be set to dynamically filter top and bottom N values by a given metric, filter to values that meet wildcard criteria, or limit the view to values that meet a specific condition. Once filters have been set up on a given sheet, they can also be applied across multiple sheets or entire data sources. Filters can also be set to occur at different times during Tableau's processing and building of a view.

Tableau's parameters are essentially user-defined variables that are always a single value but can be updated at any time. This allows users to create interactive dashboards that will update based on the value of the parameter. Parameters can be of the string, numeric, date, or Boolean type, similar to Tableau field types, and are commonly used to create dynamic calculations, filters, and reference lines.

Data Grouping

This section will take you through the methodologies to group data on Tableau Desktop, including creating sets, groups, hierarchies, and bins.

Sets

Tableau's sets are Boolean fields that effectively group values based on the membership of the set. Values that meet the conditions for membership of a set are considered to be **In**, whereas those that do not are considered **Out**. This allows the set to be used in the same way as a true/false Boolean field within Tableau. Sets are symbolized in Tableau via Venn diagram icons. Sets can only be created on dimensions.

The simplest type of set is one that has the membership defined via the inclusion or exclusion of values in a field. A set can be created by right-clicking a field, selecting **Create**, and then **Set**. The **Set Creation User interface** will then open, and values can be selected from the list shown. These values will, by default, be considered members of the set. The **Exclude** checkbox can be selected to reverse set membership, meaning that any values selected will not be members of the set.

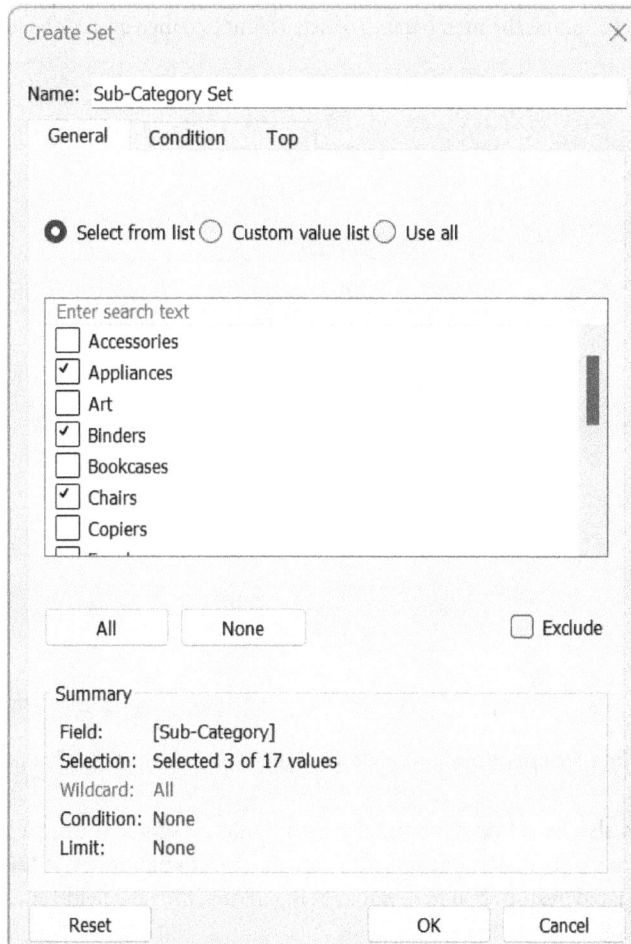

Figure 4.1: The interface to create the new Sub-Category Set set field
with Appliances, Binders, and Chairs as set members

Fixed sets can also be created by selecting data points directly and either hovering over or right-clicking one of the data points, before selecting **Create Set**. This will open a similar set creation user interface, with the member already defined as the data points selected on the visual. It is possible to use this method to create sets across multiple fields. For example, selecting data points with both a subcategory and category on the view will create a set in which the membership incorporates values from both the subcategory and category fields.

These sets are **fixed sets**, in that the membership of sets will not change even if the underlying data does.

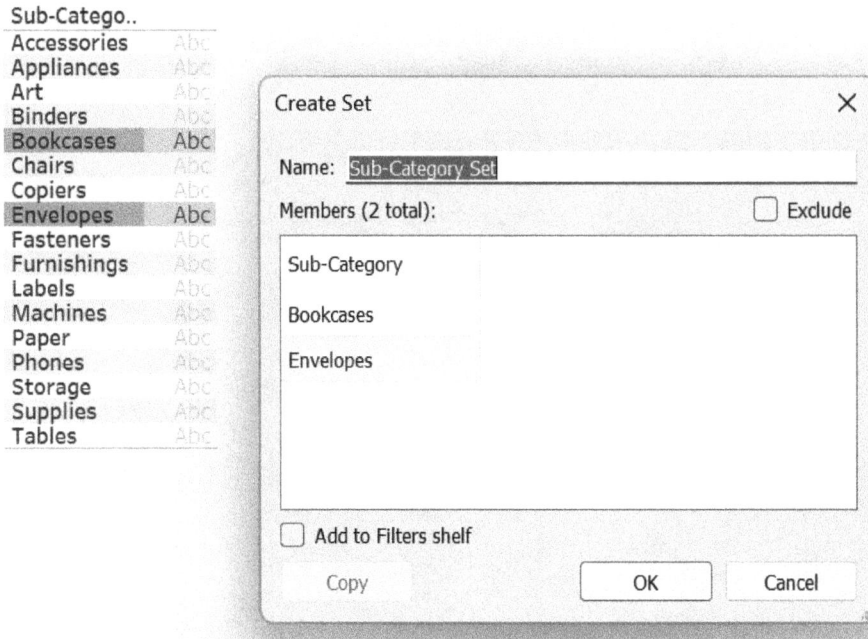

Sub-Catego..
Accessories   Abc
Appliances   Abc
Art   Abc
Binders   Abc
Bookcases   Abc
Chairs   Abc
Copiers   Abc
Envelopes   Abc
Fasteners   Abc
Furnishings   Abc
Labels   Abc
Machines   Abc
Paper   Abc
Phones   Abc
Storage   Abc
Supplies   Abc
Tables   Abc

Create Set   ✕

Name: Sub-Category Set

Members (2 total):   ☐ Exclude

Sub-Category

Bookcases

Envelopes

☐ Add to Filters shelf

Copy   OK   Cancel

Figure 4.2: The interface popup when creating a fixed set via a selection of values on the view

Set membership can also be set up dynamically with dynamic sets. Creating a set from a field and selecting membership via inclusion or exclusion constitutes a dynamic set. If **Use all** is selected and the data source is updated, resulting in new values being added, the new field values will be considered members of the set.

Dynamic sets can be made more advanced via the use of conditional logic. When creating a set by right-clicking a dimension, open the **Condition** tab. The **Condition** tab offers users the ability to set a condition by which values in a field will be considered members or not. For example, a subcategory value could only be considered a member of a set if the average sales for that subcategory are greater than 50,000. If a single, numeric-based condition is not sufficient to define set membership, a custom formula can be used.

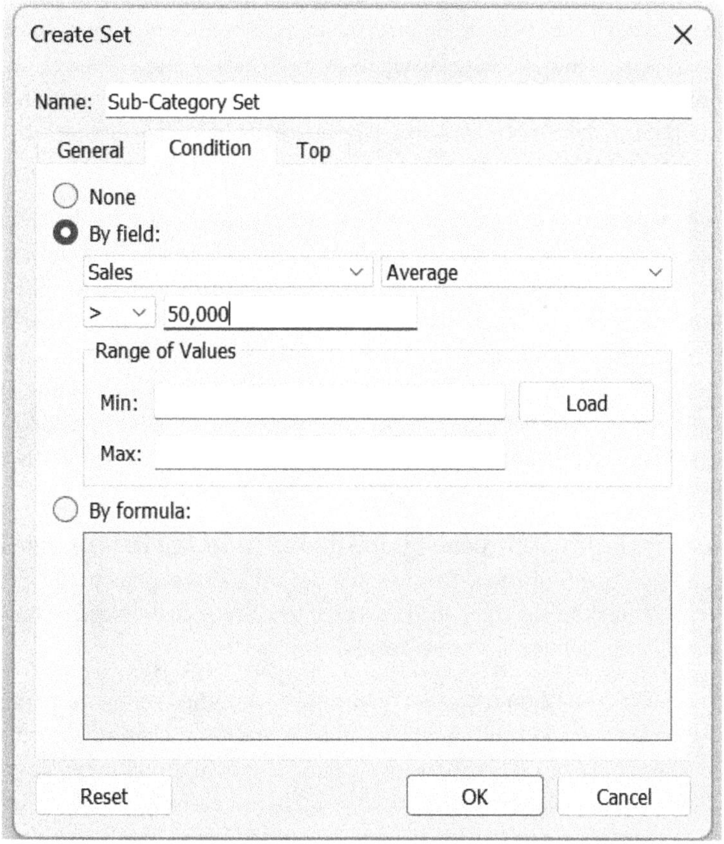

Create Set ☓

Name: Sub-Category Set

General   Condition   Top

○ None

◉ By field:

| Sales | ∨ | Average | ∨ |

| > ∨ | 50,000| |

Range of Values

Min: _____   Load

Max: _____

○ By formula:

_____

Reset   OK   Cancel

Figure 4.3: The interface configuration for a conditional set that selects sub-
category members, based on average sales being greater than 50,000

The **By formula** pane on the **Condition** tab of the set creation user interface allows users to apply Tableau's calculated field language to create Booleans that will define the membership of a set. Similar to the previous example, a subcategory value could be considered a member of a set only if the average sales are greater than 50,000 and less than 100,000. Membership conditions are not limited to numeric logic only; date logic and text-based logic can also be used – for example, a **Country** field could have a set built on it that defines membership by the country name's first letter being a **C**.

Dynamic sets can also have membership defined via a top or bottom **N**. Similar to conditional logic, top or bottom N set membership is defined using its own tab, called **Top**. The user interface is the same as the **Condition** tab, using **By Field** for basic setup or **By Formula** for more advanced membership selection. A top or bottom number (N) is selected by a given field (or calculation), and the dimension values that are in that top or bottom N are considered members of the set.

When it comes to using sets, they function in the same way as a Boolean field, where membership of a set or **In** means true. This means the set can be put on the **Filters** shelf to limit the view to only members of a set. The set can be used on the **Rows** and **Column** shelves to split a view between members and non-members of a set. The set can be placed on any item in the marks card, and they will function accordingly – for example, coloring members of a set one color and non-members another. Sets can also be used in calculated field Boolean logic where **IF [Set Name] THEN…** would result in the following logic only being applied to members of the set. Right-clicking anywhere in the visualization workspace and selecting **Show members in set** will place the set on the **Filters** shelf and limit the view to members only.

If a set is used somewhere in a view, then membership of the set can be made interactive by right-clicking the set and selecting **Show Set**. This will add the **Set Control** pane to the view, which allows users to add or remove membership of a set. The set control can be customized in the same way quick filters can (see the *Filtering* section for more details).

Two sets that have been created on the same field can be combined into a new **combined set**. The combined set can be set to include all members of both sets, only members from one set, or only members shared across both sets. This allows for advanced groupings – for example, customers that are in the top 50 in terms of sales but have a total profit below zero. To create a combined set, both sets must first be selected before right-clicking one, selecting **Create**, and then **Combine Sets**. From there, the user interface allows the user to select the membership combination and the set name.

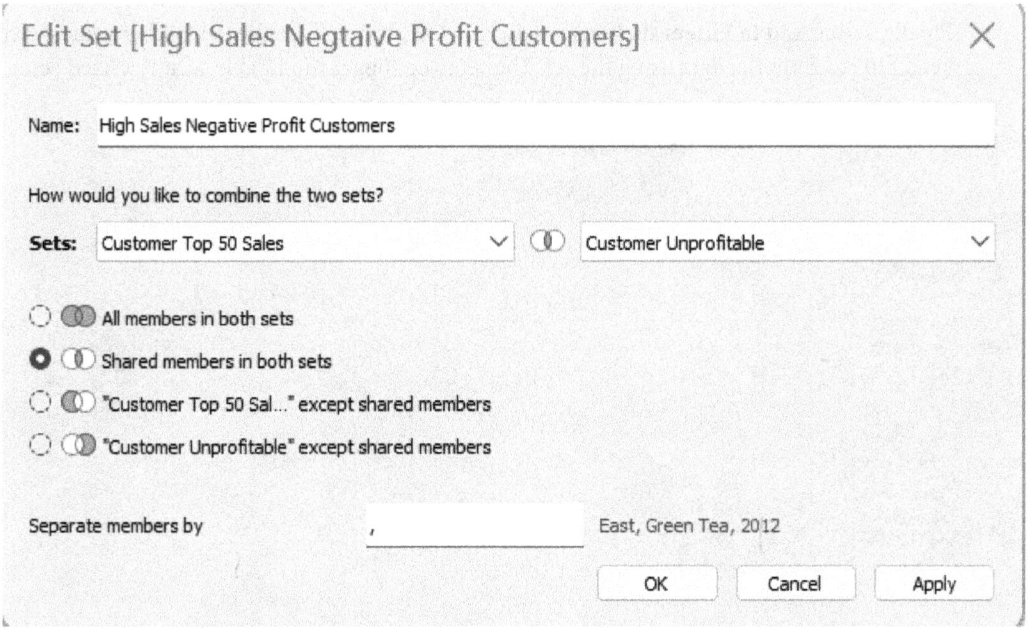

Edit Set [High Sales Negtaive Profit Customers]   ✕

Name: High Sales Negative Profit Customers

How would you like to combine the two sets?

Sets: Customer Top 50 Sales   ∨   ◖◗   Customer Unprofitable   ∨

◯ ◖◗ All members in both sets

⬤ ◖◗ Shared members in both sets

◯ ◖◗ "Customer Top 50 Sal..." except shared members

◯ ◖◗ "Customer Unprofitable" except shared members

Separate members by   ,   East, Green Tea, 2012

                                             OK   Cancel   Apply

Figure 4.4: A combined set configuration to create a set that includes only members that
are both in the top 50 in terms of total sales but have less than zero total profit

Creating a Fixed Set and Filtering Data

In this exercise, you will create a fixed set and use it to filter data in the view. Fixed sets are a speedy way to create sets and are useful for sets with a specific purpose that will not need to be updated in the future:

1. On a new sheet, drag the **Sales** measure to the **Rows** shelf and the **Profit** measure to the **Columns** shelf.

2. Next, drag the **Customer Name** field to **Detail** on the **Marks** card. Two customers at the top stand out as potential outliers that we want to exclude from our analysis (Sean Miller and Tamara Chand).

3. Drag and select both of these data points, hover over a selected point, and then select the **Create Set** option.

4. In the user interface popup, you can see that Sean Miller and Tamara Chand are considered members of the set.

5. Name the set Outliers, and select **Exclude**, as we no longer want to see these customers.

6. Finally, select **Add to Filters shelf** before clicking **OK**. You will see the two outlier values have been filtered from the data using the set. The set is no longer modifiable, as it is a fixed set.

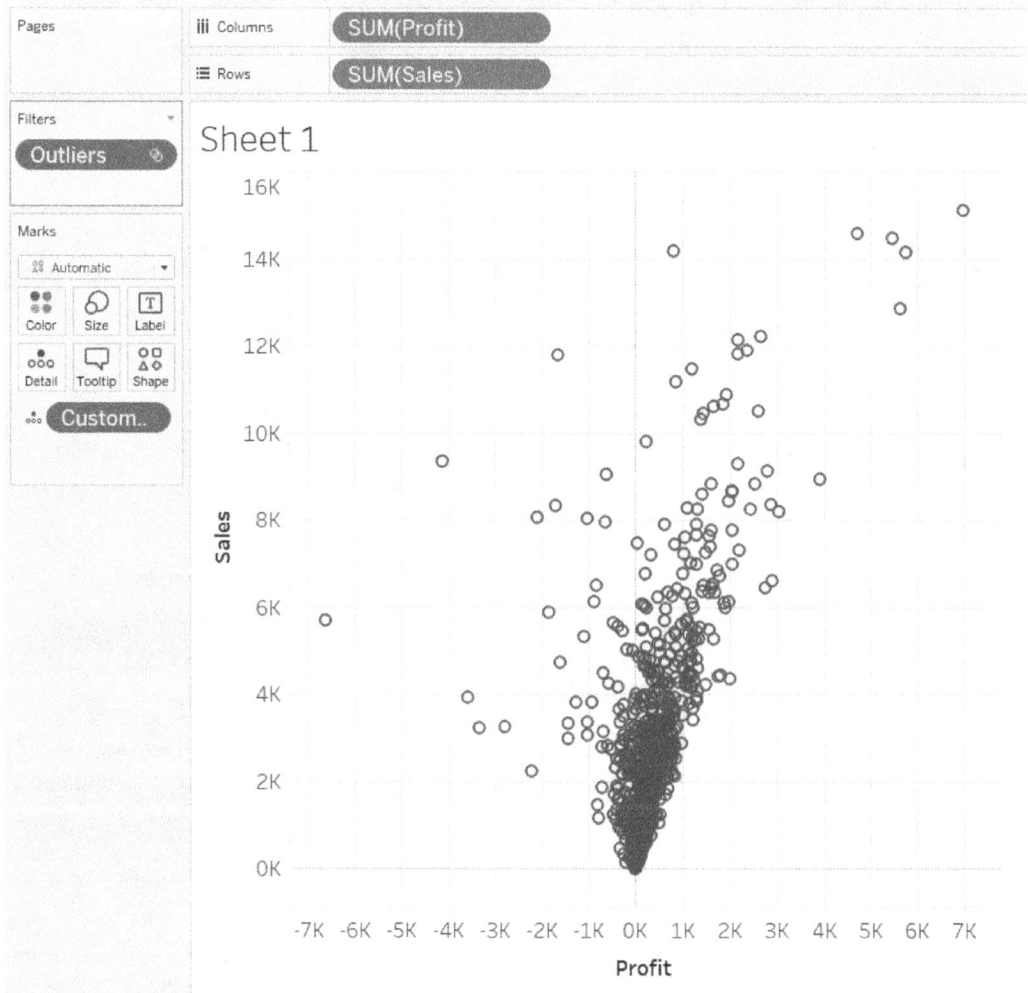

Figure 4.5: Outliers removed using a set on Customer Name

In this exercise, you created a fixed set by directly interacting with the view. You configured it to exclude the selected values, as these are outliers, and then used the set as a filter. You now know how to create a fixed set and have a greater understanding of set functionality.

Creating a Top N Set

In this exercise, you will create a conditional set that has the top 50 customers by total sales as its members. This conditional logic for set membership allows users to set rules that keep their visuals correct, as opposed to having to manually update visuals every time data is updated:

1. To create a set that shows the top 50 customers by sales, right-click the **Customer Name** field and select **Create Set**.

2. Name the set `Customer Top 50 Sales`, and navigate to the **Top** tab.

3. Select **By field**, and set the N value to `50`.

4. Set the measure to sales and the aggregation to sum, and press **OK**.

5.   Placing the newly created set on **Color** on the **Marks** card will result in the customers at the top of the chart (i.e., those with the most sales) being colored differently from those at the bottom.

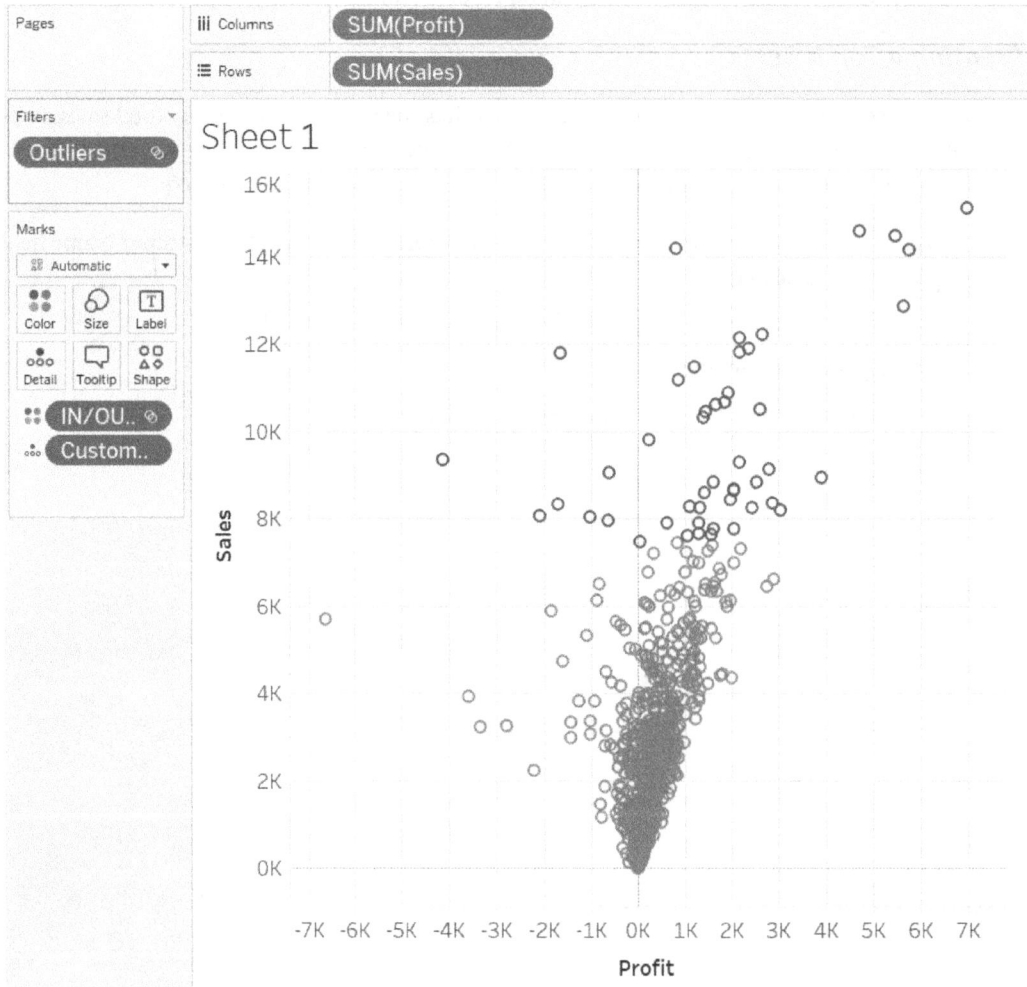

Figure 4.6: The customers in the top 50 sales colored differently from the rest

In this exercise, you have defined a set membership using a condition. This conditional set will not need its membership manually updated, as it will dynamically add and remove customers from the top 50 based on their total sales values. You have also demonstrated how sets can be used to color marks on a visual.

Creating a Conditional Set

Conditional sets can also be created with rules beyond top or bottom N members by a given metric. Rules such as whether a value for a member meets a specific threshold can also be specified. In this exercise, you will create a conditional set that has a threshold condition for membership.

1. To create a set that shows the unprofitable customers, right-click the **Customer Name** field and select **Create Set**.
2. Name the set Unprofitable Customers, and navigate to the **Condition** tab.
3. Select **By field**, and set the field to **Profit**, the aggregation to **Sum**, and the condition to less than 0.

4. Placing the newly created set on **Color** on the **Marks** card will replace the top 50 customer coloring, and customers will instead now be colored by whether or not they are profitable.

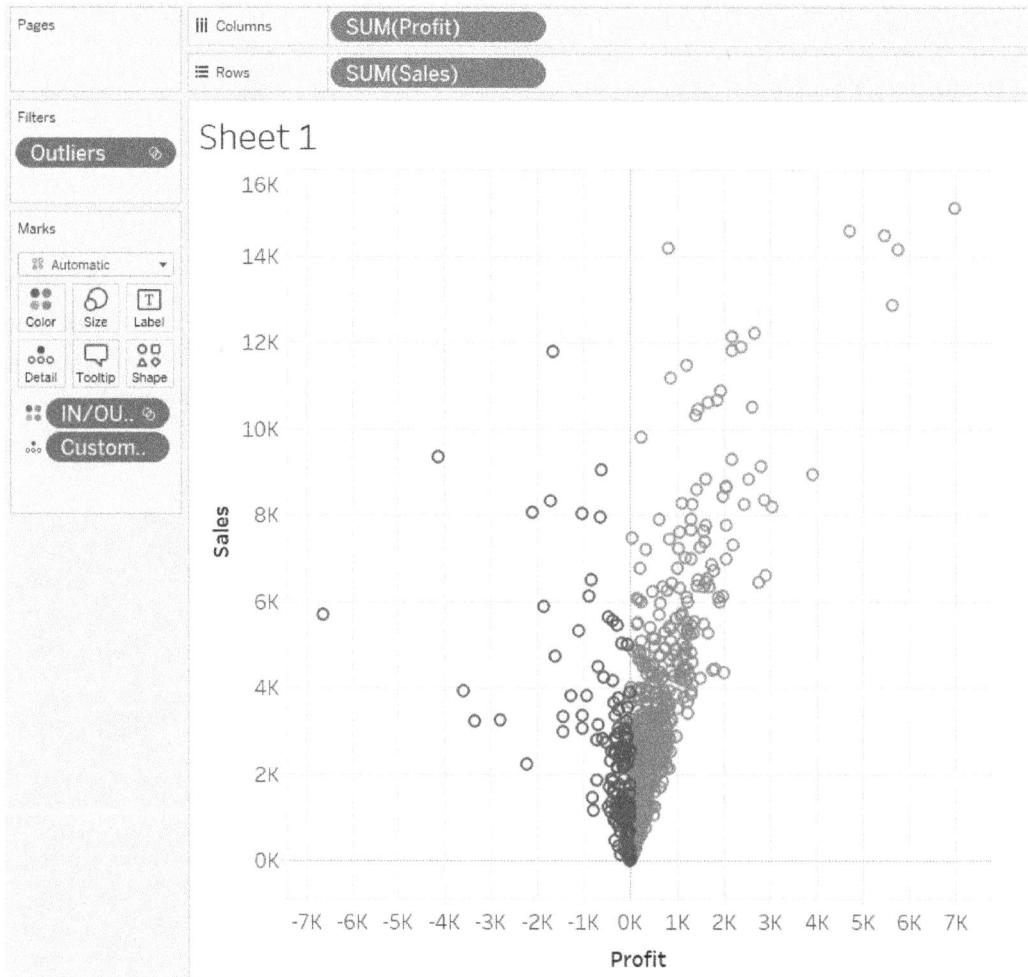

Figure 4.7: The unprofitable customers are colored differently from profitable customers

In this exercise, you defined a conditional set's membership by whether the total profit is below zero. You now know how to apply threshold conditions to define set membership.

Combining Sets for Advanced Analysis

In this exercise, you will create a combined set by merging two sets and defining how the membership across both should define membership in combination. Combined sets are useful to learn, as they allow for the functionality that comes with the intersection of multiple conditions:

1. To create a combined set, select both the **Customer Top 50 Sales** and **Unprofitable Customer** sets. Then, right-click either one, select **Create**, and then **Combined Set**.

2. Name the combined set `High Sales Negative Profit Customers`, and select **Shared members in both sets**.

3. Drop the newly created set onto **Color** to show customers that are both in the top 50 in terms of total sales but are also unprofitable.

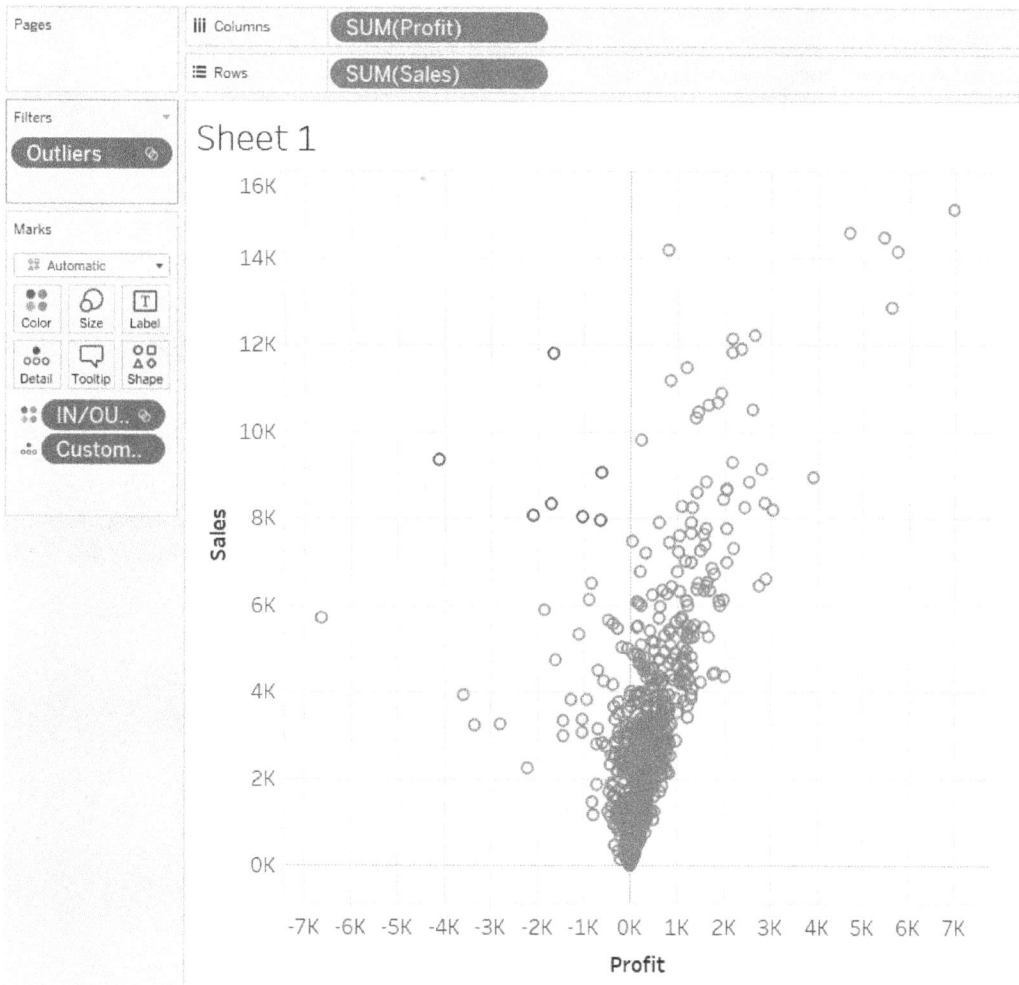

Figure 4.8: Customers with high sales but who are unprofitable are colored differently

In this exercise, you combined two sets into a single set and defined membership as only those items that meet the conditions of both sets. You now know how to configure combined sets.

Groups

Tableau's groups are newly created fields created on Tableau by manually combining the values of an existing field. Groups can be created on any data type but will always be treated as dimensions. Groups can be used to create categorical groupings of multiple lower-level values – for example, grouping countries into North, East, South, and West Europe regions. They can also be used in ad hoc analysis to answer questions such as *"What if we combined North and West Europe?"* Groups can also be used to clean up data – for example, by grouping UK and United Kingdom into a single value. Group fields are represented in Tableau with a paperclip symbol.

Groups can be created directly from the view by selecting the values to be grouped simultaneously, then either hovering over a value and selecting the paperclip icon or right-clicking a selected value, and then clicking **Group**. Creating a group directly from values in the view will add the group with the original field name, plus group as a suffix.

Groups can be edited by right-clicking the field and selecting **Edit Group**. From the user interface, the group field name can be changed and groupings can be selected. Values within the group field are shown in alphabetical order, and grouped values will be emphasized with an arrow and a paperclip symbol. Clicking the arrow will expand the group, and values can be added or removed by dragging and dropping, right-clicking, and selecting **Add to** or **Remove**, or by selecting values followed by the **Group** or **Ungroup** button below. If **Include Other** is checked, then all non-grouped values in the field will be automatically added to the **Other** grouping; this includes any new values added to the dataset when refreshed.

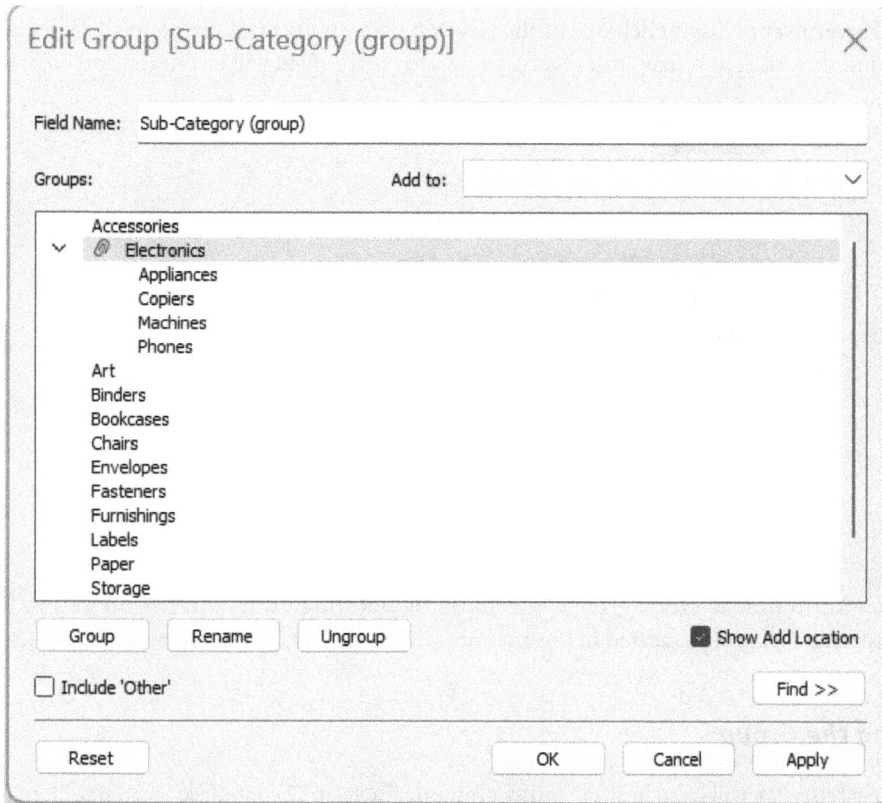

Figure 4.9: The Edit Group user interface showing the creation of a new Sub-Category (group) field that has grouped Appliances, Copiers, Machines, and Phones into a higher-level Electronics grouping

Groups can also be created from the **Data** pane by right-clicking a field directly and selecting **Create** and then **Group**. This will open the same user interface as the edit group functionality mentioned earlier, but without any groups already created.

Once a group field has been created, it will function the same way as any other discrete dimension in Tableau. The group can be used within the view as displayed values breaking down the view, colored groupings, filter options, and so on.

Creating a Group from Data Points in a View

In this exercise, you will learn how to create groups in Tableau by selecting points directly on a view. This is a quick method to create groups instead of having to scroll through lists of values:

1. On a new sheet, drag the Sub-Category field onto **Rows**, and then drag Sales onto **Columns**.

2. Hold down *Ctrl* and multi-select the bars that correspond to **Appliances**, **Copiers**, **Machines**, and **Phones**.

3. Hover over or right-click one of the selected data points until the paperclip icon is visible, and then click it. A new `Sub-Category (group)` field will be created and color the bars.

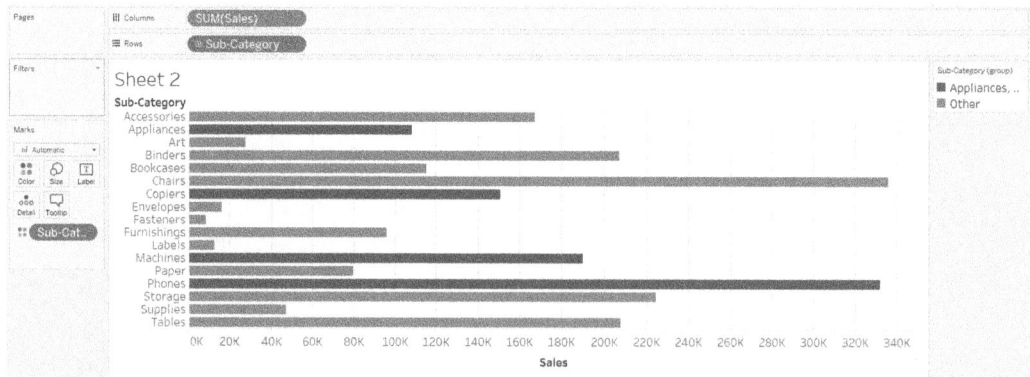

Figure 4.10: The Sub-Category grouping is created and colors the bar chart

You now know how to create groups in Tableau by selecting values directly on a view. You also demonstrated that groups created in this manner will, by default, be placed as a coloring field on the **Marks** card.

Editing the Group

In this exercise, you will learn how to modify groups in Tableau. The interface when modifying a group is the same as when creating a group directly from a field. If more than two groupings are necessary, then groups will have to be modified, as creating a group directly from the view results in a group for the selected values and an **Other** grouping:

1. Right-click `Sub-Category (group)` in the **Data** pane and select **Edit Group**.

2. Rename the field `Office Categories` and uncheck **Include Other**.

3. Rename the **Appliances, Copiers, Machines and 1 more** grouping to `Electronics` by selecting the name and clicking the **Rename** button below.

4. Phones are no longer considered part of the electronics grouping, so remove the **Phones** value from the grouping either by selecting it and dragging and dropping it followed by the use of the **Ungroup** button, or by right-clicking the value and selecting **Remove**.

5. Hold down *Ctrl* and select **Chairs**, **Furnishings**, and **Tables**, and then click **Group** to group the values together, giving the group the name `Furniture`.

6. **Bookcases** was missed as it was not included in the initial grouping process; this can be added to the already created group by either dragging and dropping, right-clicking, and selecting **Add To**, or by selecting the field and using the **Add to** dropdown at the top right.

7. The rest of the values can now be grouped into **Other** by checking **Include Other**.

8.   Click **OK**, and the bar chart will now show the updated groupings with three different colors.

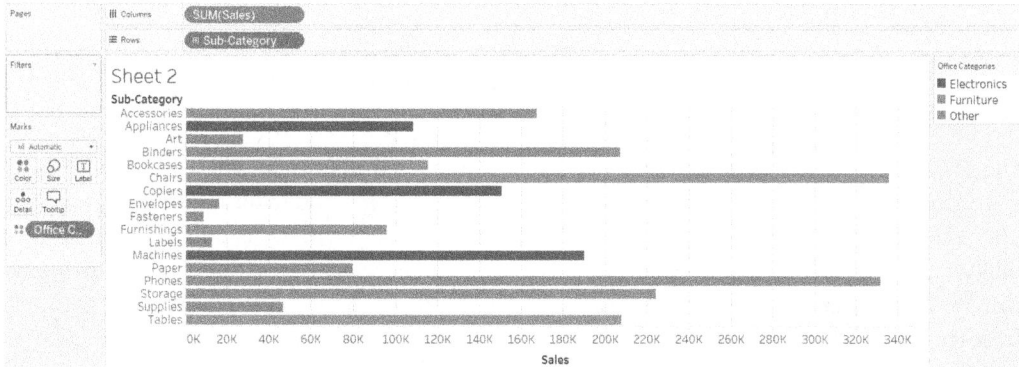

Figure 4.11: The named group field with two groupings and the other grouping coloring the bar chart

In this exercise, you modified the grouping field to create a new grouping and add and remove specific items from groupings. You now know how to both create and modify groups in Tableau.

Hierarchies

Tableau's hierarchies are a method for linking fields that are hierarchical, in that the values of one field subsume the values of another field. A common example is a geographical hierarchy where a continent field has values such as `Europe` that are a higher-level grouping of values in a `Country` field such as `United Kingdom`, `France`, `Spain`, and so on. Lower levels in the same hierarchy could be `State`, `Country`, and `Postcode`.

Other common examples of hierarchies would be levels within an organization or product groupings, but any two (or more) descriptive fields where one field is a higher- or lower-level representation of the other can be made into a hierarchy.

Hierarchies can be created in Tableau's **Data** pane by dragging and dropping the lower-level field in the hierarchy onto the higher-level field. This will cause a popup, asking the user to name the hierarchy. Selecting multiple fields, right-clicking one, selecting **Hierarchy**, and then clicking **Create Hierarchy** will have the same result. Once named, the hierarchy will be visible in the **Data** pane with a hierarchy symbol, the hierarchy name, and the relevant fields shown below, indented. The order fields shown in the hierarchy represent their order in the hierarchy; the lower the field in the **Data** pane view, the lower it is in the hierarchy.

Figure 4.12: Dragging and dropping the Postal Code field on top of the
City field creates a hierarchy from City to Postal Code

Multiple fields can be added or removed from a hierarchy by dragging and dropping or right-clicking, selecting **Hierarchy**, and then **Add to Hierarchy**. The hierarchy order can also be updated by dragging and dropping fields.

Hierarchies can be deleted by right-clicking and selecting **Remove Hierarchy**. This will cause the fields to return to the **Data** pane ungrouped.

The advantage of creating a hierarchy is that it allows users to drill from high-level data to low-level data in the view. Adding a hierarchy to the view will result in the first field in the hierarchy being added. The pill for that field will have a small plus symbol in a box displayed to the left, the drill. Clicking the plus symbol will add the next-level field in the hierarchy to the view. The original field will now have a minus symbol in a box, which can be used to roll up the hierarchy and remove the lower-level field from the view. If the hierarchy has more than two fields, then the second-level field will have the plus symbol, allowing the user to display the third field in the hierarchy, and so on.

Date fields in Tableau are not displayed as hierarchies in the **Data** pane, but when they are added to a view as a discrete dimension, they have the hierarchy functionality. Dragging a date onto the view will, by default, display the year. If the plus icon is clicked, the year and the quarter will be shown. The date hierarchy will then continue when the plus icon is clicked with the month and, finally, the day.

Creating a Product Hierarchy

In this exercise, you will create a hierarchy in the **Data** pane. Hierarchies allow users to drill from one level to another in the view, so it is important to understand how they can be created:

1. On a new sheet, observe the **Data** pane. The `Sample - Superstore` data source, by default, provides the user with two hierarchies, **Location** and **Product**.

2. Start by removing the **Product** hierarchy. Right-click **Product** and select **Remove Hierarchy**. The hierarchy will be removed, and the fields will be selected.

3. Drag the `Sub-Category` field on top of the `Category` field to recreate the hierarchy.

4. Name the hierarchy `Product Hierarchy`.

5. Drag **Product Name** into **Product Hierarchy** below **Sub-Category**.

6. Right-click **Manufacturer**, and select **Hierarchy**, and then add it to the **Product Hierarchy** hierarchy.

7. **Manufacturer** should be above **Product Name** in the hierarchy, so drag it to the correct place.

Figure 4.13: Product Hierarchy recreated from scratch

In this exercise, you removed a hierarchy, created a new hierarchy, added two fields to a hierarchy using different methods, and rearranged the fields in a hierarchy. You are now proficient in hierarchy creation in Tableau.

Using a Hierarchy in a View

Now that you know how to create and manage hierarchies, you need to learn how to utilize them in a view. It is important to understand how hierarchies are utilized so that you can use them yourself as an end user of dashboards, and so that you know how to implement them with the end user in mind:

1. Drag the `Product Hierarchy` field to the **Rows** shelf, and then drag `Sales` to the **Columns** shelf.

2. Hover over the **Category** field on **Rows**, and then click the plus icon to show the **Sub-Category** level in the view.

3. Click the plus icon on **Sub-Category** to show **Manufacturer**, and then repeat on **Manufacturer** to show **Product Name**.

4. Select the minus symbol on any field above **Product Name** in the hierarchy to drill up to that level.

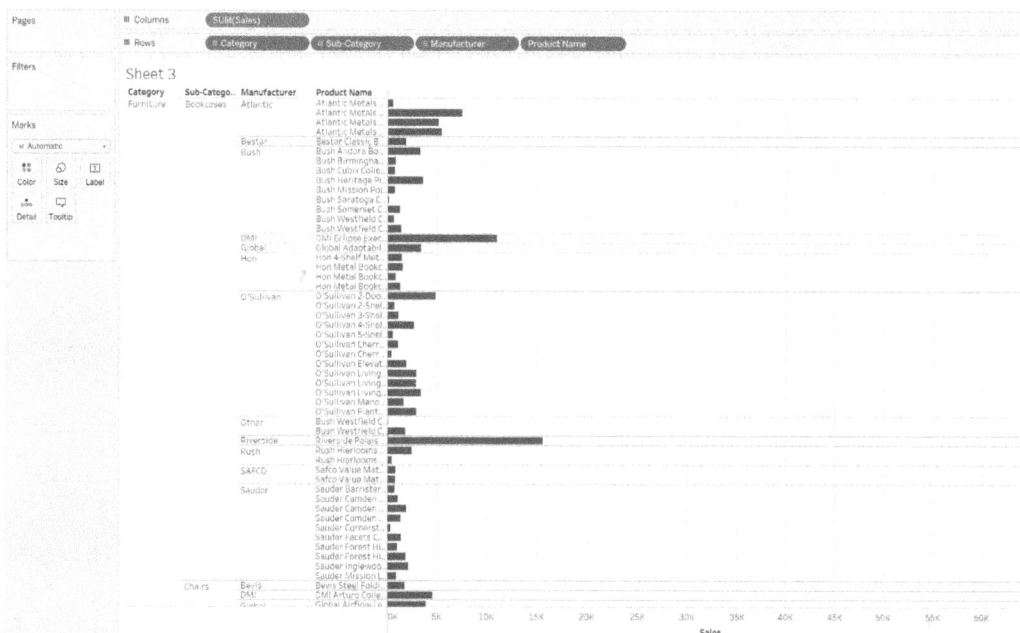

Figure 4.14: A hierarchy added to the view and the levels drilled down and up

In this exercise, you added a hierarchy to a view and drilled both up and down the hierarchy. You now know how hierarchies function for end users in Tableau.

Bins

In Tableau, when a measure is plotted against a discrete field, the categorical values of the discrete field can be described as bins for the aggregated values of the numeric field. Tableau's bin field functionality allows for the binning, or grouping, of a numeric field against a numeric field by creating discrete categories from the ranges of values within a numeric field. Bins are essentially intervals in a numeric field by which the values within it are grouped. For example, when creating bins of size 500 on a sales field, values 0 to 499 would be grouped in bin 0, and the next bin value would be 500, which holds any sales values between 500 to 999, and so on. Bin fields are represented with a histogram symbol.

To create a bin field, right-click the measure you want to create bins on, select **Create**, followed by **Bins**. This will open up a popup in which the new bin field name can be set, as well as the bin size (the size of the interval between groupings). By default, a bin size will be suggested based on the **Number of Bins = 3 + log2(n) * log(n)** formula (although if the calculation is not completed quickly enough, the bin size may default to 10; **Suggest Bin Size** can be selected to force the calculation). The default bin size can be manually overwritten to any number. The range of values for the minimum, the maximum, the difference between the minimum and maximum, as well as the number of distinct rows are displayed in the user interface to aid users in bin creation.

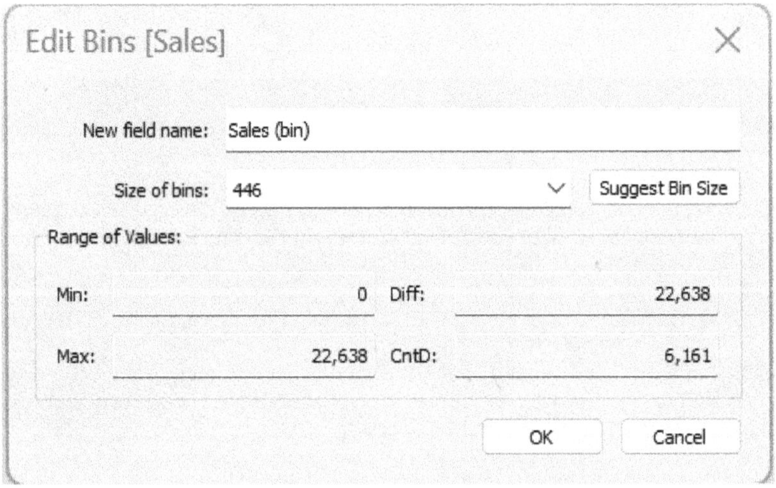

Figure 4.15: The Edit Bins [Sales] user interface allowing you to name the bin field and setting the bin size

Bins can be used in the same way as any other field on the view. By default, bins are discrete dimensions but can also be converted to continuous measures.

Creating a Sales Bin and Counting the Customers in Each

In this exercise, you will create a bin field based on the sales measure. You will learn how to both create bins and specify the bin size, and we will also demonstrate how bins work in a view:

1. On a new sheet, right-click the **Sales** field and select **Create**, followed by **Bins**.

2. Name the bin field Sales (binned by 500), set the bin size to 500, and then press **OK**.

3. To see how many customers fall into each sale bin grouping, drag the **Sales (binned by 500)** field to the **Columns** shelf. Then, hold right-click down on the **Customer Name** field in the **Data** pane, and then drag it to the **Rows** shelf.

4. Select **CNTD** in the aggregation election popup. The view now shows the number of distinct customers in each total sales bin.

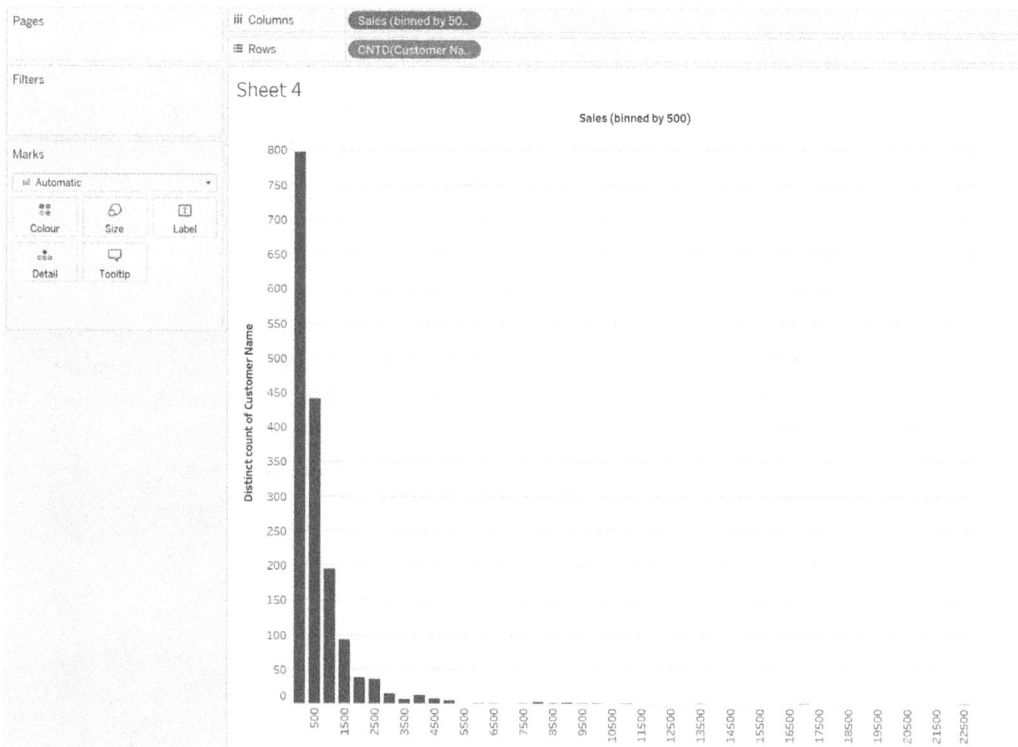

Figure 4.16: The bin field created and the number of customers in each bin displayed

In this exercise, you learned how to create bins and specified a bin size. You also demonstrated that bins hold records that fall within intervals and are a great method for visualizing counts of dimensions along a measure.

Filtering

This section will walk you through Tableau Desktop's filtering functionality, including the types of filters that are available, how to configure filters, how to apply filters across multiple sheets, as well as Tableau's order of operations.

Types of Filters

Filtering the data displayed in a Tableau view is as simple as placing the field you want to filter on the **Filters** shelf and then selecting from the options available which values to show and which not to show.

The options available when filtering will depend on the type of field used. While filtering using a measure, the first decision to make is what style of filtering should be applied to that measure. The initial popup when filtering with a measure asks **How do you want to filter on [FIELD NAME]?**. The first option available is **All Values**, and this will filter the data at a row level. This means that if the data model is at the invoice level and the sale value measure filter is set to show all values less than 50, only invoices with a value less than 50 are shown.

Alternatively, data can be filtered at an aggregated level, meaning that the filter options available will depend on the level of data of the view. For example, if the invoice level data has a country field and this is on the view, and **Sum** is selected as the aggregation level for the sale value measure filter, then the options for filtering will be between the country with the least sales value and the country with the most. The other levels of aggregation available are **Average**, **Median**, **Count**, **Count (Distinct)**, **Minimum**, and **Maximum**, as well as calculations of **Standard Deviation**, **Standard Deviation (Population)**, **Variance**, and **Variance (Population)**.

Filter Field [Sales]   ✕

How do you want to filter on [Sales]?

#	**All values**
#	Sum
#	Average
#	Median
#	Count
#	Count (Distinct)
#	Minimum
#	Maximum
#	Standard deviation
#	Standard deviation (Population)
#	Variance
#	Variance (Population)
#	Attribute

Next >   Cancel

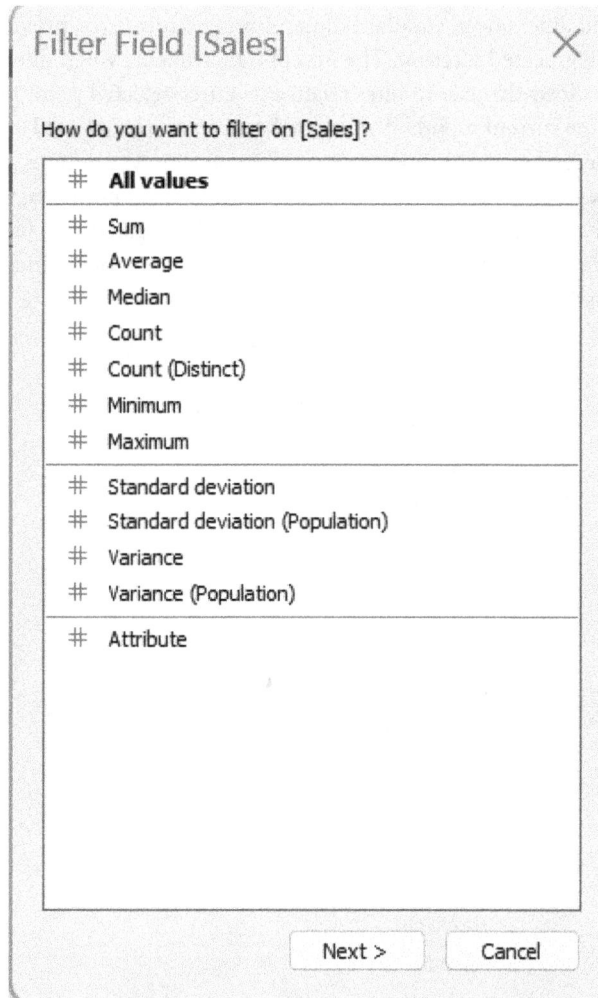

Figure 4.17: The Measure filter aggregation methods available

Using a dimension filter differs from a measure filter in that it does not depend on the level of aggregation and the fields in the view; it is simply a case of selecting which values within a dimension should have their corresponding data included or excluded in the view. Dimension filters are commonly used in the same way as measure filters by dragging and dropping the dimension field onto the **Filters** shelf. Dimension measures can also be created straight from the view by selecting the data points you want to keep or exclude, right-clicking, and then selecting either **Keep Only** or **Exclude**. This will add a field to the **Filters** shelf that then acts as the desired filter. If one dimension is needed for the filtering, then that dimension field will be added as the filter; if multiple dimensions are needed, then a set will be created across both fields and utilized to filter members in or out, based on the user's selection.

Filtering on a date field differs from standard dimensions and measures, providing options for both aggregated and non-aggregated filtering. The first option available when filtering on a date field is **Relative Date**, which allows the user to filter relative to a user-selected point in time – for example, the last three years or the current month. A range of dates can also be selected – for example, between January 1, 2020 and December 31, 2021. Dimension/discrete filter type options are also available, such as selecting specific exact dates, years, quarters, months, days, week numbers, weekdays, month and year combinations, or month, year and day combinations. Aggregated date filtering is also possible via filtering, by a count of dates by dimensions on the view, a count of distinct dates, the minimum date, or the maximum date.

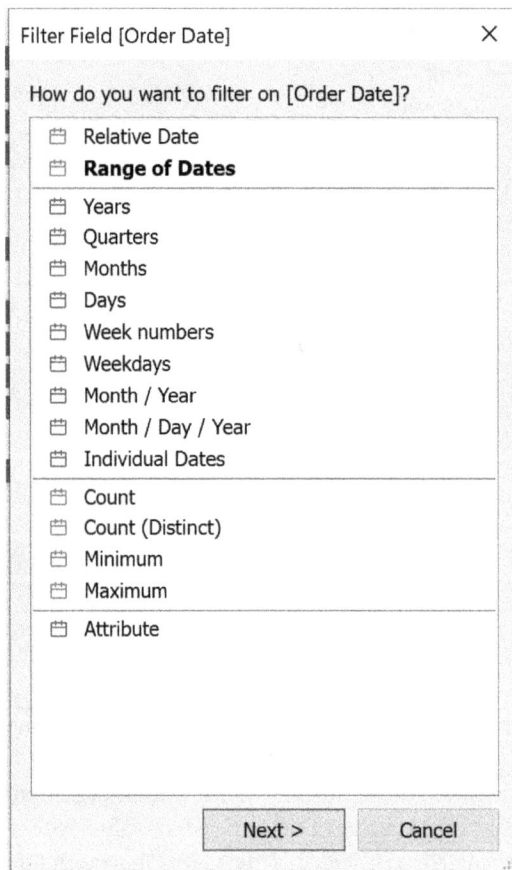

```
Filter Field [Order Date]                              ✕

How do you want to filter on [Order Date]?

  🗓  Relative Date
  🗓  Range of Dates

  🗓  Years
  🗓  Quarters
  🗓  Months
  🗓  Days
  🗓  Week numbers
  🗓  Weekdays
  🗓  Month / Year
  🗓  Month / Day / Year
  🗓  Individual Dates

  🗓  Count
  🗓  Count (Distinct)
  🗓  Minimum
  🗓  Maximum

  🗓  Attribute




                         [ Next > ]      [ Cancel ]
```

Figure 4.18: The date filter methods available

Table calculation filters can also be added to the view if a table calculation has been written and saved in a calculated field. Table calculation filters are added to the **Filters** shelf and have the usual configuration options available, depending on whether the field is a continuous measure or discrete field. Table calculation filters differ in that they do not filter the underlying data; they instead wait until the view has been evaluated, including any table calculations, and then filter based on how they are configured directionally over the view.

Filtering Directly from the View

In this exercise, you will learn how to filter records in a Tableau view by selecting them directly. You will learn how to keep selected records only and also to exclude selected records. This methodology is often useful for excluding outliers:

1. On a new sheet, add the **Sub-Category** dimension to **Rows** and the **Sales** measure to **Columns**.

2. Drag the cursor over the chart to select the top three bars, **Accessories**, **Appliances**, and **Art**.

3. Right-click one of the bars and select **Keep Only**. A **Sub-Category** filter will be created on the **Filters** shelf that includes the **Accessories**, **Appliances**, and **Art** sub-categories only.

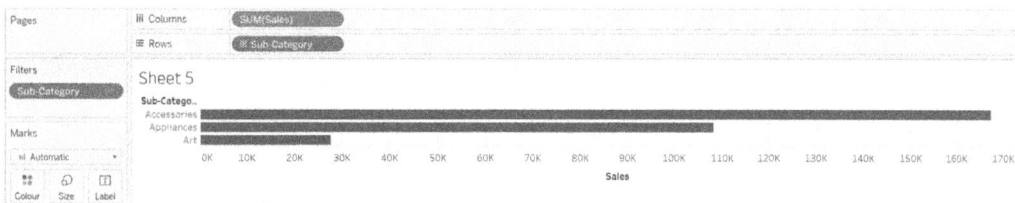

Figure 4.19: A Sub-Category filter created directly from the view by selecting data points

4. Repeat the exercise, but select **Exclude** instead of **Keep Only**. Note how the **Sub-Category** filter created now excludes instead of includes the **Accessories**, **Appliances**, and **Art** sub-categories.

In this exercise, you filtered by selecting records in the view, directly followed by **Keep Only** or **Exclude**. You now know the difference between **Keep Only** or **Exclude** and can filter any data points you observe on a Tableau chart.

Filter Configuration

Filter configuration options differ based on the type of filter used. Once a filter has been configured, it can be shown on the view by right-clicking the filter on the **Filters** shelf and selecting **Show Filter**.

The configuration options available for a measure filter, once the level of aggregation for the filter has been selected, are to set a range of values, a minimum value, a maximum value, or a special filter that can include or exclude `null` values. For the first three options, null values are excluded by default but can be included by checking the **Include Null Values** checkbox. Values can be typed in or the slider in the user interface can be used. By default, only relevant values are shown based on data available in the view, but if you would like to show a full range of filter options, the **Show** dropdown can be switched from **Only Relevant Values** to **All Values in Database**. Once the filter has been configured, it can be shown on the view as a slider (this can be changed from **At least** to **At most**).

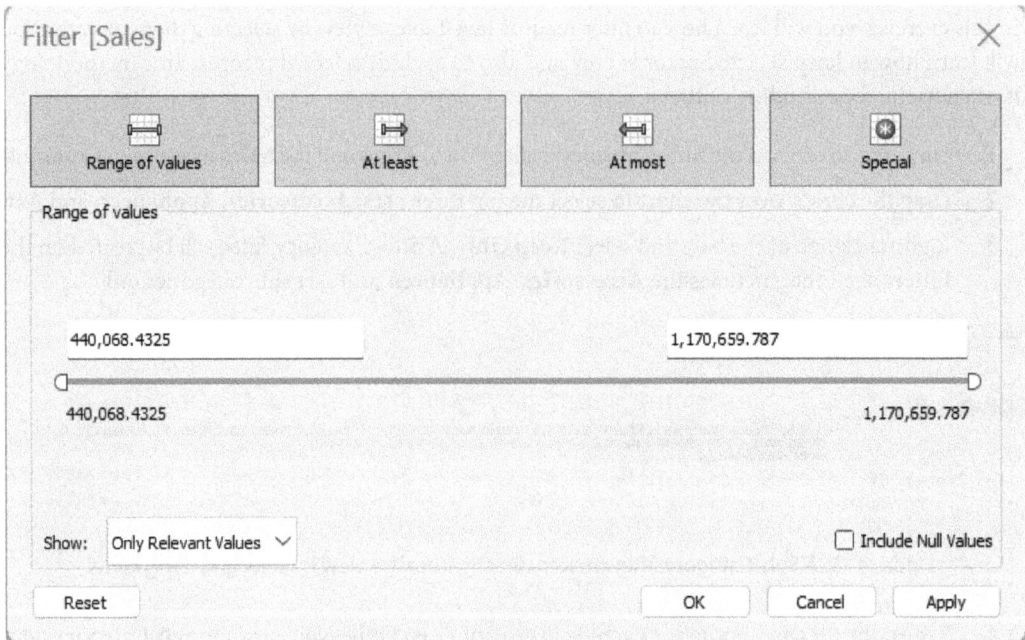

Figure 4.20: The measure filter configuration

Dimension filter configuration offers the simple include and exclude functionality for values within a dimension on the **General** tab of the **Filter configuration** popup. From this tab, all the values of a field are listed and can be checked or unchecked to specify whether they will be included in the view, but if the **Exclude** checkbox is selected, they will be excluded instead.

Dimension filters can also have more advanced logic applied. The second tab in the **Dimension filter configuration** popup allows the user to input a **Wildcard** value that the filter can check each dimension value for. The filter can be set to check whether the text input into the **Wildcard match value** box is contained anywhere within a dimension value, whether the dimension value starts with the wildcard text, whether it ends with it, or whether there is an exact match. This could be useful, for example, when trying to filter phone numbers to a specific area code. In that case, the wildcard match input would be the area code (e.g., +44) and the type would be **Starts with**. Again, the option to exclude values instead of include them is available, and whether or not to include all values if there are no matches to the wildcard text is also possible.

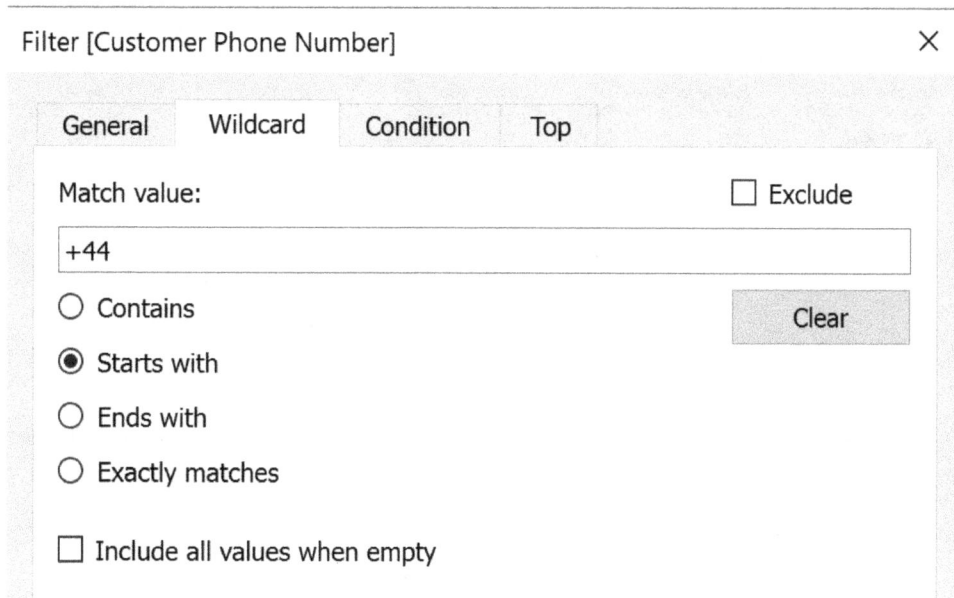

Figure 4.21: Wildcard filter configuration to limit data in the view to UK phone numbers only

The **Condition** tab in the dimension filter configuration popup allows users to include or exclude based on a set piece of logic. The **By field** section of the configuration allows for filtering on whether another field in the dataset when aggregated is equal to, not equal to, less than, greater than, less than or equal to, or greater than or equal to a given value. More complex logic can also be written using Tableau's calculated field language in the **By formula** section. This allows for multiple conditions to be set but requires a Boolean (`true` or `false`) output.

The **Top** tab in the dimension in the filter popup for dimensions allows users to filter a dimension to values that are in the top or bottom N (the selected number) by a given measure. The **By field** section facilitates selecting whether to show the top or bottom N, as well as by how many and by what measure and level of aggregation. The **By formula** section provides the same functionality but for a measure created by written Tableau calculated field logic.

After selecting **Show Filter** on a dimension filter, the filter will be shown in the view providing the functionality to select or deselect values in that dimension, in order to include or exclude them from the view. The interactive filter can be set to display as either a single value or multiple value (selection) list or dropdown, a single value slider, multiple values from a custom list, or a wildcard match search box.

Date filter configuration depends on the type of date filter selected. Relative date filters have their own user interface window for the selection of the relative date period. Discrete date types function in the same way as dimension filters, and aggregated date filters work in the same way as measure filters.

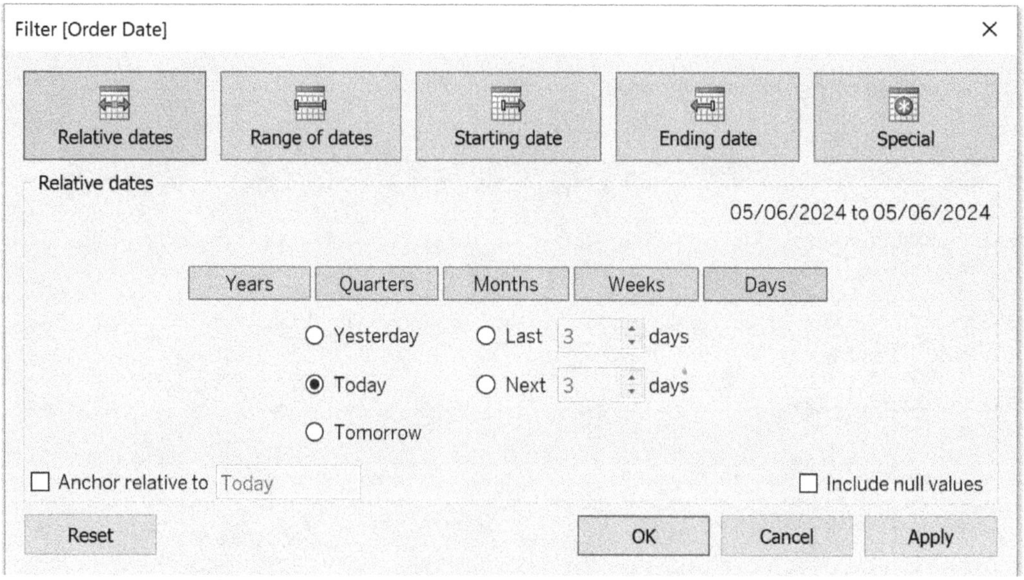

Figure 4.22: Relative date filter configuration allows you to select an anchor
date, as well as the period relative to that date, to filter the data to

Table calculated fields used as filters can be used as either continuous measures or discrete filters. Continuous table calculation filters have the same configuration as measures. Discrete table calculation measures have a similar configuration to dimension filters in terms of the inclusion and exclusion of values but with a wildcard. Conditional and top/bottom N filters are not permitted. For a table calculation filter to be applied to totals in the view, the filter must be right-clicked, and **Apply to Totals** must be selected. The filter values available will depend on the table calculation configuration.

Creating a Wildcard Filter

In this exercise, you will create a wildcard filter in Tableau Desktop. There are multiple ways to set up dynamic filtering, and this example of wildcard filtering allows for values to be kept or excluded based on the presence of specific substrings:

1. On a new sheet, drag **Ship Mode** to the **Rows** shelf.

2. Drag **Sales** to the **Columns** shelf to create a bar chart showing the total sales by each ship mode.

3. To limit the view to only ship modes with the word **Class** as a suffix, right-click **Same Day** in the view, and then click **Exclude**. This creates a **Ship Mode** filter that excludes same-day delivery.

4. If the data were to update and a new method of delivery was brought into the dataset that was neither **Same Day** nor ended in **Class**, it would show in the bar chart. If the exclusion filter was edited to become an inclusion filter with **First Class**, **Second Class**, and **Standard Class** selected, and then a new ship mode ending in **Class** was added to the dataset, it would not show on the bar chart.

5. To create a dynamic filter that shows any ship mode ending with the word **Class**, edit the existing **Ship Mode** filter on the **Filters** shelf by right-clicking the field and selecting **Edit Filter**.

6. In the first tab in the filter configuration popup, select the **Use all** radio button at the top right to remove the exclusion logic.

7. Open the **Wildcard** tab and type **Class** into the **Match value** box.

8. Select **Ends with**, and then press **OK**. The bar chart now filters out **Same Day**, as it does not end in the word **Class**, but it will also filter out any new records in the data that do not end in **Class**.

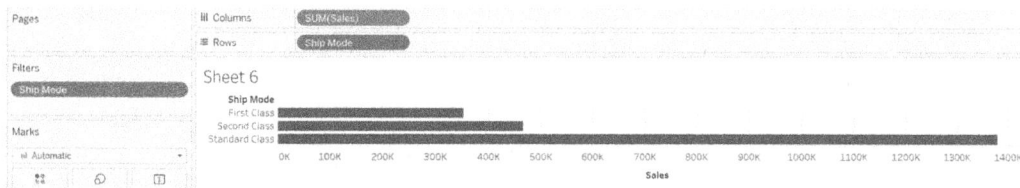

Figure 4.23: Ship Mode filtered to modes with the word Class as a suffix

In this exercise, you successfully created a wildcard filter that excludes all values that end with **Class**. You will now be able to generalize what you have learned and apply wildcard filters for any substring and to values that start and end with the specified substring.

Creating a Conditional Filter

In this exercise, you will create a conditional filter that filters out values based on set rules. These filters are useful, as even if the data is updated, new values can be kept or filtered based on applied logic, removing the need for manual maintenance:

1. On a new sheet, drag **Sub-Category** onto **Rows** and **Sales** onto **Columns** to create a **Sub-Category by Sales** bar chart.

2. To create a filter that limits the view to sub-categories with less than 100,000 sales, drag the **Sub-Category** field onto the **Filters** shelf.

3. Navigate to the **Condition** tab in the filter configuration popup and select **By field**.

4. Set the field to **Sales** and the aggregation to **Sum** to ensure that each sub-category is filtered based on its total summed sales value.

5. Set the condition to **less than** and the value to **100,000**, and then press **OK**.

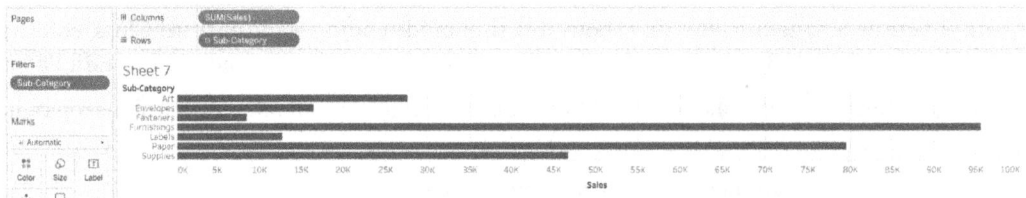

Figure 4.24: The bar chart filtered to sub-categories with less than 100,000 sales

In this exercise, you configured a conditional filter. If the data needs to be updated, the conditional filter will still apply, and the rules will filter out any new data that does not meet the condition. Conditions can be made more complex with multiple clauses, using the **Calculated field** box in the configuration.

Creating a Top N Filter

In this exercise, you will learn how to create a top N filter. This is similar to conditional and wildcard filters, in that updated data will be filtered in accordance with the logic setup:

1. On a new sheet, drag **State/Province** onto **Rows** and **Sales** onto **Columns** to create a total sales by state/province bar chart.

2. To filter to the top 20 states/provinces, drag **State/Province** onto **Filters** and navigate to the **Top** pane.

3. Select **Top**, and change the N value to **20**.

4. Ensure that the measure is **Sales** and the aggregation is **Sum**, and then press **OK**.

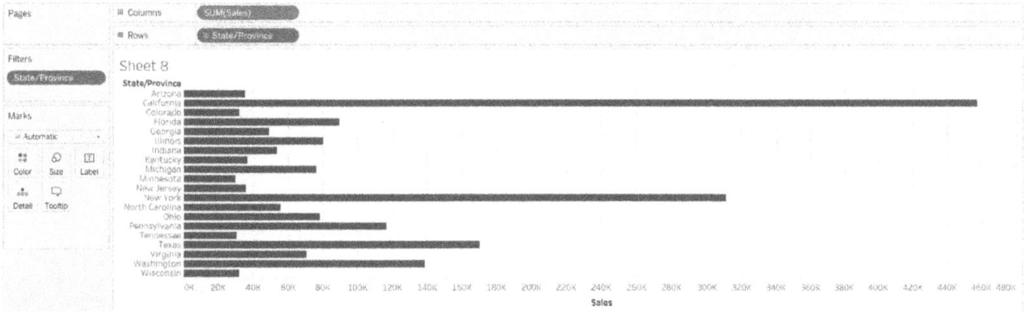

Figure 4.25: The top 20 states/provinces by sales bar chart created

You successfully configured a top N filter that limits the view to the top 20 states/provinces. Any new states/provinces that come in will be filtered based on whether their total sales value is in the top 20 or not. The same configuration can be used to filter for bottom N records.

Creating a Date Filter

In this exercise, you will create a relative date filter. Relative date filters provide a wide range of customizable filtering to end users so are a useful filter type to learn:

1. On a new sheet, drag **Order Date** onto **Rows** and expand the hierarchy so that the year, quarter, month, and day are shown.

2. Now, drag **Order Date** to the **Filters** shelf and select **Relative Date** in the popup.

3. Check **Anchor relative to**, and set the date to 1 January, 2020 using the calendar provided.

4. To show data for January 2020, only select the **Months** tab, and then select **Anchor Month**. This filters the data to the anchor month only (only days with data will be shown), press **OK** to apply the filter.

5. Right-click **Order Date** on **Filters** and select **Show Filter** to display the relative date filter on the view.

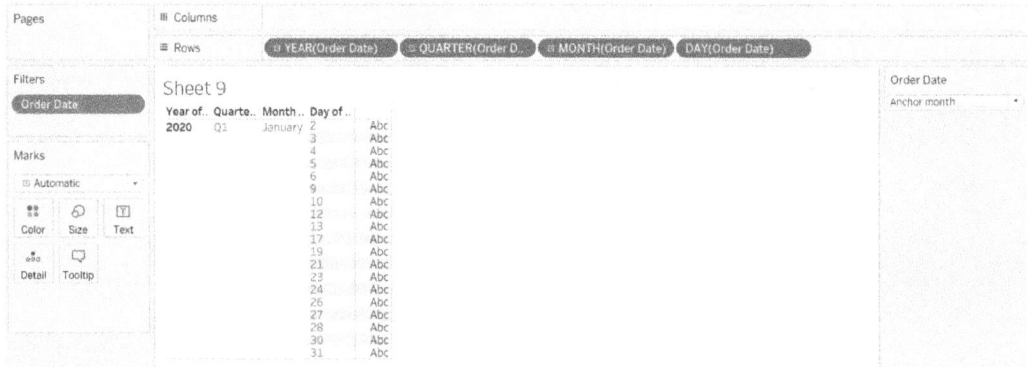

Figure 4.26: Data showing for January 2020 only using relative state filter logic

6. To update the relative date filter logic to show the data for the previous year (2019), click the **Order Date** relative filter on the view (at the top right). In the filter configuration, navigate to **Years**, and then select **Previous year**.

In this section, you added a relative date filter, saw the options available, changed the filter, and observed the effects in the view. You now know how to use a relative date filter as an end user and know when they are suitable to provide to end users as a developer.

Creating a Table Calculation Filter

In this exercise, you will create a table calculation filter. Table calculations run based on the current view configuration and are the last filter to occur in Tableau's order of operations. The fact that table calculation filters occur last in the order of operations provides them with some unique functionalities, so they are worth learning how to implement:

1. On a new sheet, drag **Sub-Category** to **Rows** and **Sales** to **Columns** to create a **Sub-Category by Sales** bar chart.

2. To filter to the top five bars positionally, the index table calculation function can be used as a filter. Create a calculated field by clicking the carrot at the top right of the **Data** pane and selecting **Create Calculated Field**.

3. Call the field **Index**, type INDEX() into the calculated field logic interface, and then press **OK**.

4. The Index function returns the positional number of each item in the view – for example, the **Accessories** index value is 1 whereas **Tables** is 17 (**Index** can be dragged onto **Columns** to provide a visual representation of each item's positional value).

5. To only show the top five bars in the view, drag the **Index** calculation to **Filters**, select **At most**, set the maximum value to 5, and then press **OK**.

6. Use the sort ascending and sort descending icons above the **Columns** shelf to update the sort order of the sub-categories on the view. Observe how the sub-categories being filtered update, based on the sorting, because table calculations filter last in the order of operations.

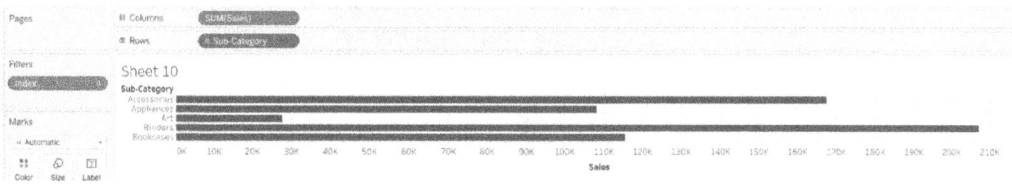

Figure 4.27: Only the top five bars (positionally) in the view are visible

In this exercise, you used the INDEX table calculation that calculates the position of each item in the view. This was used as a filter to show the top five records positionally. You now know how table calculation filters can be useful for creating dynamic views that update with user interaction.

Applying Filters to Single or Multiple Sheets

When filters are created, by default, they only apply to the view that they were created on. All non-table calculation filters can be applied across multiple sheets simultaneously by right-clicking the filter on the **Filters** shelf, selecting **Apply to Worksheets**, and then either **All Using Related Data Sources**, **All Using This Data Source**, or **Selected Worksheets**.

The **All Using Related Data Source** option automatically applies the filter to any sheet in the workbook that uses a data source that is the same as the field being filtered or related data sources (via blending). Similarly, the **All Using This Data Source** option applies the filter to all sheets in the workbook that are from the same data source as the field being filtered. If a new sheet is created that uses the same data source (or a related one), then the filter will be automatically applied. The **Selected Worksheets** option allows the user to apply a filter to sheets of their choosing by selecting those worksheets in the apply filter popup. The filter that is applied will overwrite any existing filters of the same field on that sheet.

By default, filters do not have an icon, but when applied to multiple sheets, an icon will be displayed depending on which methodology has been used. Filters applied to **All Using Related Data Source** show a colored database icon connected to a white database icon. Filters applied to **All Using This Data Source** have a white database icon. Filters applied to **Selected Worksheets** have a stacked sheets icon.

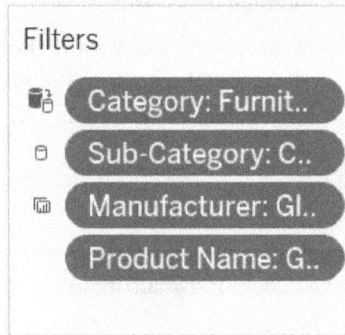

Figure 4.28: Filter icons, with Category applied to All Using Related Data Source, Sub-Category applied to All Using This Data Source, Manufacturer applied to Selected Worksheets, and Product Name applied to the current worksheet only

Applying a Filter to Multiple Worksheets

In this exercise, you will apply a filter to multiple worksheets. This is a useful technique, as it allows users to carry filter options throughout a workbook. It is also incredibly useful when building dashboards that combine many charts, as a single filter can be applied across multiple sheets/charts:

1. Create a new worksheet, double-click or right-click the sheet name, and select **Rename**.

2. Call the sheet Category List, and then drag the **Category** dimension onto **Rows**.

3. Right-click the sheet again and select **Duplicate** to create a copy of the sheet, called Category List (2).

4. On **Category List (2)**, drag the **Category** field onto the **Filters** shelf and ensure that only **Furniture** is selected before pressing **OK**. The view now only displays **Furniture**.

5. To apply the filter to the original **Category List** sheet, right-click the filter, select **Apply to Worksheets**, and then select **Selected Worksheets**.

6. In the apply filter pop-up box, scroll down, check **Category List**, and then press **OK**.

7. Navigate to the **Category List** sheet, which also now only displays **Furniture** and has the **Category** field on the **Filters** shelf, with the selected worksheets icon visible.

Pages	⋮⋮⋮ Columns
	☰ Rows · · · · · ⊞ Category
Filters	
Category: Furniture	Category List
	Category
	Furniture Abc
Marks	

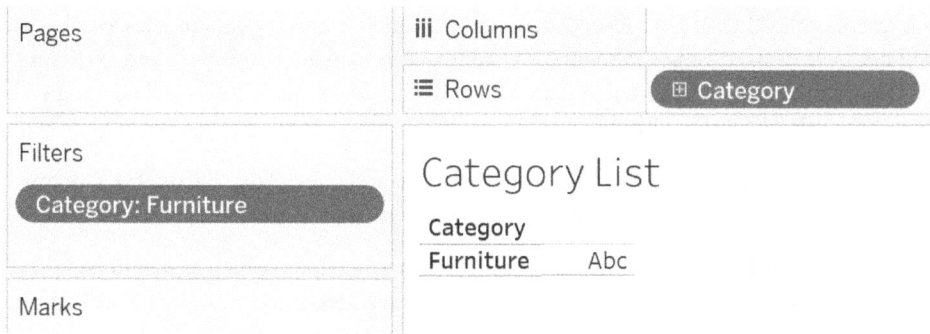

Figure 4.29: The Category filter applied to the selected worksheet

You successfully applied a filter across multiple sheets in a Tableau workbook. You are now able to specify filters that act across a whole workbook or on specific selected sheets, a skill that will come in handy when developing dashboards that contain multiple sheets.

Filtering and the Order of Operations

Filters in Tableau can be configured to occur at different times during Tableau's execution, and this will influence what data is returned to the view. Tableau's order of execution is referred to as the order of operations.

The first filter to occur in Tableau's order of operation is the extract filter. Extract filters are filters that limit the data that is stored in a Tableau extract. They occur during the generation or refreshing of a Tableau extract. Extract filters can be added to an extract during creation by selecting **Add** in the **Filters** section of the **Configuration** pane, and then selecting and configuring the field to be filtered.

The second filter to occur in Tableau's order of operation is the data source filter. Data source filters are similar to extract filters in that they limit data directly from the data source, preventing it from reaching the view, but they do not directly reduce the data contained in the data source. Data source filters can be applied by selecting **Add filters** in the top-right corner of the data source page, or by right-clicking a data source in the top left of the worksheet view and selecting **Edit Data Source Filters**. In the data source filter popup, a field can be selected and configured.

The third filter in Tableau's order of operations is the context filter. Context filters are normal filters, created by placing fields on the **Filters** shelf (as discussed in the chapter previously) but have been specifically designated to occur higher up in the order of operations. Sets, fixed level of detail calculations, conditional filter logic, and top N filter logic are all executed before standard include and exclude dimension filtering. Adding a filter to **Context** places it before the aforementioned executions in the order of operations.

An example of context filtering is a level of detail calculation that fixes the sum of sales total by category and will not have its fixed values influenced by adding **Sub-Category** to **Filters** and excluding some values. However, if the filter is added to context, the fixed category sales totals will be updated by the filtering of the sub-categories within.

To add a filter to context, right-click it on the **Filters** shelf and select **Add to Context**. Context filters can be identified by their color; they are gray as opposed to blue or green.

The fourth filter in Tableau's order of operations is the dimension filter, which as mentioned previously occurs after sets and fixed level of detail calculations are executed, and conditional filter logic and top N filter logic are executed before inclusion and exclusion filtering.

The fifth filter to occur in the order of operations is the measure filter, which is executed after include and exclude level of detail calculations are executed and also after blending has been executed.

The final filter in Tableau's order of operations to be executed is the table calculation filter, which is dependent on the view to work. Forecasts, table calculations, clusters, and totals are all calculated before table calculation filters (unless **Apply to totals** is checked for the filter). Only trend lines and reference lines are calculated after table calculations.

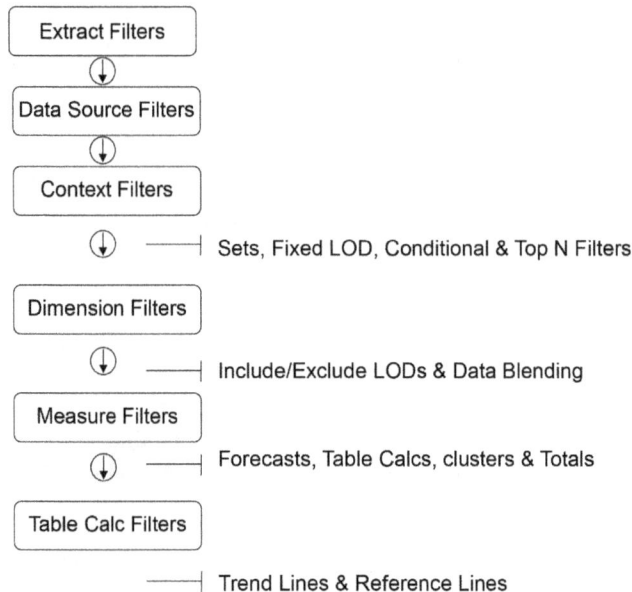

Figure 4.30: Tableau's order of operations – filters are shown in the boxes
with other operations to the right

Creating a Data Source Filter

Data source filters are used to pre-filter data in Tableau Desktop. In this exercise, you will create a data source filter and observe its effects. Data source filters can be a great way to speed up workbooks that rely on huge datasets by reducing the amount of data that has to be loaded right at the start of the order of operations:

1. On a new sheet, place **Sub-Category** on **Rows** to create a list of sub-categories in the view.

2. Right-click the **Sample - Superstore** data source at the top left and select **Edit Data Source Filters**.

3. To reduce the data source to data related to **Furniture** only, select **Add**, then **Category**, and check **Furniture** for inclusion only.

4. After pressing **OK** and **OK** again, you will return to the view and see that only sub-categories that are **Furniture** remain in the list of sub-categories.

5. Dragging the **Category** field to **Filters** will only provide **Furniture** as a filter option because the other two categories were pre-filtered at the data source earlier in the order of operations.

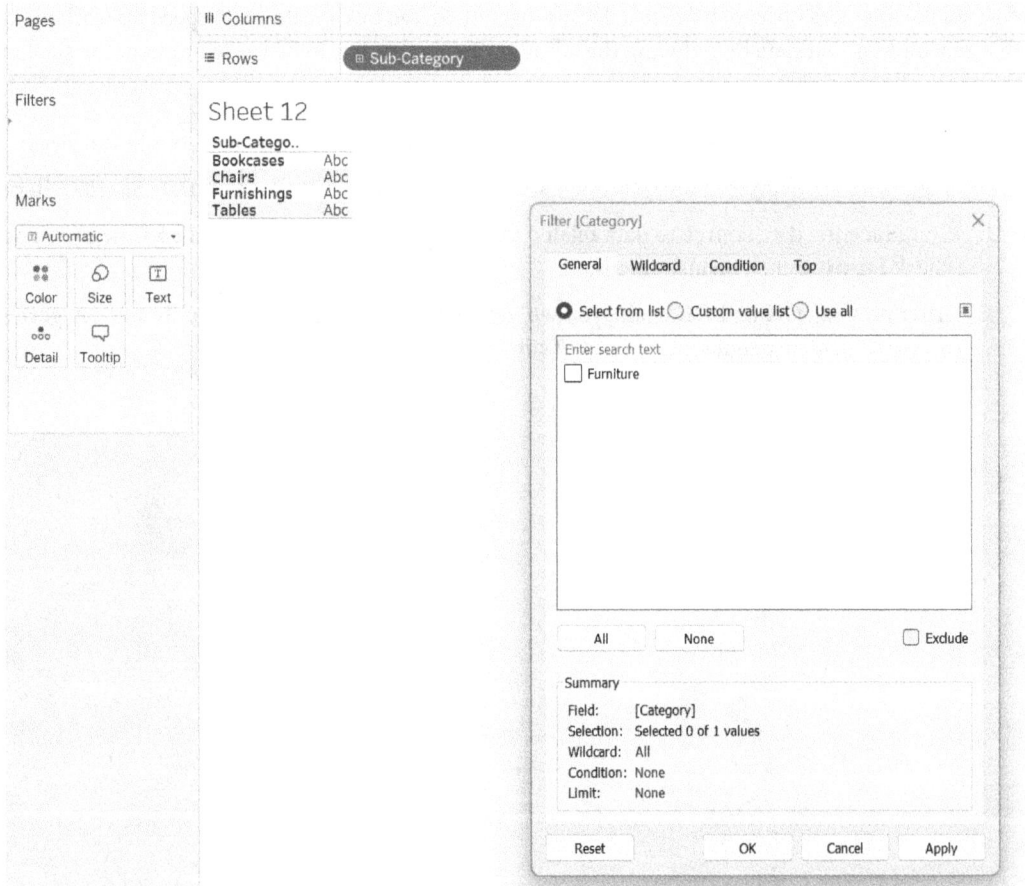

Figure 4.31: The Category data source filter added, pre-filtering the data to Furniture only

In this exercise, you applied a data source filter and observed how it limits the data available in the view and even the data variable in normal filtering, as the data was pre-filtered at the start of the order of operations.

Before continuing to the next exercise remove the **Category** data source filter previously added to ensure all data is available going forward.

Context Filtering Exercise

In this exercise, you will learn how to add filters to context. Adding a filter to context moves it up in the order of operations and ensures a filter works even on level of detail calculations (calculations that fix data at a specific level of detail regardless of normal filtering):

1. On a new sheet, put **Sub-Category** on **Rows** and **Sales** on **Columns** to create a bar chart showing the total sales by each sub-category.

2. Click the sort descending icon above **SUM(Sales)** on **Columns** to sort the sub-categories from most to least sales.

3. Create a sub-category filter that limits the view to the top five sub-categories by sales (**Chairs**, **Phones**, **Storage**, **Tables**, and **Binders**).

4. Now, add **Category** to the **Filters** shelf, select **Office Supplies**, then press **OK**. The bar chart now only contains two sub-categories, **Storage** and **Binders**. This is because in the order of operations, the top five **Sub-Category** filter reduces the sub-categories before the **Category** filter reduces to **Office Supplies**.

5. To show the top five **Office Supplies** sub-categories, the **Category** filter must be moved up in the order of operations. To do this, right-click the **Category** filter and select **Add to Context**. The view now shows the top five **Office Supplies** sub-categories.

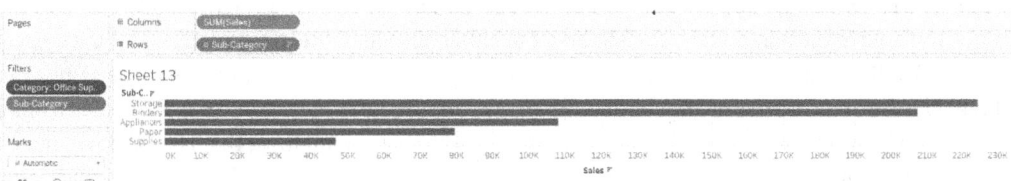

Figure 4.32: The Category filter has been added to context and
now occurs before the sub-category top N filter

In this exercise, you added a filter to context to move it up in the order of operations. As a result, the filter has occurred before a top N filter. The result of this is shown in the visual and demonstrates the importance of understanding Tableau's order of operations. You now know how to ensure a normal filter occurs before a top N filter, as well as how to ensure that a normal filter applies to level of detail calculations.

Parameters

This section covers Tableau's parameters, including the different types of parameters, how to use them in calculated fields, how to use them to filter, and how to use them to customize reference lines.

Parameter Creation

Tableau's parameters are essentially single-value variables that can be updated by the user to increase interactivity. An example of a parameter would be creating a percentage parameter that enables the user to select a value between 0% and 100%. The parameter could then be used in a calculated field to create a targeted percentage growth value for next year's sales, based on the current year's sales and a user-input percentage growth. This would then allow the end user to play with the parameter and see the impact of different percentage growth targets.

Parameters can be created using either the values of existing fields or user-defined values. To create a parameter from the values in an existing field, right-click the desired field, select **Create**, and then select **Parameter**. This will open up the parameter configuration window. The parameter configuration window allows the user to give the parameter a name, select the data type, and display the format of the data, the current value of the parameter, and how to set the value of the parameter when the workbook opens.

Parameters can be string, float, integer, Boolean, date, or datetime types, similar to Tableau fields. The values available to the parameter can then be set to either all values, meaning anything (of the correct data type), a list of allowable values, or a range of acceptable values to choose from. The values can also be fixed or set to dynamically update when the workbook opens by pulling the values from a selected field. By default, when creating a parameter from an existing field, the data type will match that of the field, and the values in that field will make up the list of allowable values from the parameter.

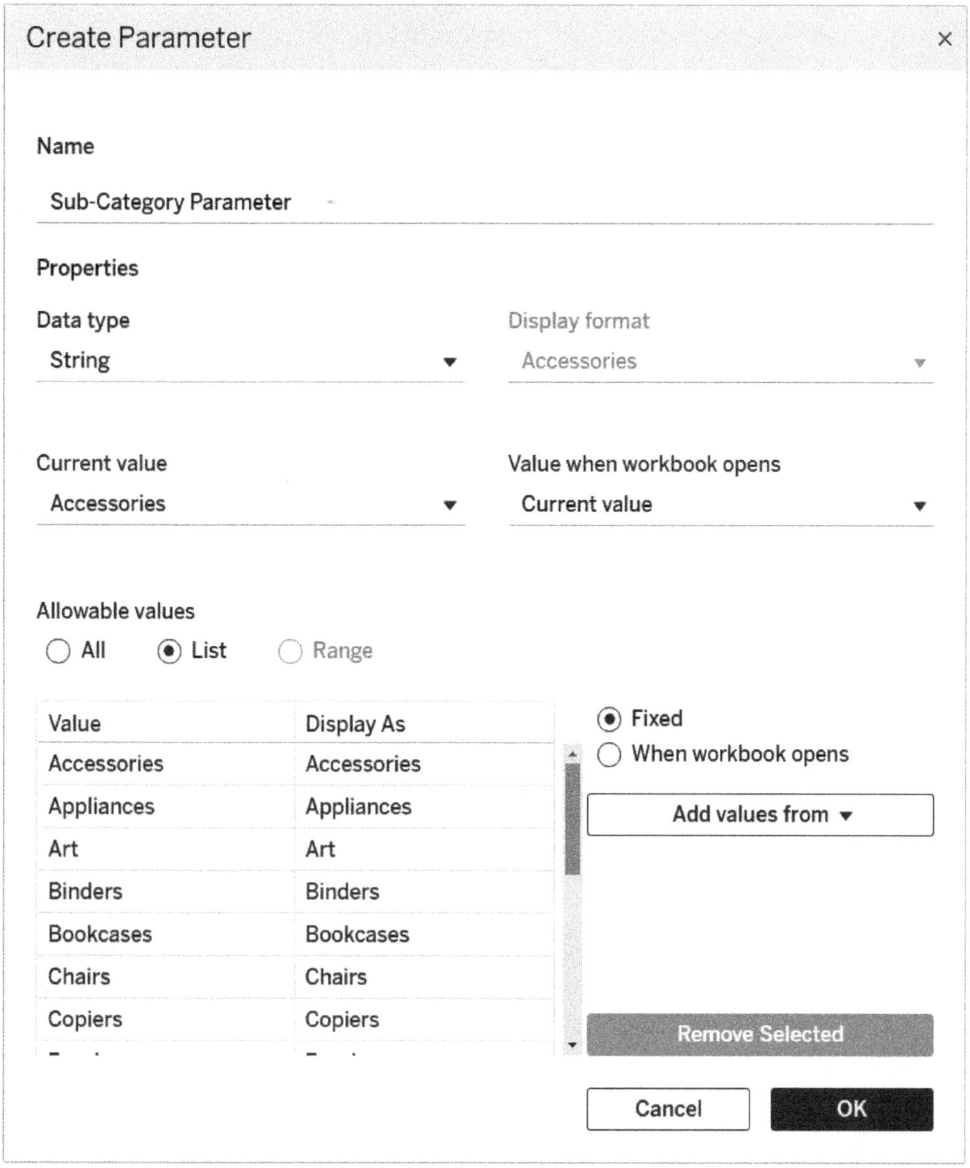

Figure 4.33: A parameter created from the Sub-Category field set to a string type by default, with the allowable values taken as a list from the Sub-Category field

To create a parameter from scratch, without any predefined values, click the carrot at the top right of the **Data** pane and select **Create Parameter**.

Once a parameter has been created, it needs to be applied to an element in Tableau via a calculation, filter, or reference line. Once it has been made functional, the parameter can then be displayed to the end user. To display the parameter configuration for end users, the parameter must be right-clicked and **Show Parameter** must be selected. Parameters can be found at the bottom of the **Data** pane.

Creating a String Parameter from an Existing Field

In this exercise, you will create a parameter using values from an existing field. Parameters are often created using the values from existing fields, as it allows those values to become static reference points that are not technically part of the data source itself. These reference points can then be used in a variety of ways, such as in calculated fields or reference lines.

To complete this exercise in Tableau Desktop, connect to Tableau's readily available training data source, the Sample - Superstore data source, which can be found in the **Saved Data Sources** section of the home screen:

1. On a new sheet, right-click the **Sub-Category** field in the **Data** pane and select **Create Parameter**.
2. The parameter will, by default, be called **Sub-Category Parameter** and will take its values as a list from the **Sub-Category** field.
3. Switch the current value to Art to change the current parameter value.
4. Click **OK** to create the parameter, which can now be seen at the bottom of the **Data** pane.
5. Right-click it and select **Show Parameter** to show the parameter control. The value is Art, as set during the configuration, but this can be changed using the drop-down control.
6. If the data source is updated, any new **Sub-Category** field values will not, by default, be included in the parameter. To ensure the parameter values update dynamically with the data source, right-click the parameter and select **Edit**.
7. In the configuration window, select **When workbook opens**, and then select the Sample - Superstore data source and the **Sub-Category** field.
8. The values in the parameter will now be refreshed from the **Sub-Category** field whenever the workbook is opened.

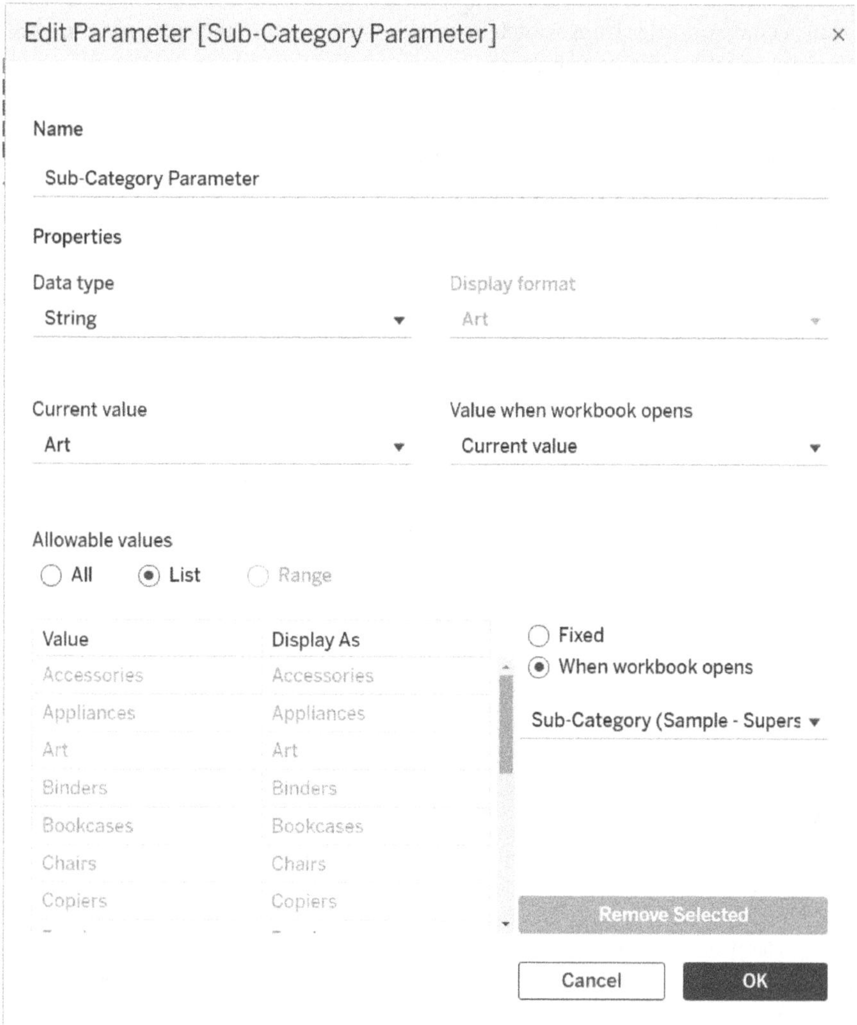

Edit Parameter [Sub-Category Parameter]   ✕

Name

Sub-Category Parameter

Properties

Data type   Display format

String   ▼   Art   ▼

Current value   Value when workbook opens

Art   ▼   Current value   ▼

Allowable values

○ All   ⦿ List   ○ Range

Value	Display As	
Accessories	Accessories	○ Fixed
Appliances	Appliances	⦿ When workbook opens
Art	Art	Sub-Category (Sample - Supers ▼
Binders	Binders	
Bookcases	Bookcases	
Chairs	Chairs	
Copiers	Copiers	

Remove Selected

Cancel   OK

Figure 4.34: The Sub-Category parameter created and edited to
update the values when the workbook opens

In this exercise, you created a string parameter using the values from an existing field, showed the parameter on the view, and configured the parameter such that if the data source is updated and new values are added, the parameter will be updated to incorporate these new values.

Creating an Integer Parameter

In this section, you will create an integer parameter. Integer parameters are useful for things such as setting top and bottom N reference points that can be used for filtering:

1. Create a new parameter from scratch by selecting the carrot in the top-right corner of the **Data pane** and selecting **Create Parameter**.

2. Name the parameter Top N, as it will be used as a customizable top N value for filtering.

3. Set the data type to Integer and make the allowable values a range, with a minimum of 1, a maximum of 50, and a step size of 1.

4. The current value, which will be the default, should be set to 10.

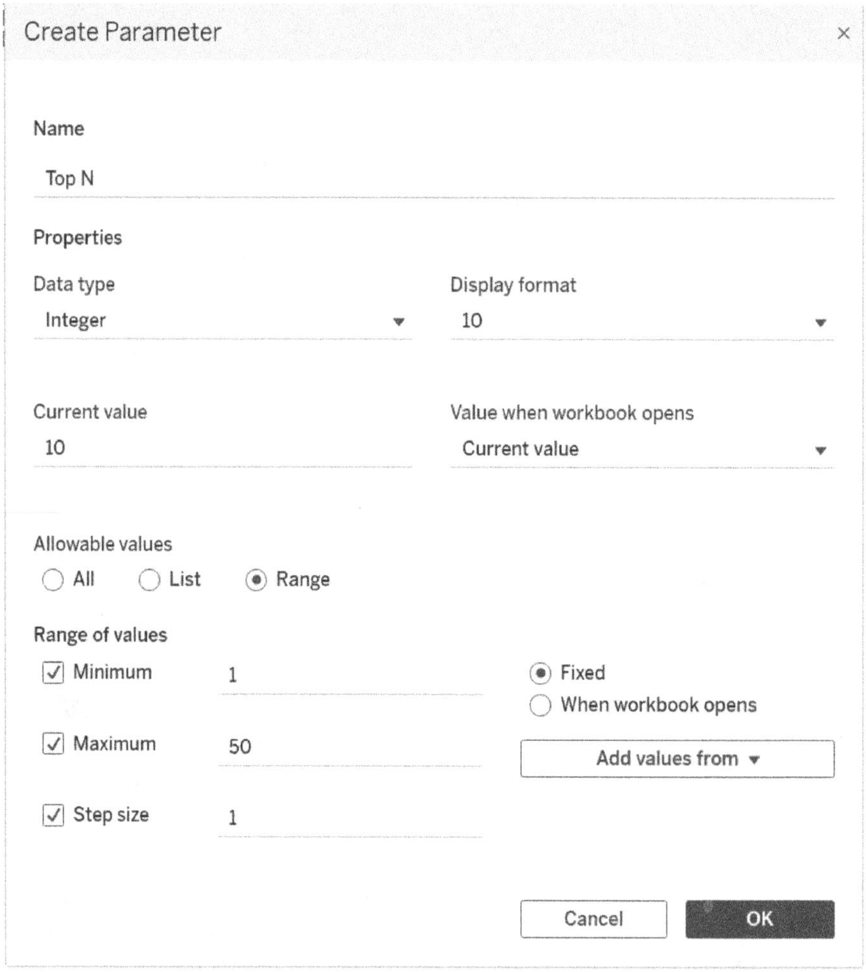

Create Parameter   ×

Name

Top N

Properties

Data type   Display format

Integer   ▼   10   ▼

Current value   Value when workbook opens

10   Current value   ▼

Allowable values

○ All   ○ List   ⦿ Range

Range of values

☑ Minimum   1   ⦿ Fixed
                                         ○ When workbook opens

☑ Maximum   50   ┌──────────────────────────┐
                                         │   Add values from ▼   │
                                         └──────────────────────────┘

☑ Step size   1

                                         Cancel   OK

Figure 4.35: The integer parameter created from scratch, with allowable
values between 1 and 50 and the default set to 10

In this exercise, you created a parameter from scratch instead of from a field, and you defined the values yourself. You now know how to create and configure an integer parameter.

Creating a Date Parameter

In this exercise, you will create a date parameter. Date parameters are useful for setting reference dates that can be customized by the user for things such as target dates, reference lines, and calculations, such as sales from a specified date:

1. Create a new parameter from scratch by selecting the carrot in the top-right corner of the **Data pane** and selecting **Create Parameter**.

2. Call the parameter `Select Date`, as it will be used to provide users with a customizable date with which they can set a reference line.

3. All date values should be allowed. Press **OK** to create the date parameter.

4. To set a default date for the parameter dynamically on today's date, create a calculated field called `Today`, write the calculations function, `TODAY()`, and then press **OK**.

5. The `TODAY()` function returns today's date so that it can be used to dynamically set the default date in the date parameter.

6.   Right-click the **Select Date** parameter and select **Edit**. In the **Edit Parameter** dialog box, set the **Value when workbook opens** option to the **Today calculated field**. Then press **OK**.

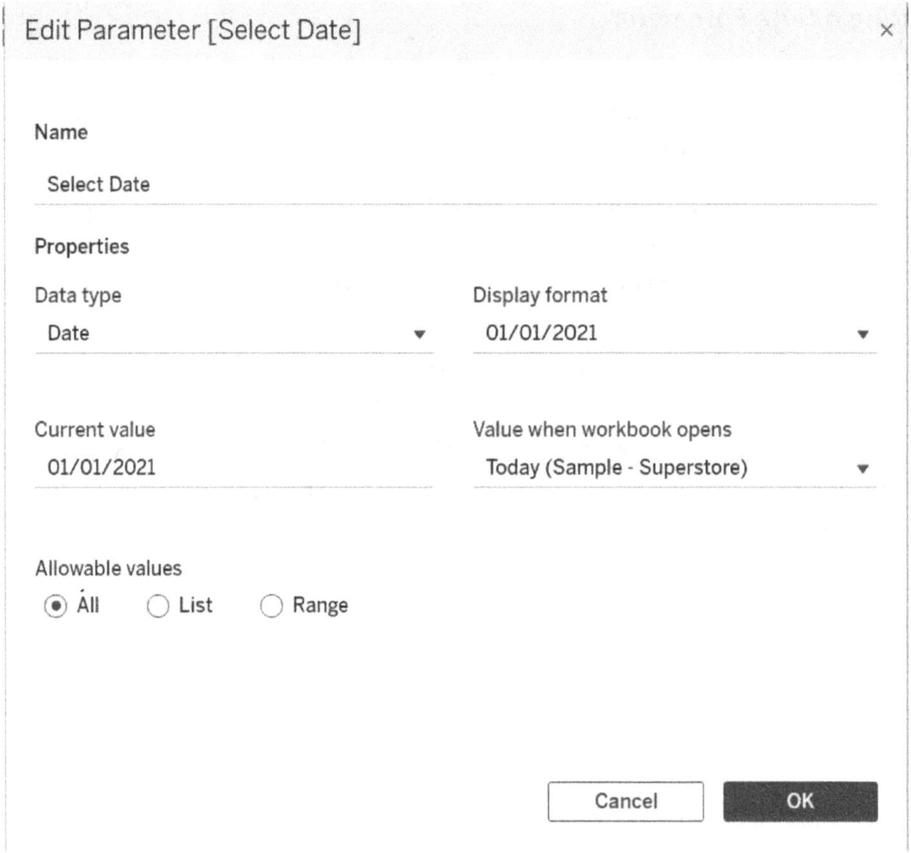

Edit Parameter [Select Date]   ×

Name

Select Date

Properties

Data type   Display format

Date   ▼   01/01/2021   ▼

Current value   Value when workbook opens

01/01/2021   Today (Sample - Superstore)   ▼

Allowable values

⦿ Áll   ○ List   ○ Range

                                                        Cancel   OK

Figure 4.36: The date parameter created that has a default value dynamically set to today's date

In this exercise, you learned how to create a date parameter and also understood how to set the default value for a parameter when the workbook is opened, based on a calculated field.

Parameters in Calculated Fields

Parameters can be used in calculated fields as customizable reference values – for example, a float parameter could be used in a calculation to set a limit/threshold to compare sales values against. Parameters are referenced in calculated fields directly by name. So, in the previously mentioned example, the calculated field could look something like SUM([Sale]) > [Float Parameter]. This would create a Boolean field that would return true or false, depending on whether the summed sales in the view are above or below the user-defined parameter value, which can in turn be modified as and when needed by end users via the parameter control. The calculated field can then be used in the same way as any other calculated field – it can be placed as a field on the view, it could be used to color data points where the condition is met, or it could be used as a Boolean filter.

Using a Parameter in a Calculated Field

In this exercise, you will use a parameter in a calculated field to color a chart based on the user's selection. Parameters allow users to select reference values that can then be used in calculations.

To complete this exercise in Tableau Desktop, connect to Tableau's readily available training data source, Sample - Superstore, which can be found in the **Saved Data Sources** section of the home screen:

1. If you have not completed the *Creating a String Parameter from an Existing Field exercise* in the *Parameter Creation* section, do so now to create a **Sub-Category** string parameter.

2. Once the **Sub-Category** parameter has been created, create a new sheet, and then put **Sub-Category** on **Rows** and **Sales** on **Columns** to create a **Total Sales by Sub-Category** bar chart.

3. Show the parameter control by right-clicking the **Sub-Category parameter** and selecting the **Show parameter**.

4. The parameter control is now visible on the right-hand side, and the dropdown can be used to select a sub-category. However, selecting a sub-category does not change anything on the view.

5. To add logic that checks whether a sub-category is selected in the parameter, first create a new calculated field. Type into the calculated field [Sub-Category] = [Sub-Category Parameter], then name the calculated field Selected Sub-Category, and press **OK**.

6. Drag the **Selected Sub-Category** calculated field onto **Color** on the **Marks** card, and the selected sub-category in the parameter control should now be colored.

7.   Changing the sub-category control to **Binders**, for example, will cause the colored bar to switch to the selected sub-category.

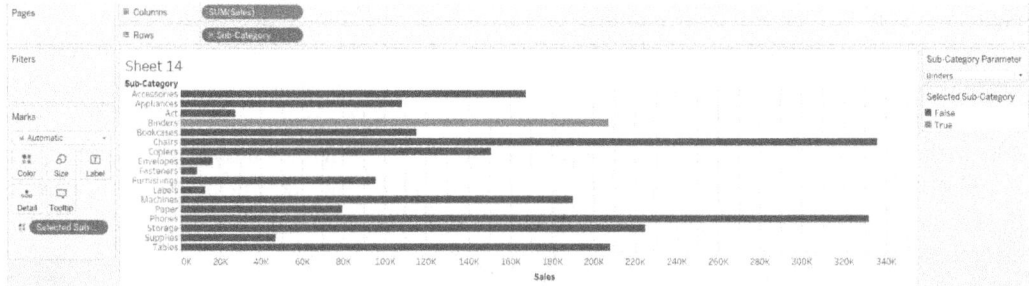

Figure 4.37: The Sub-Category parameter used in a calculated field, which
is applied to Color to highlight the selected sub-category

In this exercise, you used a parameter in a calculated field to color the bars in a chart, based on the selected sub-category. This basic demonstration of the combination of parameters and calculated fields demonstrates the increased user interactivity that can be made possible via parameters.

Filtering with Parameters

Parameters can be used in top and bottom N filters, providing users with dynamic and customizable top and bottom filtering. When using a dimension filter, the **By field** section in the **Top** tab allows the user to select the top or bottom by a given number. The number can be a typed value but can also be set to a parameter in the workbook. This allows end users to decide, using the parameter control, how many items to show in the top or bottom list.

Filtering with a Parameter

In this exercise, you will use an integer parameter in a top N filter to enable end users to select a top N to filter by. This, again, is a great skill to learn in Tableau as a developer, as it allows you to provide more interactivity to reports for end users.

To complete this exercise in Tableau Desktop, connect to Tableau's readily available training data source, Sample - Superstore, which can be found in the **Saved Data Sources** section of the home screen:

1.   If you have not completed the *Creating an Integer Parameter* exercise in the *Parameter Creation* section, do so now.

2.   You now have a sheet with a **Sub-Category by Sales** bar chart from the previous exercise, with a selected sub-category highlighted and an integer parameter called **Top N**.

3. To create a filter that allows users to show a customizable number of top sub-categories by sales, drag the **Sub-Category** field onto filters.

4. Navigate to the **Top tab**, select **By field**, and then switch the default **10** value to the **Top N** parameter.

5. The view will now be filtered to the top 10 sub-categories by sales only because the default/current value of the **Top N** parameter is 10.

6. To customize the **Top N** value, right-click the **Top N** parameter and select **Show Parameter**.

7. The **Top N** parameter is now visible as a slider and the **Top N** value can be changed (to five, for example) using this slider.

8. Updating the **Top N** parameter value updates which sub-categories are displayed on the view, and this can be made more obvious by sorting the sub-categories descending by sales.

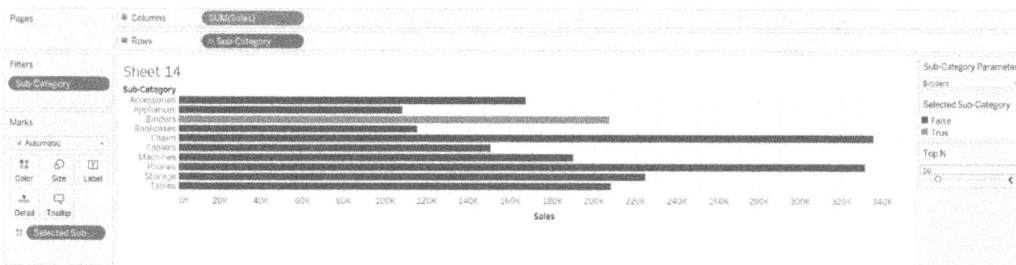

Figure 4.38: The Top N parameter is used to filter the number of sub-categories displayed in the view

In this exercise, you learned how to provide users with the ability to set a number of top items to show on a chart. This can also be applied to bottom N filtering. Parameters can also be applied to other conditional filters – for example, an end user could set a numeric threshold to be a value that acts as a threshold for filtering on a view.

Customizing Reference Lines Using Parameters

Reference lines on charts can be placed dynamically and customized using parameters. When adding a reference line to a chart, a value must be specified to determine where on the axis to place the line. The value can be either an aggregated measure, a hardcoded constant value, or a parameter. If a parameter is selected, then the reference line can be moved on the relevant axis by changing the value in the parameter control.

Customizing a Reference Line Using a Parameter

In this section, you will apply a parameter to a reference line, allowing end users to move the reference line along the axis based on their selection. This, again, provides greater interactivity to end users.

If you have not completed the *Creating a Date Parameter* exercise in the *Parameter Creation* section, do so now:

1. On a new sheet, add **Sales** to **Rows**, hold down right-click on the **Order Date** field, and then drag it to **Columns**.

2. In the **Aggregation selection configuration** popup, choose **MONTH(Order Date)** to display the continuous months in the data by total sales.

3. Add a reference line to the date axis by right-clicking the **Month of Order Date** axis and selecting **Add Reference Line**.

4. In the value selection, choose the **Select Date** parameter created earlier, and then click **OK**.

5. A line will now be added to the chart on today's date (this will go beyond the line, as the superstore data stops at December 2022).

6. To customize the reference line, right-click the **Select Date** parameter, and then click **Show Parameter**.

7. Click on the parameter control and change the date (to January 1, 2021, for example), and the reference line will move to that date.

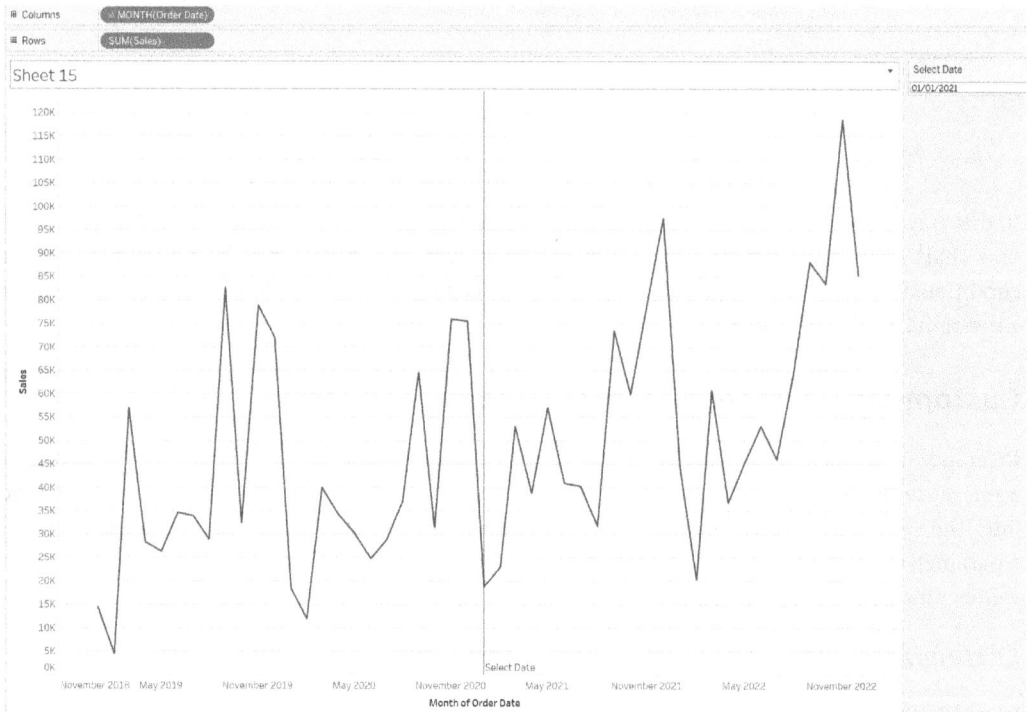

Figure 4.39: The customizable reference line created using a date parameter

In this exercise, you created a reference line that takes its value from a parameter input. You changed the parameter input and observed the line move accordingly along the axis.

Summary

In Tableau, grouping data, filtering, and enhancing user interactivity with customizable values is made easy with inbuilt functionality.

Grouping in Tableau can be done at the field level by grouping hierarchical fields into the hierarchical field type. This enhances functionality in the view, allowing users to drill up and down levels of hierarchy. Values within a field can easily be manually grouped together into a new field using Tableau group functionality, and the new field can be used in exactly the same way as the original field. Values within fields can also be defined based on conditions, resulting in a new Boolean-type field called a set. Sets can then be combined together to create more complex interlocking conditions for values within a field. Fields can also be created on numeric data by binning the values into consistent ranges, allowing users to see the distribution of values across a field.

There are a multitude of filter options available in Tableau, both in terms of type and the order in which the filter occurs. Types of filters include numeric measure filters that allow for minimum and maximum values, as well as specific range filtering. Dimension fields allow for basic inclusion and exclusion filtering, as well as filtering by wildcard matches in the text, a specific user-defined condition, and the top or bottom values by a given aggregated field. Dates can be filtered in the same way, as both measures and dimensions, but they also provide a relative date littering option in which an anchor date can be set. Table calculations can also be used as fillers, with the filter values available dependent on the table calculation configuration. Filters can be set to occur at different points in Tableau's order of operations, with extract filters running first, followed by data source filters, followed by context filters, and then dimension, measure, and table calculation filters in succession.

Tableau's parameters provide the ability for end users to set specific values that Tableau can then use in calculation logic, top or bottom N filtering, or reference lines. Parameters can be limited to a specific set of available values for end users, any value between a given range, or any value possible.

Exam Readiness Drill – Chapter Review Questions

Apart from a solid understanding of key concepts, being able to think quickly under time pressure is a skill that will help you ace your certification exam. That is why working on these skills early on in your learning journey is key.

Chapter review questions are designed to improve your test-taking skills progressively with each chapter you learn and review your understanding of key concepts in the chapter at the same time. You'll find these at the end of each chapter.

> **How to Access these Resources**
>
> To learn how to access these resources, head over to the chapter titled *Chapter 9, Accessing the Online Practice Resources*.

To open the Chapter Review Questions for this chapter, perform the following steps:

1. Click the link – `https://packt.link/TDA_CH04`.

    Alternatively, you can scan the following **QR code** (*Figure 4.40*):

Figure 4.40: QR code that opens Chapter Review Questions for logged-in users

2. Once you log in, you'll see a page similar to the one shown in *Figure 4.41*:

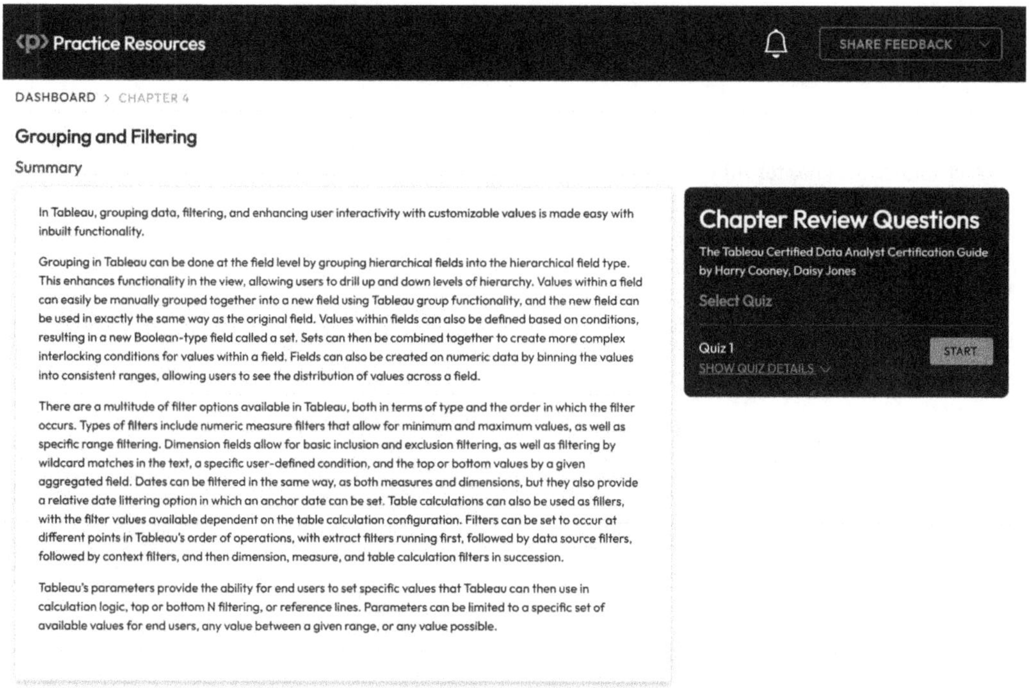

Figure 4.41: Chapter Review Questions for Chapter 4

3. Once ready, start the following practice drills, re-attempting the quiz multiple times.

Exam Readiness Drill

For the first three attempts, don't worry about the time limit.

ATTEMPT 1

The first time, aim for at least **40%**. Look at the answers you got wrong and read the relevant sections in the chapter again to fix your learning gaps.

ATTEMPT 2

The second time, aim for at least **60%**. Look at the answers you got wrong and read the relevant sections in the chapter again to fix any remaining learning gaps.

ATTEMPT 3

The third time, aim for at least **75%**. Once you score 75% or more, you start working on your timing.

> **Tip**
>
> You may take more than **three** attempts to reach 75%. That's okay. Just review the relevant sections in the chapter till you get there.

Working On Timing

Target: Your aim is to keep the score the same while trying to answer these questions as quickly as possible. Here's an example of how your next attempts should look like:

Attempt	Score	Time Taken
Attempt 5	77%	21 mins 30 seconds
Attempt 6	78%	18 mins 34 seconds
Attempt 7	76%	14 mins 44 seconds

Table 4.1: Sample timing practice drills on the online platform

> **Note**
>
> The time limits shown in the above table are just examples. Set your own time limits with each attempt based on the time limit of the quiz on the website.

With each new attempt, your score should stay above **75%** while your "time taken" to complete should "decrease". Repeat as many attempts as you want till you feel confident dealing with the time pressure.

5
Charts

Introduction

This chapter will cover chart creation in Tableau Desktop. The first section will include the different types of charts available and how to create them. How to conduct spatial analysis with geographic charts will then be covered, followed by the advanced analytical capabilities available in Tableau when building a chart.

The primary function of Tableau is to help analyze data and, to do this, charts can be built to visualize the data. Decisions can be made with the insights provided. In this chapter, the aim is to go through the main types of charts that can be built in Tableau, give a step-by-step guide on how to build them, take a look at the minimum requirements for a chart, and provide tips to help make the charts stand out.

These are the topics that will be covered in the chapter:

- How to build the following charts:

  - Area chart

  - Bar chart

  - Box plot

  - Bullet graph

  - Heatmap

  - Gantt chart

  - Highlight table

  - Histogram table

  - Line chart

  - Packed bubble chart

- Pie chart

- Scatter plot

- Text table

- Treemap

- Combination chart

- Geographic charts

- Analytical functionalities

For each chart that has been built, the data must always be prepped for analysis. Before building anything in Tableau, ensure that the following has been completed:

1. Have Tableau Desktop open.

2. Connect to the desired data source.

3. Ensure that the data types are assigned correctly, and the required fields are available for analysis.

Once these steps have been completed, the developer can proceed to build Tableau charts.

Area Chart

Area charts are very similar to line charts, except that the space between the line and the base of the axis is shaded in color, as shown in Figure 5.1. This is used to display changes over time and allows the stacking of multiple lines to show the breakdown of the value.

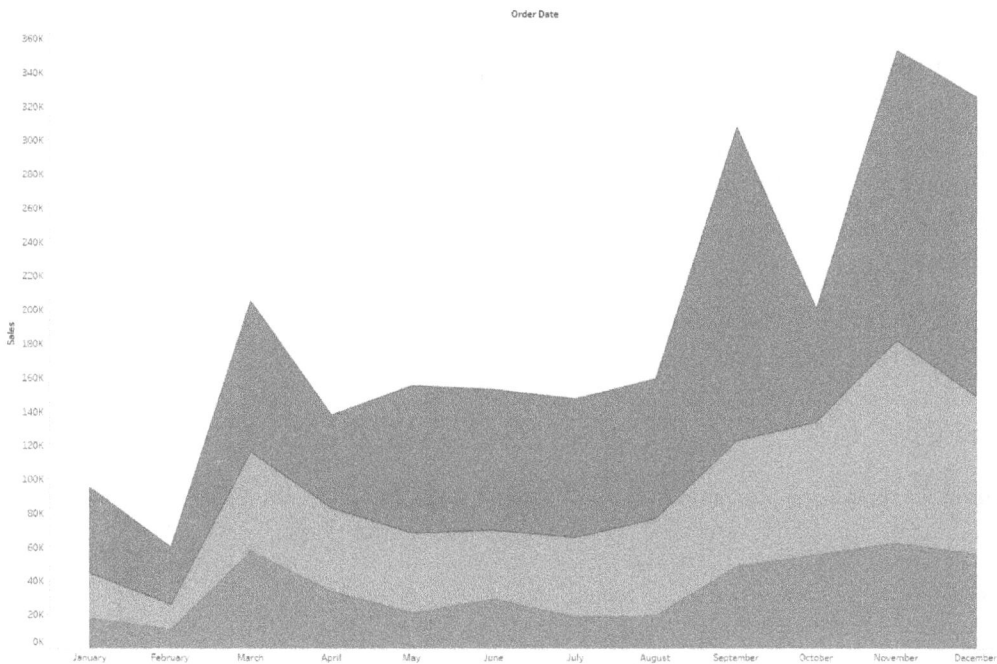

Figure 5.1: Example of an area chart

To build an area chart, you will need the following:

Columns	1 Dimension
Rows	1 Measure
Marks	Area

Table 5.1: Requirements to build an area chart

Here are the steps you need to follow to build an area chart:

1. Start by choosing dimensions and measures. For the chart shown in *Figure 5.1*, **Superstore Data**, **Sales**, **Order Date**, and **Category** were used.

2. Drag the dimension(s) that represents the categories or time intervals to the **Columns** shelf.

3. Drag the measure(s) to visualize onto the **Rows** shelf.

4. You can now start creating the area chart. Add **Order Date** to **Columns**, using it as a discrete dimension of the months.

5. Then, bring in the measure of **Sales Value** into **Rows**.

6. With the dimensions and measures placed on the shelves, Tableau will automatically generate a chart type based on the selected fields. By default, it might create a bar chart or line chart.

7. To change the chart type to an area chart, click on the drop-down option in the **Marks** card and select an **Area** chart.

8. An additional step is to add a second dimension so that the area chart will be broken down into the categories presented, such as the **Category** field.

Now that the area chart has been built, try building another chart without looking at the instructions. The next chart is the bar chart.

Bar Chart

One of the most used and popular charts in Tableau is the bar chart. This chart can easily compare categories in a straightforward way to convey fields in a particular order.

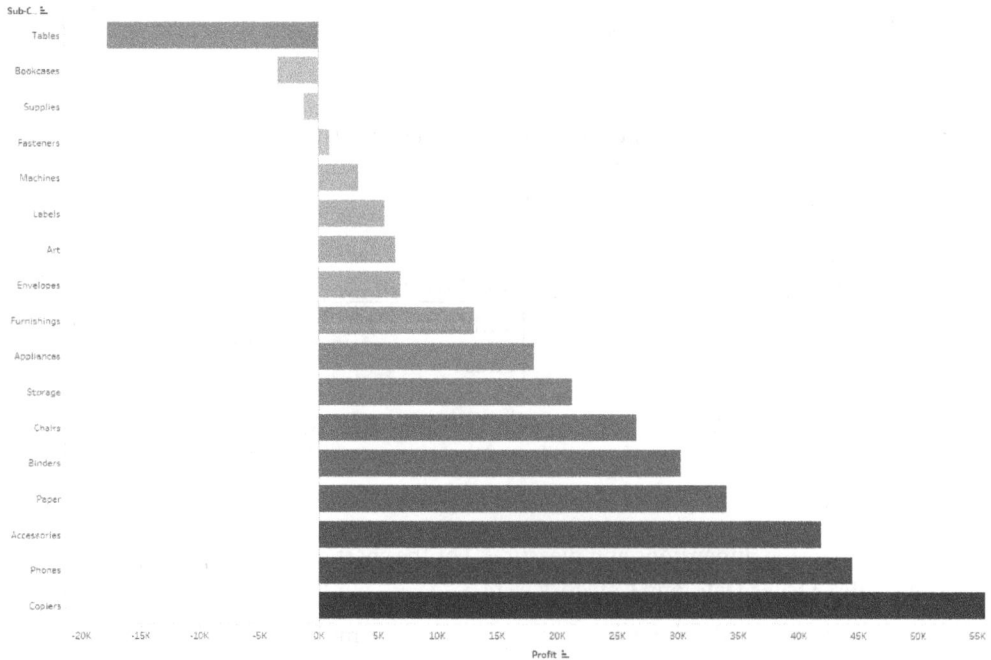

Figure 5.2: Example of a bar chart

Items needed to build this chart are as follows.

For horizontal bars, the following are needed:

Columns	1 Measure
Rows	1 Dimension
Marks	Bar
Detail	Measure or dDimension for colors

Table 5.2: Requirements to build a horizontal bar chart

For vertical bars, the following are needed:

Columns	1 Dimension
Rows	1 Measure
Marks	Bar
Detail	Measure or dDimension for colors

Table 5.3: Requirements to build a vertical bar chart

These are the steps to build a bar chart:

1. From the **Data** pane on the left, select the dimension(s) and measure(s) needed for the bar chart. For *Figure 5.2*, the data source used is **Superstore Data**, the dimension used is **Sub Category**, and the measure is **Profit**.

2. Note that, depending on the design requirement for the report, building a horizontal or vertical bar will impact where the dimensions and measures go in the **Field** shelf.

3. Drag the dimension(s) that represents the categories or groups needed to analyze onto the **Columns** shelf.

4. Drag the measure to compare by onto the **Rows** shelf.

5. Tableau will automatically generate a chart type based on the selected fields. It is likely that, by default, it will create a bar chart.

6. In the case where Tableau creates a line or text chart, all that is needed to change the chart type is to select the **Bar** option in the dropdown of the **Marks** card.

Bar charts are the most commonly used chart type in Tableau, so it is important to take time to explore how different measures can change the view.

Box Plot

Also known as a box-and-whisker plot, this chart allows a user to see the distribution of values across an axis. The box in the middle highlights 50% of the data, split into two with the high and low parts. The 'whiskers' extend to the smallest and largest data points that are within 1.5 times the interquartile range (IQR) from the lower and upper quartiles, respectively. In some variations, the whiskers extend to the actual minimum and maximum values in the dataset.

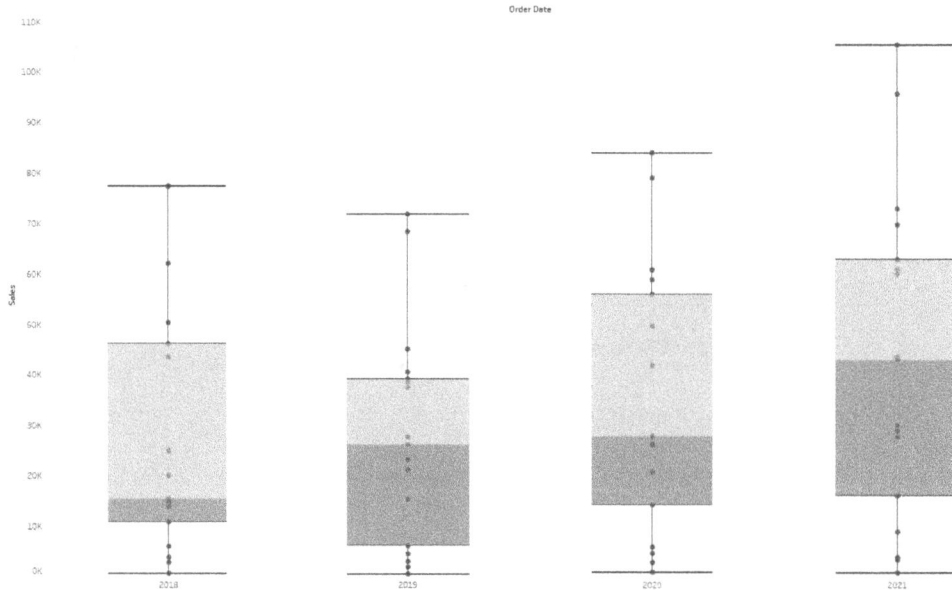

Figure 5.3: Example of a box plot

Items needed to build this chart are as follows:

Columns	1 Dimension
Rows	1 Measure
Marks	Circle
Detail	1 Dimension

Table 5.4: Requirements to build a box plot

It is time to build a box plot. The following are the steps to recreate *Figure 5.3*:

1. From the **Data** pane on the left, select the dimension(s) and measure(s) you want to use in your box plot. In *Figure 5.3*, which uses **Superstore Data**, the dimension is **Month** of **Order Date**, and the measure used is **Sales**.

2. Drag the dimension(s) that represents the categories or groups needed to analyze the data onto the **Columns** shelf.

3. Drag the measure(s) required to visualize onto the **Rows** shelf.

4. With the dimensions and measures placed on the shelves, Tableau will automatically generate a chart type based on the selected fields.

5. To change the chart type to a box plot, click on the **Show Me** button in the top-right corner of the screen.

6. From that pane, find and click on the **Box Plot** icon. This will convert the existing chart into a box plot.

You can include highlights or colors in the box plot so the data points can be easily spotted within the chart. Try using different dimensions to see how the chart changes to display the information.

Bullet Graph

A bullet graph is a version of a bar chart that allows a clear way to compare a primary metric to one or more other metrics.

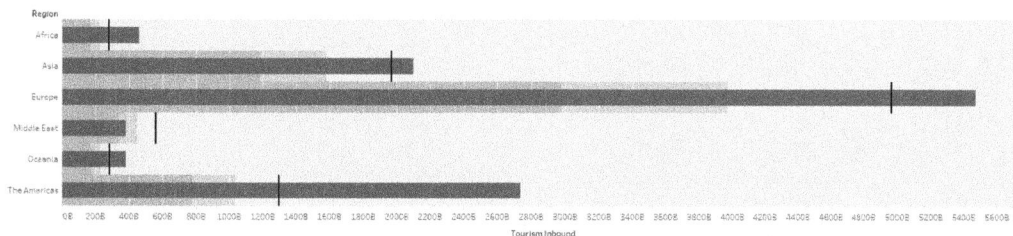

Figure 5.4: Example of a bullet graph

Items needed to build this chart are as follows:

Columns	Measure
Rows	Dimension
Marks	Automatic
Detail	Measure

Table 5.5: Requirements to build a bullet graph

Follow these steps to build a bullet graph:

1. From the **Data** pane on the left, select the dimensions and measures needed for the area chart. For *Figure 5.4*, the data source is **World Indicators Data**, and the **Tourism Inbound** and **Tourism Outbound** fields have been used as measures and **Region has been used as a dimension**.

2. The simplest way to build a bullet graph is to select the desired measures by holding down Ctrl and selecting **Tourism Outbound** and **Tourism Inbound**.

3. Then, click on the **Show Me** tab in the top-right corner and select ing the **Bullet Graph** image. This will create a single-line bullet graph.

4. To break this down into the other categories, drag **Region** to **Rows** and there will be a bar per category.

5. If needed, the reference line can be edited by right-clicking on the axis and selecting **Edit Reference Line**.

Try building this chart again without looking at the instructions to memorize how to build it. Explore different fields to see how the chart changes and how the averages adapt.

Tip: An additional calculation can be created and added to color to or highlight bars where it is lower than the target. For this example, creating the take a look at the following calculation:

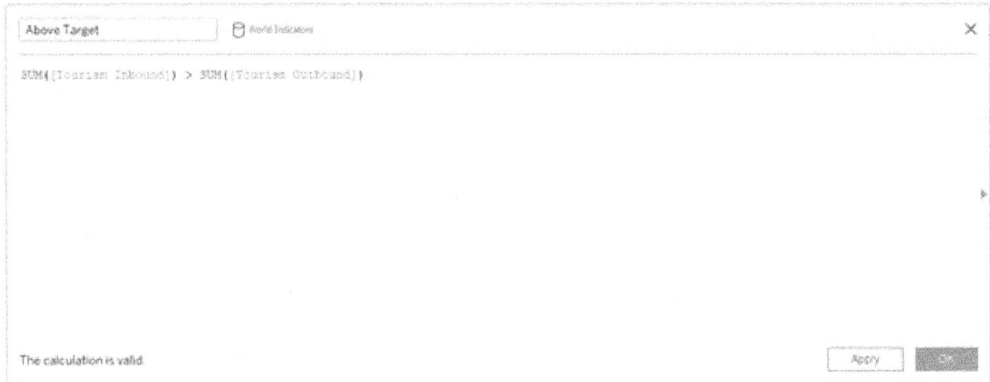

Figure 5.5: Calculation window for Boolean

Use this Boolean in **Color** and adjust the colors so that the bars lower than **Tourism Outbound** can be highlighted for analysis.

Heatmap

Heatmaps are otherwise known as density maps and are used to mark multiple data points that are layered. These are then color-coded to show the density of the marks in a specific chart. This can help visualize a report to show trends or geographical locations where there are denser points of data.

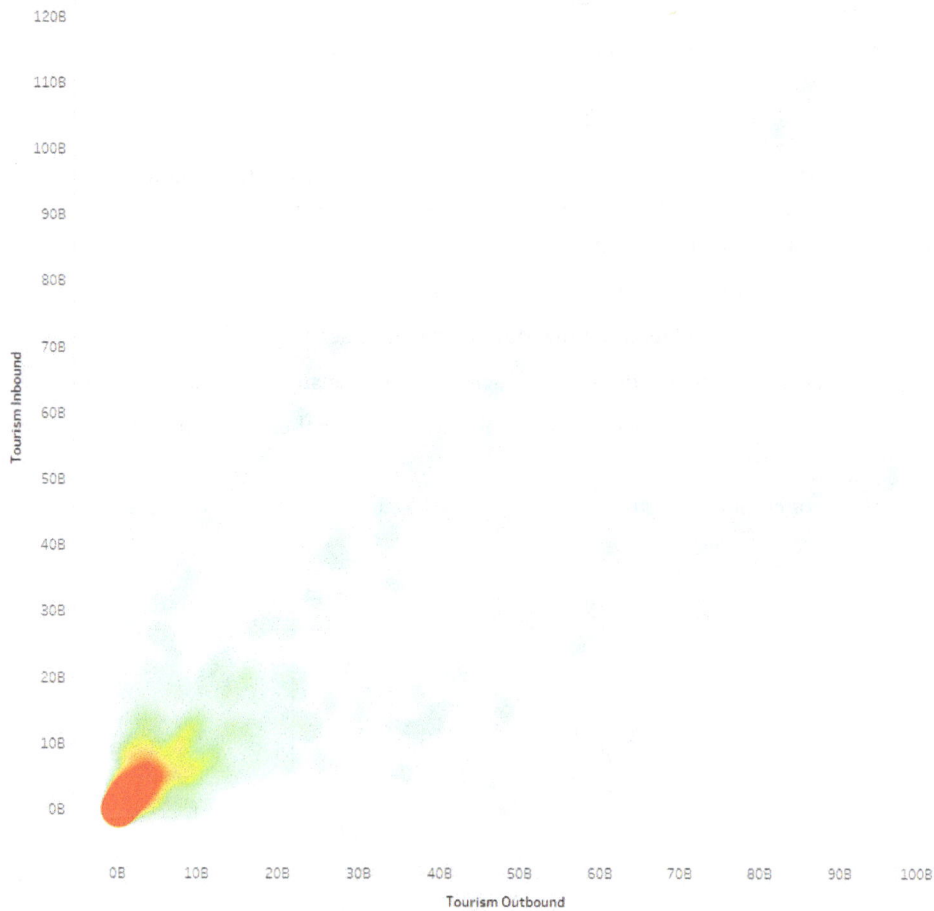

Figure 5.6: Example of a heatmap

Items needed to build this chart are as follows:

Columns	Measure
Rows	Measure
Marks	Density
Detail	Dimension

Table 5.6: Requirements to build a heatmap

The following steps will describe how to build a heat map:

1. From the **Data** pane on the left, select the dimension or measure that you want to visualize using the density chart. In the preceding example, **Sales** and **Profit** have been used as measures and **Product ID** as the dimension.

2. Drag the measures to the **Columns** and **Rows**, one in each.

3. Drag the dimension into **Detail** on the **Marks** card and a scatter graph will be created.

4. To change the chart type to a density chart, click on the **Density** option in the **Marks** card.

5. This will merge the points together and paint the area where there are a high number of data points a deep color.

6. The choices of colors can be changed in the **Marks** card, where Tableau has a default set of color palettes for density.

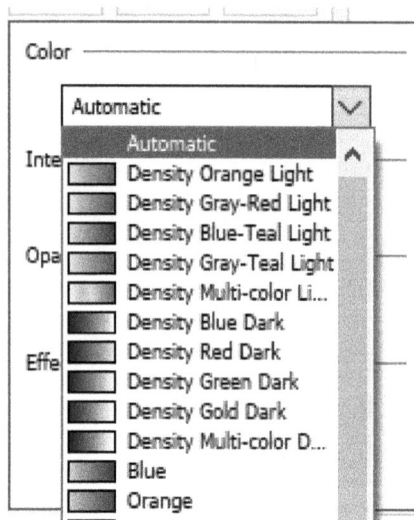

Figure 5.7: Density color choices

Density charts can also be used in maps, so take some time to explore using density in a map. You can use **UFO Sightings Data** to see how this looks.

Gantt Chart

Gantt charts help show the duration of time spent on certain dimensions. Each separate bar or mark will show the duration of time that was spent. These can be used to assess how long a project should take and in which order to take on tasks.

Figure 5.8: Example of a Gantt chart

Items needed to build this chart are as follows:

Columns	Date field
Rows	Dimension
Marks	Gantt
Detail	Dimension

Table 5.7: Requirements to build a Gantt chart

Take some time to identify what fits in the requirements to build a Gantt chart and then follow the next steps to create one:

1. From the **Data pane** on the left, select the dimension(s) and measure(s) required. For the chart shown in *Figure 5.8*, there will likely need to be a calculation to work out the timeframe for the project. For our example, the fields being used are **Order Date**, **Category**, **Ship Mode**, and an additional calculation called **Days to Ship**.

2. The **Days to Ship** calculation is created using a DATEDIFF function that calculates the number of days between the order date and the shipping date. To calculate days, make sure to type in day. Other time intervals can also be month, quarter, year, or even minute.

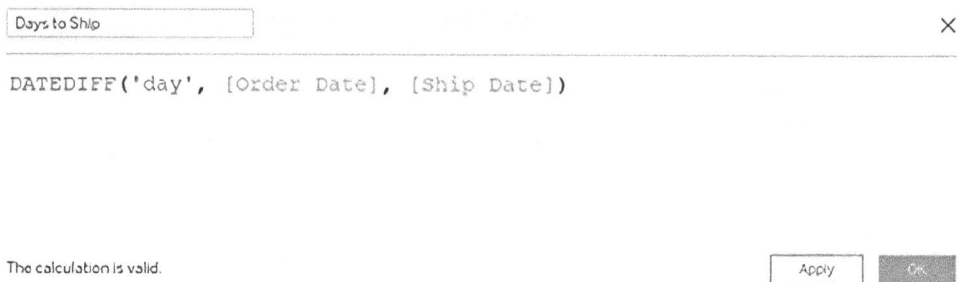

Days to Ship   ×

```
DATEDIFF('day', [Order Date], [Ship Date])
```

The calculation is valid.   Apply   OK

Figure 5.9: Calculation window for Gantt chart

3. 3. Drag the dimension(s) to the **Rows** shelf.

4. Drag the dimension(s) that represent the time intervals or durations onto the **Columns** shelf.

5. When the fields are placed on the shelves, Tableau will likely build a bar graph or Gantt chart to represent the fields.

6. Drag the calculated **Days to Ship** field onto **Size** in the **Marks** card, changing it to a Gantt chart. Change the aggregation to AVG to get the average **Days to Ship**.

Remember that the DATEDIFF function can be used to calculate the minutes or hours so the user has control of the granularity of the Gantt lines.

Highlight Table

Highlight tables are used to compare categorical values in a table format. This format is a clean way to instantly show any outliers or specific areas on which the developer should focus in the data.

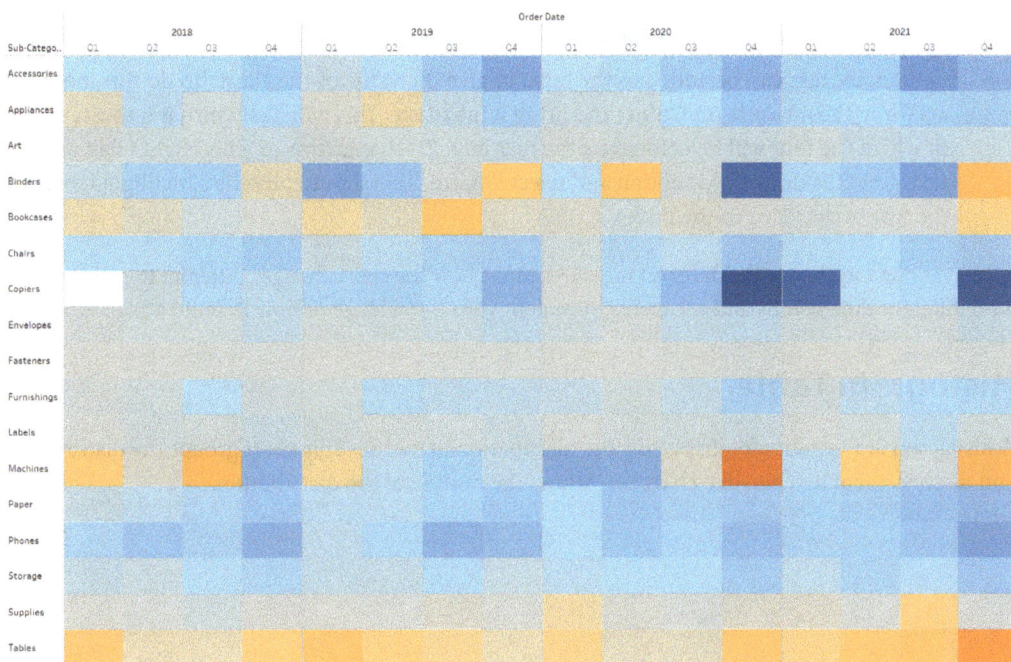

Figure 5.10: Example of a highlight table

Items needed to build this chart are as follows:

Columns	Dimension
Rows	Dimension
Marks	Square
Color	Measure

Table 5.8: Requirements to build a highlight table

Once the fields have been identified to build a highlight table, follow these steps to build the chart shown in *Figure 5.8*:

1. Drag the dimensions that are needed to the **Rows** and **Columns** shelves. For *Figure 5.10*, **Order Date** and **Sub Category** were used from **Superstore Data** and **Profit** was the measure used.

2. Drag the measure(s) onto **Color** in the **Marks** card.

3. Optionally, text can be added to the table to give the value of the chart. To do this, hold Ctrl and drag the measure on **Color**, and bring it into **Text**. This will likely turn it into a text table in which the font will be colored by the measure. To change this to a highlight table with the text, go to the drop-down option and select **Square**. This should provide a highlight table with the text of the value in the boxes.

Take a moment to change the different measures to see how the chart changes. Attempt to recreate them with different dimensions and measures to cement your knowledge of how to build a highlight table.

Histogram Table

A histogram is a variation of displaying the distribution of trends. This chart groups the distributions and displays them in a bar-like chart that shows the outline of the distribution.

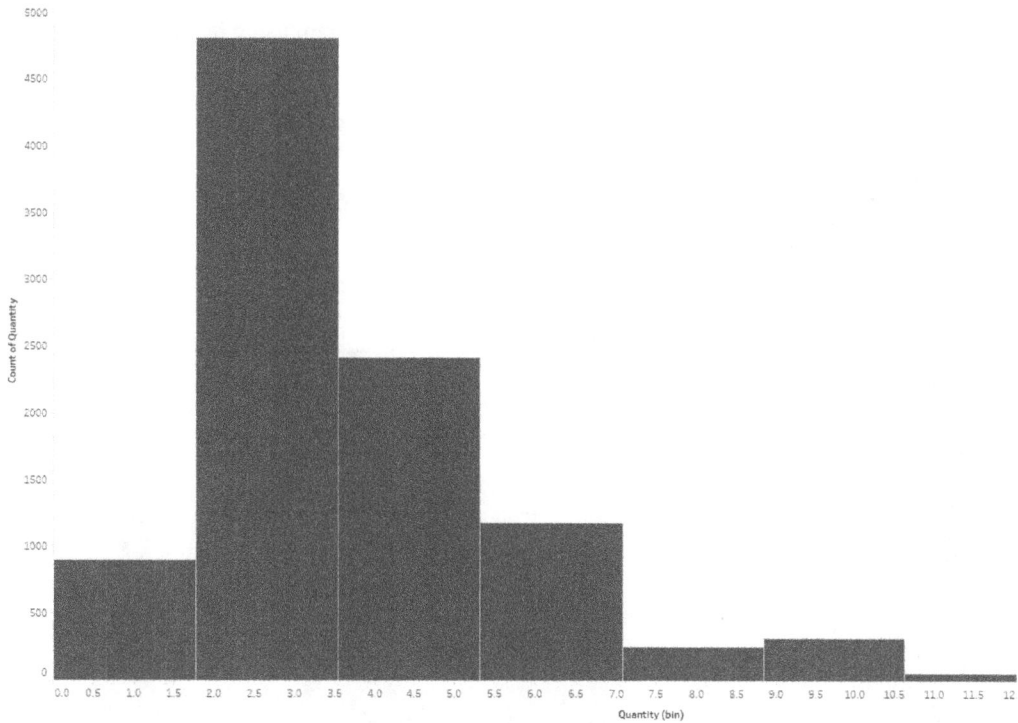

Figure 5.11: Example of a histogram

Items needed to build this chart are as follows:

Columns	Bin
Rows	Continuous measure (aggregated as either CNT or CNTD)
Marks	Automatic

Table 5.9: Requirements to build a histogram

Begin by connecting **Superstore Data** in Tableau to build a histogram using the following steps:

1. Select the measure to be turned into bins, right-click, and select **Create** > **Bins**. A window will open that will allow the user to edit the size of the bins, as shown in *Figure 5.12*:

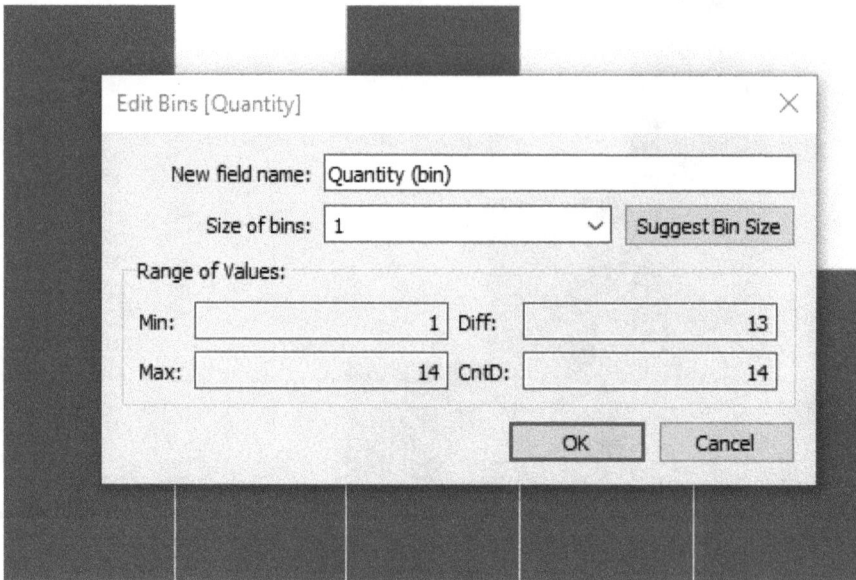

Figure 5.12: Edit Bins window

2. Drag the **Bin** field into **Columns**.
3. Drag the other measure (the measure you want to analyze in relation to the binned field) onto the **Columns** shelf, making sure to aggregate it to CNT or CNTD. In the preceding example, you will need to aggregate the measure into the CNT of **Quantity**.
4. The shape of the bar chart presented will be similar to a histogram., to To make the bars stick together, change the **Bin** field into a continuous field.
5. A dimension can be included to break down the chart into the categories selected. Simply drag the dimension into **Color**.
6. By default, Tableau will automatically determine the bins for your histogram. However, you can adjust the **Bin** size to customize the granularity of the histogram. Right-click on the **Bin** field in the **Data** pane and select **Edit**.
7. This will reopen the window when first building the calculation and the developer can edit the **Bin** sizes there.

A histogram can be broken down with a categorical field to see the relationship between the dimension and the quantity specified.

Line Chart

A line chart connects several data points across a timeline to create a line, to showing trends over time.

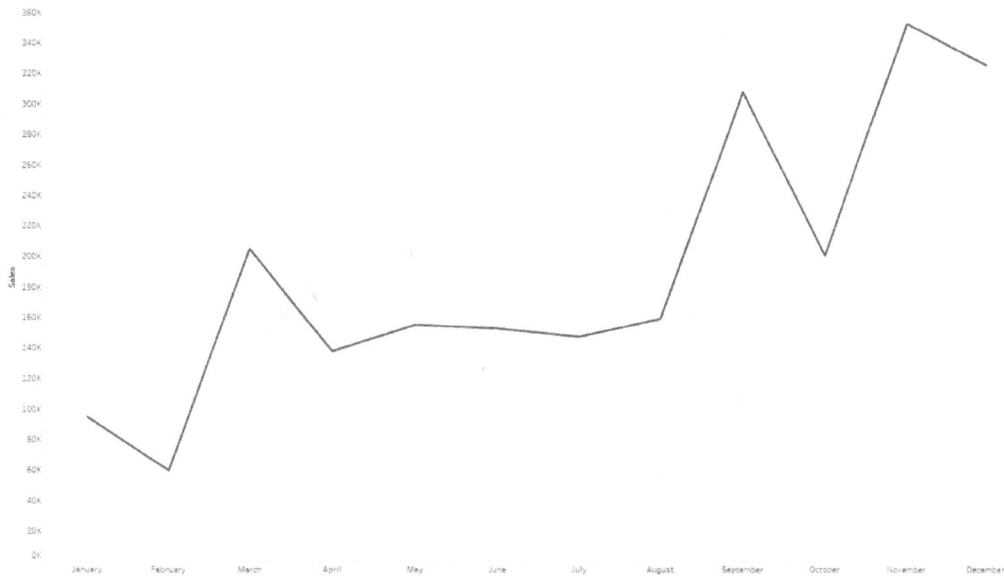

Figure 5.13: Example of a line chart

Items needed to build this chart are as follows:

Columns	Date Field, Continuous
Rows	Continuous measure
Marks	Line

Table 5.10: Requirements to build a line chart

Now it is time to start building a line chart. Use the following steps to create one:

1. From the **Data** pane, select the required fields for analysis. The preceding chart uses **Superstore Data** with **Order**, **Date**, and **Profit**.
2. Drag the dimension that represents the time periods onto the **Columns** shelf.
3. Drag the measure to analyze the trend onto the **Rows** shelf.
4. It is likely that as soon as the fields are placed on the shelves, Tableau will automatically create a line chart. If it does not, simply change the marks to **Line** in the **Marks** card and this will create a line chart.

A line chart along with a bar chart is one of the most commonly used charts in Tableau. A developer can include an additional dimension such as **Category** to **Color** to add more lines to compare the performance of each category.

Packed Bubble Chart

Packed bubble charts are used to display categorical values by size. You can then define the distribution of the measures across the dimensions.

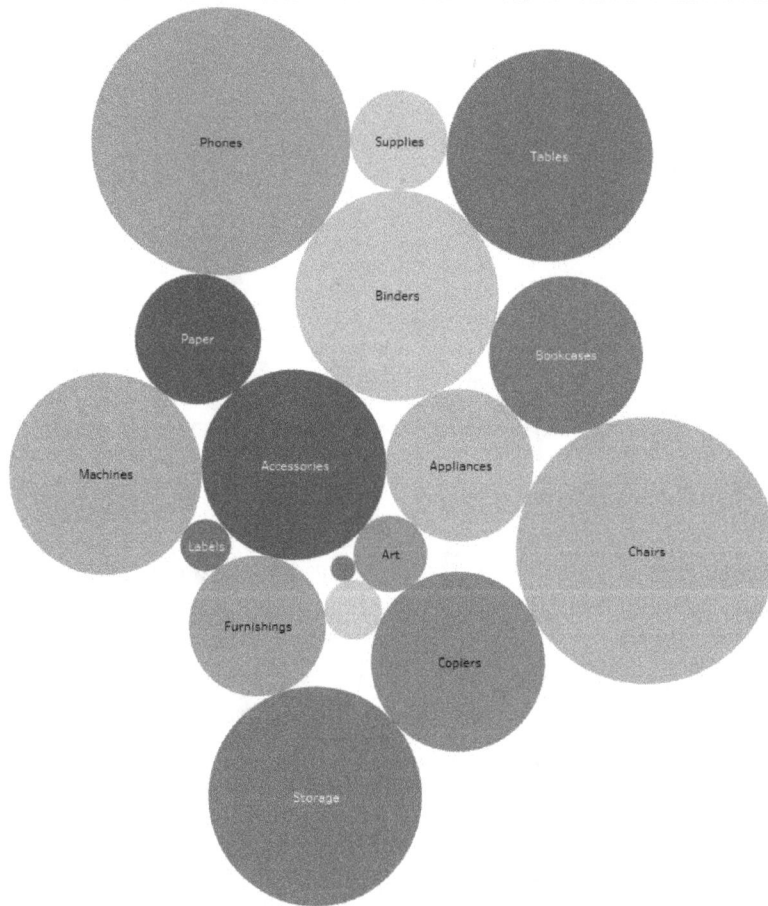

Figure 5.14: Example of a packed bubble chart

Items needed to build this chart are as follows:

Color	Dimension
Size	Measure
Marks	Circle

Table 5.11: Requirements to build a bubble chart

It is time to follow the next steps to build a packed bubble chart:

1. From the **Data** pane, identify the fields required.

2. Drag the categorical dimension onto the **Detail** tab in the **Marks** card.

3. Drag a value measure to **Size**. Tableau will likely transform this chart into a treemap.

4. To change the chart type to a packed bubble chart, click on the **Show Me** button in the top-right corner of the screen.

5. In the **Show Me** pane, find and click on the **Packed Bubble** chart icon. This will convert your existing chart into a packed bubble chart.

A packed bubble chart is a creative chart to visualize differences by using size. Take some time to explore it using other measures and fields to see how the chart changes.

Pie Chart

A pie chart is commonly used to show the distribution of a whole. Each slice will be a categorical dimension and the size of the slice will be based on how much of a whole that category has contributed to it. From this, a user can easily and instantly see which category has made the largest contribution to the whole value.

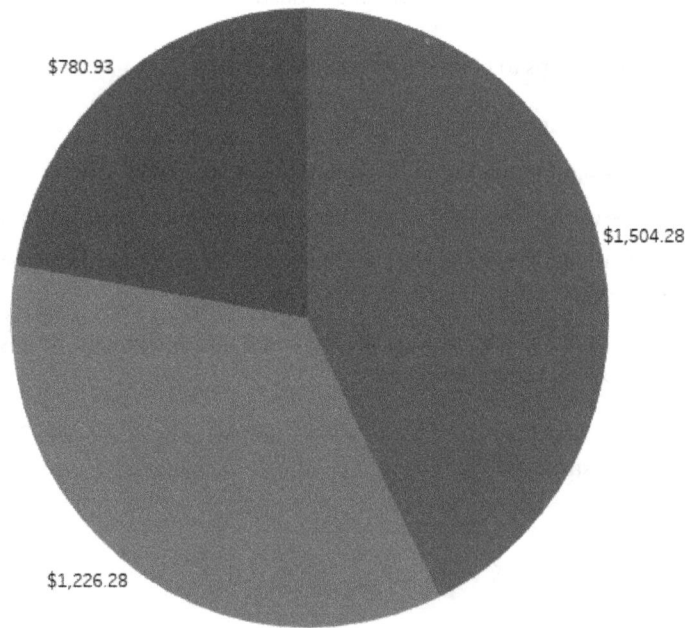

Figure 5.15: Example of a pie chart

Items needed to build this chart are as follows:

Color	Dimension
Angle	Measure
Marks	Pie

Table 5.12: Requirements to build a pie chart

Once the fields required have been identified, open Tableau and follow these steps to build a pie chart:

1. From the **Data** pane on the left, select the dimension(s) and measure(s) needed for analysis. For *Figure 5.15*, **Category** and **Quantity** from **Superstore Data** were used.

2. Drag the dimension that represents the categories or color blocks into **Color** on the **Marks** card.

3. Drag the measure to represent the size into **Size** on the **Marks** card. With the dimensions and measures placed on the shelves, Tableau will likely build a treemap.

4. To change the chart type to a pie chart, select **Pie** from the dropdown in the **Marks** card; this will transform the chart into a pie chart.

Using a pie chart can be useful for dashboard actions so that a user can select a slice to filter the other charts.

Scatter Plot

Scatter plots are used to visualize a relationship between two numerical values, explaining the correlation of the data points as well as highlighting any outliers that exist.

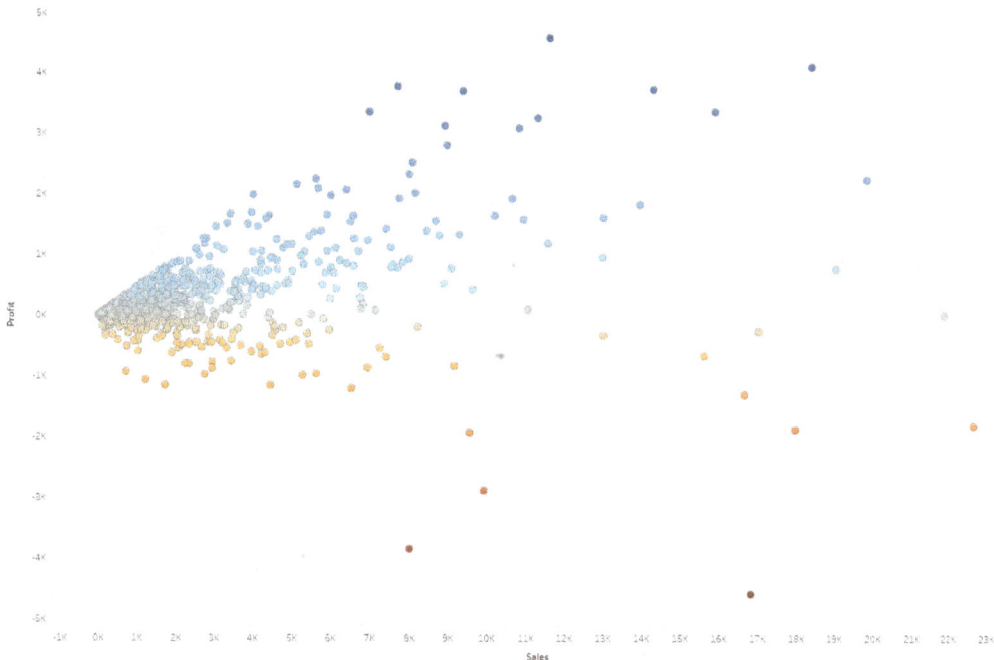

Figure 5.16: Example of a scatter plot

Items needed to build a scatter plot are as follows:

Columns	1 Measure
Rows	1 Measure
Marks	Automatic
Detail	Dimension

Table 5.13: Requirements to build a scatter plot

To build a scatter plot, connect to **Superstore Data** and follow these steps:

1. Select the measures to compare and identify the dimension to provide details of the comparison.

2. Drag one measure onto the **Columns** shelf.

3. Drag the other measure onto the **Rows** shelf.

4. Automatically, Tableau will create an x- and y-axis chart with a singular point in the top-right corner. This is because Tableau will always aggregate unless specified by the user.

5. To break down the aggregation, bring a categorical dimension into the **Detail** tab; Tableau will transform the chart into a scatter plot.

6. This kind of chart can help pinpoint any outliers in the data, but too many points can look congested and hard to see. An additional ways to help separate the points is by including color in the marks of one of the measures.

7. Borders and opacity can also be used to help separate the data points. Click on **Color** in the **Marks** card, and at the bottom, lower the opacity and include a border to see each point. If the border selection is grayed out, make sure to select **Circle** from the drop-down options.

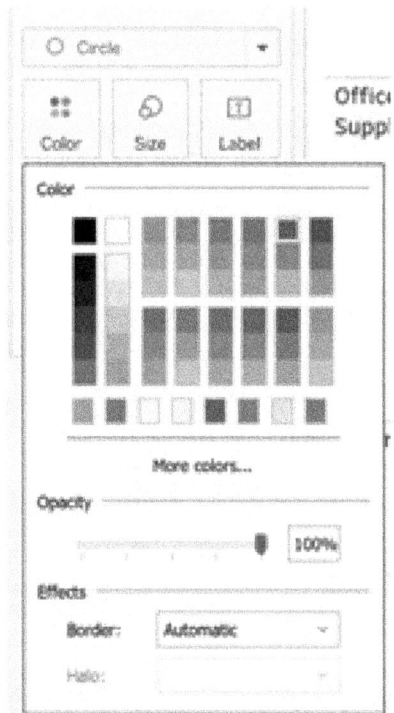

Figure 5.17: Color, Opacity, and Effects options

Take the time to explore using different measures to see whether any patterns emerge from the data.

Text Table

A text table is a simple visual that displays data to the granular level specified. What is key about text tables is their use of discrete fields in both **Columns** and **Rows**. Continuous fields will require an axis, but a text table will be an aggregated value of the dimensions specified.

Category	Sub-Category	Sales	Profit	Profit Ratio	Discount	Quantity
Furniture	Bookcases	114,880	-3,473	-3.02%	48	868
	Chairs	328,449	26,590	8.10%	105	2,356
	Furnishings	91,705	13,059	14.24%	132	3,563
	Tables	206,966	-17,725	-8.56%	83	1,241
Office Supplies	Appliances	107,532	18,138	16.87%	78	1,729
	Art	27,119	6,528	24.07%	60	3,000
	Binders	203,413	30,222	14.86%	567	5,974
	Envelopes	16,476	6,964	42.27%	20	906
	Fasteners	3,024	950	31.40%	18	914
	Labels	12,486	5,546	44.42%	25	1,400
	Paper	78,479	34,054	43.39%	103	5,178
	Storage	223,844	21,279	9.51%	63	3,158
	Supplies	46,674	-1,189	-2.55%	15	647
Technology	Accessories	167,380	41,937	25.05%	61	2,976
	Copiers	149,528	55,618	37.20%	11	234
	Machines	189,239	3,385	1.79%	35	440
	Phones	330,007	44,516	13.49%	137	3,289

Figure 5.18: Example of a text table

Items needed to build this chart are as follows:

Columns	Dimension(s)
Rows	Dimension(s)
Marks	Text

Table 5.14: Requirements to build a text table

The following steps will show how to build a text table. Open Tableau and connect to **Superstore Data** to practice building this chart:

1. 1. Drag the dimensions that are needed to the **Rows** and **Columns** shelves. For *Figure 5.18*, **Sub Category**, **Category**, and various measures such as **Sales**, **Profit**, **Discount**, **Quantity**, and **Profit Ratio** have been used.

2. In the **Data** pane, select the measures required while holding down *Ctrl*, and then drag the measures over the **Abc** box on the table.

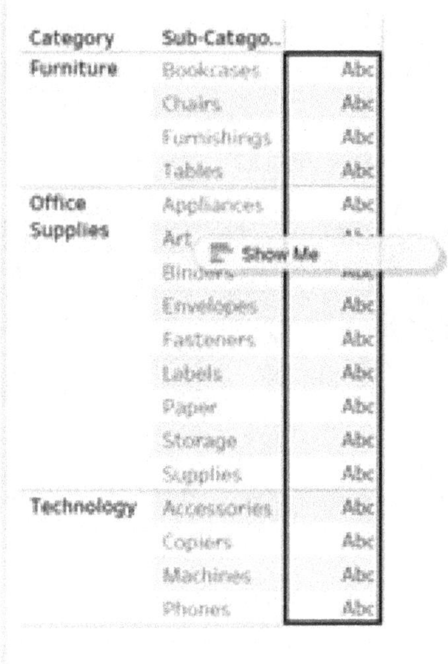

Figure 5.19: Example of Abc

3. Below the **Marks** card, a box named **Measure Values** should appear. This holds the measures being used. Drag any additional measures to that box to include in the text table. The order can be rearranged by moving the fields. A small orange arrow should appear to show you where the field is being placed.

A text table is a relatively simple chart to build but can show a vast amount of information. However, it is important to note that this kind of detail takes a lot of time to run, so this is not the most efficient chart to use with large quantities of data. Try using other charts as a dashboard action to filter the table for a detailed result.

Treemap

A treemap is used to show the percentage of a whole broken down by data points included in the view. This area will be blocked in individual squares sized depending on the percentage/value that it contributes to the whole.

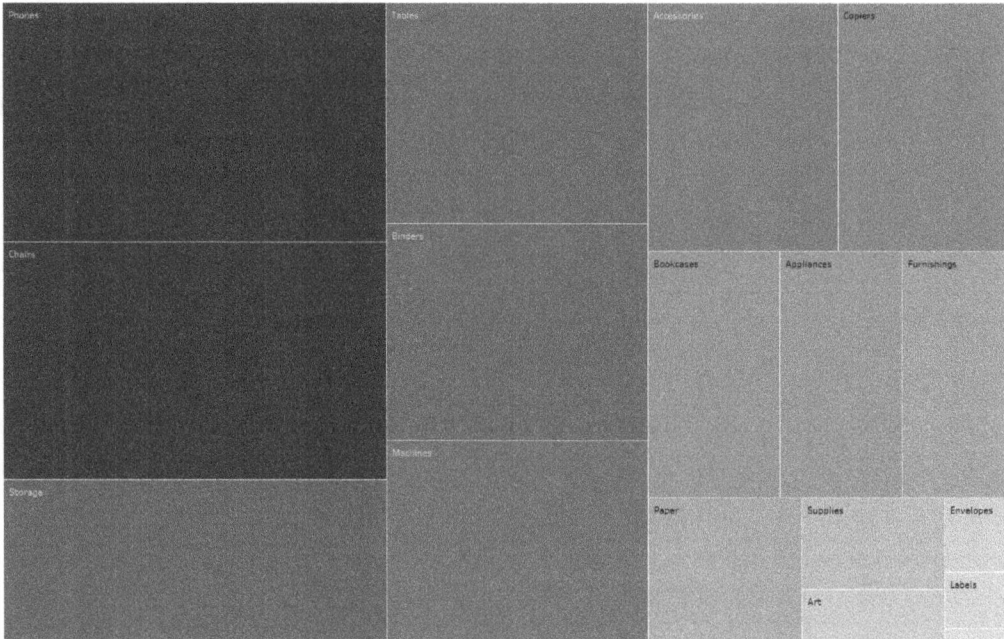

Figure 5.20: Example of a treemap

Items needed to build this chart are as follows:

Size	Measure
Detail	Dimension
Marks	Automatic
Color	Measure – Can be the same measure as in Size to further illustrate the data points

Table 5.15: Requirements to build a treemap

The following steps show how to build a treemap. Open Tableau and connect to **Superstore Data** to create *Figure 5.20*:

1. Identify the fields required to build this chart. For *Figure 5.20*, **Sub Category** and **Sales** have been used.

2. Drag the dimension that represents the categories or color blocks into **Color** on the **Marks** card.

3. Drag the measure to represent the size into **Size** on the **Marks** card.

4. Tableau will automatically build a treemap based on the configuration.

Try using other dimensions to break down the measure and observe the changes. It is important to note that this chart will fill the sheet automatically, so if there are too many dimensions included, the result will look skewed and hard to read.

Combination Chart

In combination charts, two charts share a dual axis to create one singular chart. These are useful to compare two measures and can allow flexibility in the types of charts that can be used to compare.

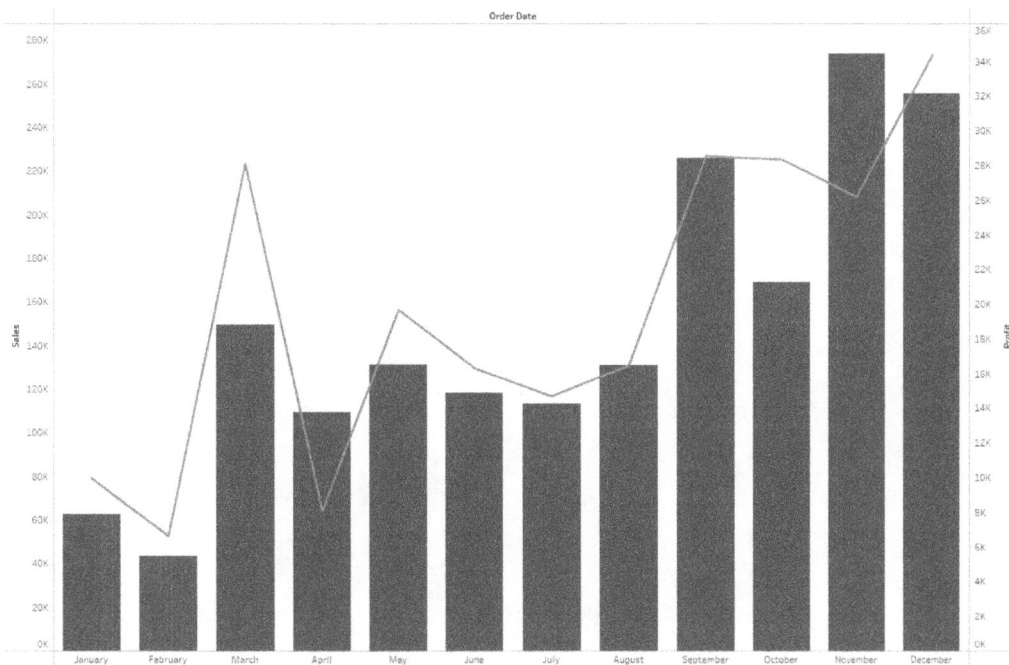

Figure 5.21: Example of a combination chart

Items needed to build this chart are as follows:

Columns	Dimension
Rows	Two measures that are **Dual Axis**
Marks	Depending on the axis

Table 5.16: Requirements to build a combination chart

Now it is time to build a combination chart. Follow the next steps to see how this chart is built:

1. Choose the two measures desired to compare and a dimensional field to break down the view.

2. Drag the dimension to **Columns**.

3. Drag the measures next to each other in the **Rows** shelf.

4. This should generate two stacked axes. To convert this into a combination chart, right-click on one of the measures and select **Dual Axis**, as shown in *Figure 5.22*:

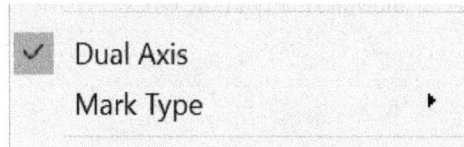

Figure 5.22: Axis selection

5. This will create one chart with two axes. Note that, in the **Marks** card, there will be two additional tabs for each measure. This is the section where the type of chart can be chosen.

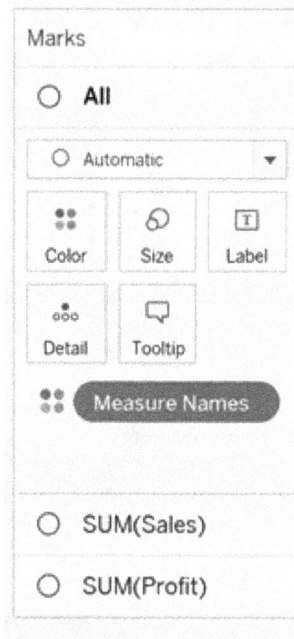

Figure 5.23: Marks card

6. To recreate the previous chart, select the **Sales** tab and change the dropdown to **Bar**.

7. Then, go to the **Profit** tab and change the dropdown to **Line**.

8. The first tab in the **Marks** card, called **All**, will adjust the formatting of all the charts. Select **Color** to change the dimension colors.

9. A developer can also synchronize the axes so that they match if a more accurate comparison is required. Right-click on one of the axes and select **Synchronize Axes**.

Geographic Charts

Spatial or geographic data in Tableau is often best visualized on a map. Maps are a great visualization in Tableau for identifying geographic trends or comparing specific spatial areas/regions in a way that can be much more intuitive than simply presenting comparisons of the geographic names in other chart types.

When using a geographic field in Tableau, the field will have a **Geographic Role** option selected. For example, a **Country** field will have the **Country/Region** geographic role selected. It is important to select the correct geographic role for a field as this is how Tableau will try to map the data. Names of places are the most common field type. They are essentially string fields that have been given a geographic role that tells Tableau how to map them. These geographic roles can range from airport names to counties, postcodes, and cities.

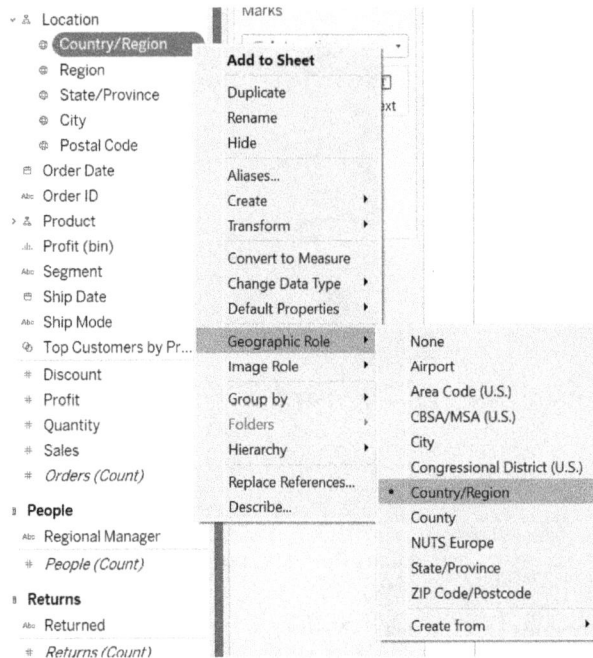

Figure 5.24: The Country field with the Country/Region geographic role applied

Tableau will often automatically identify and assign geographic fields, but if this does not happen by default, users can right-click the field, followed by **Geographic Role**, and then select the correct assignment. When a geographic field is present in a data source, Tableau will automatically generate **Latitude** and **Longitude** fields. Map charts are created in Tableau by placing the **Latitude** field on **Columns** and the **Longitude** field on **Rows**. The type of map and the data points plotted depend on which geographic fields are placed on the **Marks** card and what type of mark is selected.

Longitude and latitude do not have to be generated by Tableau. If the data source has its own numeric **Latitude** and **Longitude** fields, these can be configured as geographic types and set to **Latitude** and **Longitude** roles.

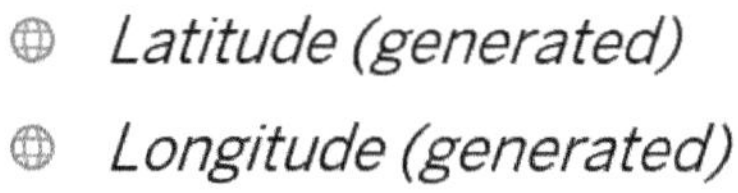

⊕ *Latitude (generated)*

⊕ *Longitude (generated)*

Figure 5.25: Latitude and Longitude fields generated by Tableau will have a (generated) suffix

When hovering over any map chart in Tableau, a toolbar appears at the top left that allows users to zoom in and out of the map (also possible with the mouse scroll wheel) and pin the map to a specific view. The toolbar also provides options for moving on the map and selecting multiple data points. There is also a search bar allowing users to jump to specified geographic locations.

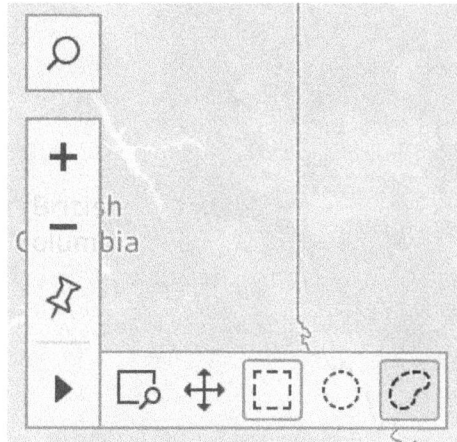

Figure 5.26: Tableau map toolbar allowing location search, zooming, location pinning, panning, and selection of data points via square, circle, or custom shape

Symbol Maps

Symbol maps in Tableau allow users to plot specific data points on a map by their longitude and latitude with a symbol denoting the position of each point. This chart type is great for showing the occurrence of a metric in specific locations.

Usually, symbol maps represent points on the map as circles but, for additional context, discrete fields can be used to change the symbol of a point based on the value of the field. An example of this could be mapping disease outbreak locations with each disease being represented by a different symbol. Similarly, different colors can be used to represent different dimension values.

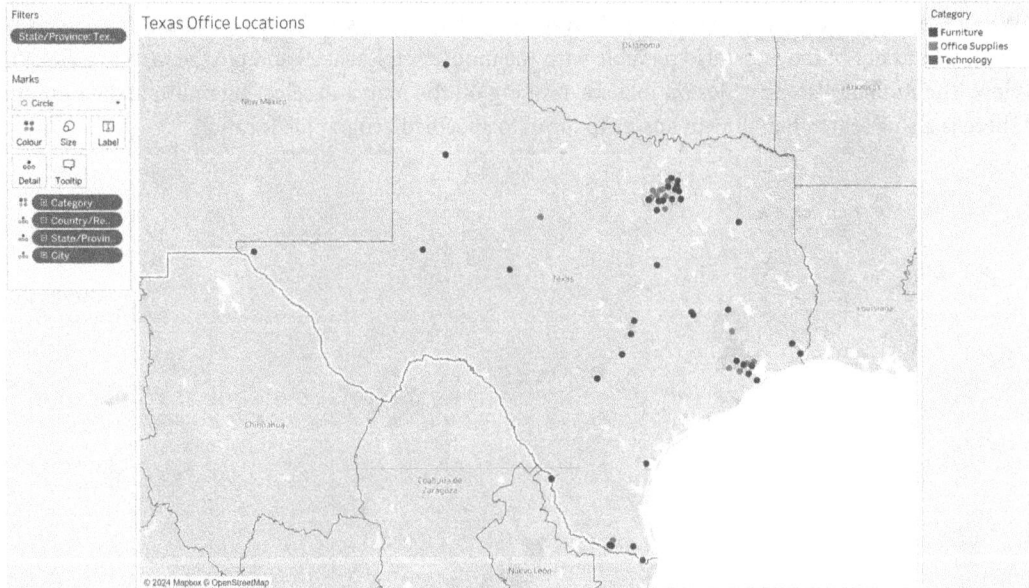

Figure 5.27: Symbol map showing office locations in Texas colored by Category

It is often useful to add context to symbol maps by sizing the symbol by a measure. An example of this could be a point on the map representing each city that has sales with the size of the circle representing the number of sales. This allows for a comparison of sales by city but also allows the user to see geographically where the sales cluster. The opacity of the symbols on the map can also be modified to ensure that all symbols are visible even if they overlap.

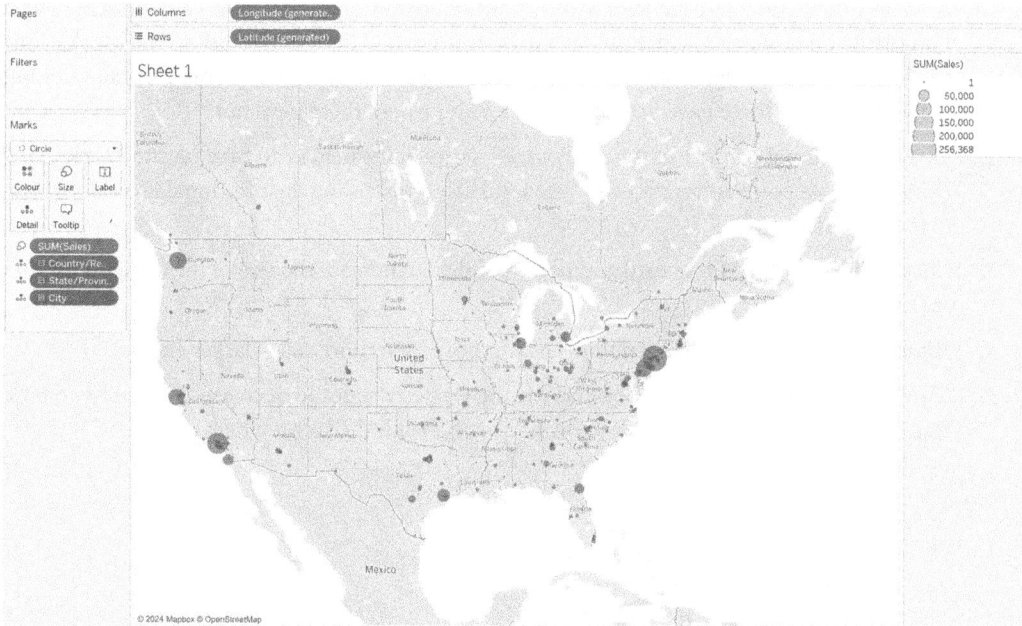

Figure 5.28: Sales mapped per city using a symbol map in Tableau

To create a symbol map, **Longitude** must be placed on **Columns**, **Latitude** on **Rows**, and the field containing the geographic points to map should be on **Detail** on the **Marks** card. A shortcut to achieve this is to double-click the geographic field that has the points to be mapped. By default, a symbol map is created with this configuration, and the **Circle** mark type is used. This can be updated to the **Symbol** type and a dimension can be placed on **Symbol** on the **Marks** card to break up data points by that field. A measure can be placed on **Size** to add additional context and **Color** can also be used to categorize the points or to add an additional measure to the analysis.

Creating a Symbol Map in Tableau

In this exercise, you will create a symbol map in Tableau. This will enable you to create symbol maps to plot individual data points on a map, size them based on measures, and color them if needed. Follow these steps:

1. On a new sheet, double-click the **Postal Code** field to automatically create a symbol map with **Longitude** on **Columns**, **Latitude** on **Rows**, and **Postal Code** on **Detail** on the **Marks** card. Each postcode in the data source is now mapped as an individual point. This chart allows users to identify where there have been orders/sales and where there have not.

2. To add additional context to the analysis, drag the **Sales** field onto **Size** on the **Marks** card. Each postcode point is now sized by the total sales, providing an additional layer of context to the analysis.

3. Click on **Size** on the **Marks** card and use the slider to increase the size of the circles until they are more visible.

4. Some postcodes are now hidden underneath those with more sales (larger circles).

5. Click on **Color** on the **Marks** card and use the **Opacity** slider to decrease the opacity to 50%. Data points previously hidden can now be seen beneath the larger circles.

6. For a final piece of additional context, drag the **Profit** field onto **Color**. Postcodes with a negative profit can now be easily identified from postcodes with a positive profit.

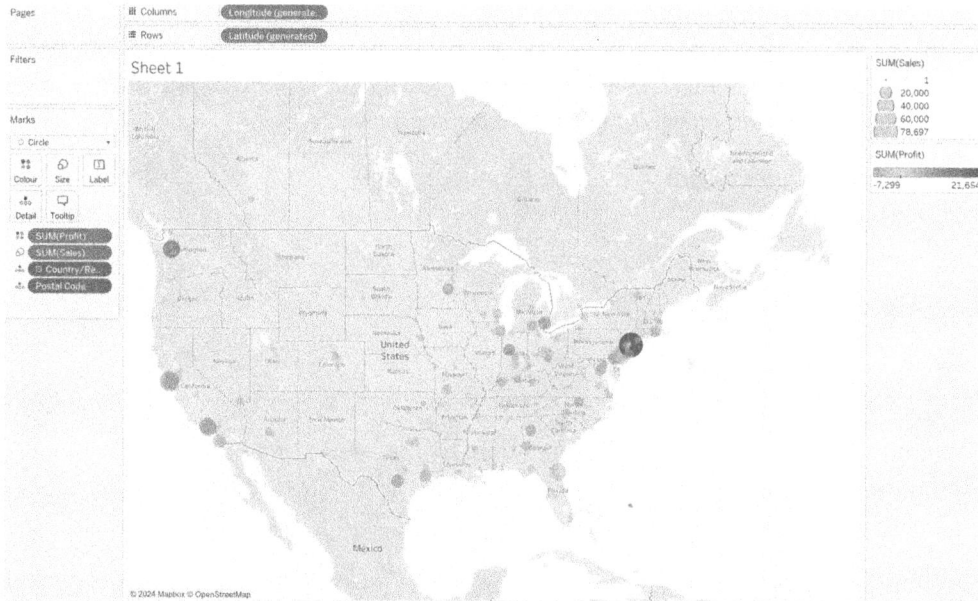

Figure 5.29: Symbol map created showing every postcode where there were
sales in the USA sized by the total sales and colored by total profit

In this exercise, you created a symbol map that plots cities in North America, sizes the data points by **Sales**, and colors the data points by **Profit**. You can now plot data points on a map in Tableau and provide additional context via sizing and color. If you would like more practice with symbol maps, try switching the mark type to **Shape** and then play with the various options on the **Marks** card.

Density/Heat Maps

Density maps or **heatmaps** in Tableau use color intensity to visualize clusters of data. Crime rates could be plotted for a city with the color on the chart being more intense the higher the crime rate in that area. Density maps solve the problem of overlapping data points in symbol maps, as the concentration of points in a specific area results in a more dense/intense coloring. Density maps are best used when the data source has a significant geographical overlap in data points.

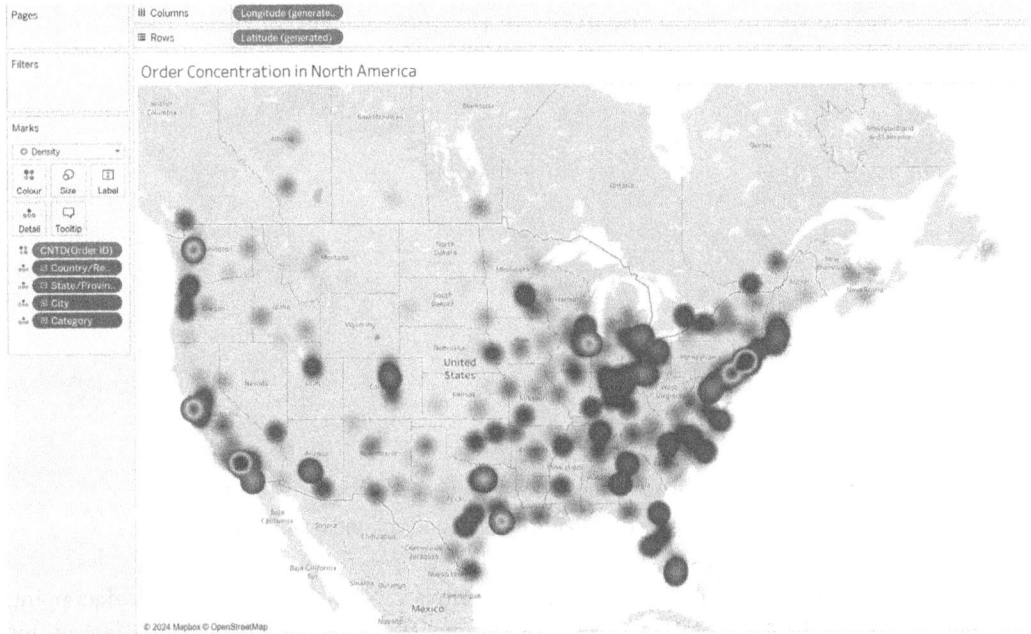

Figure 5.30: Density map showing the concentration or density of orders across North America

While locations such as counties and cities result in generated **Longitude** and **Latitude** fields, such a high level of geographic breakdown is not best suited for density maps as there will be no continuous coloring for easy identification of trends. Specific data points such as postcodes or, preferably, longitude and latitude coordinates are better, as the intensity of the color will be mapped continuously across the map.

To create a density map in Tableau, the **Latitude** and **Longitude** fields should be on **Rows** and **Columns**, respectively, and the dimension that will identify each individual data point should be placed on **Detail** on the **Marks** card. The mark type should be **Density** and the intensity and scheme of the coloring can be set using **Color** on the **Marks** card. The coloring will indicate the density of data points in a given area, but a measure can also be placed onto **Color** on the **Marks** card to instead display the density of that metric.

Creating a Density Map in Tableau

In this exercise, you will create a density map in Tableau. It is useful to learn how to make density maps in Tableau as they are a great way to visualize comparative concentrations of data. Follow these steps:

1. On a new sheet, double-click the **Postal Code** field to automatically create a symbol map with **Longitude** on **Columns**, **Latitude** on **Rows**, and **Postal Code** on **Detail** on the **Marks** card.

2. Drag the **State/Province** field to the **Filters** shelf and, in the **Configuration** popup, select **California** and click **OK**.

3. The map now displays every postcode in California in the dataset as individual points. Change the mark type to **Density** to instead display the frequency of orders in every postcode in California by the intensity of the coloring.

4. Drag the **Sales** field onto **Color** on the **Marks** card to instead show the number of sales in each postal code in California by the intensity of the coloring.

5. The intensity can be made more or less pronounced by selecting **Color** on the **Marks** card and then increasing or decreasing the intensity percentage using the slider.

6. A color scheme can also be selected in the **Color** dropdown; for example, the **Temperature Diverging** color scheme runs from green through yellow to red based on the intensity of data points in the measure being visualized.

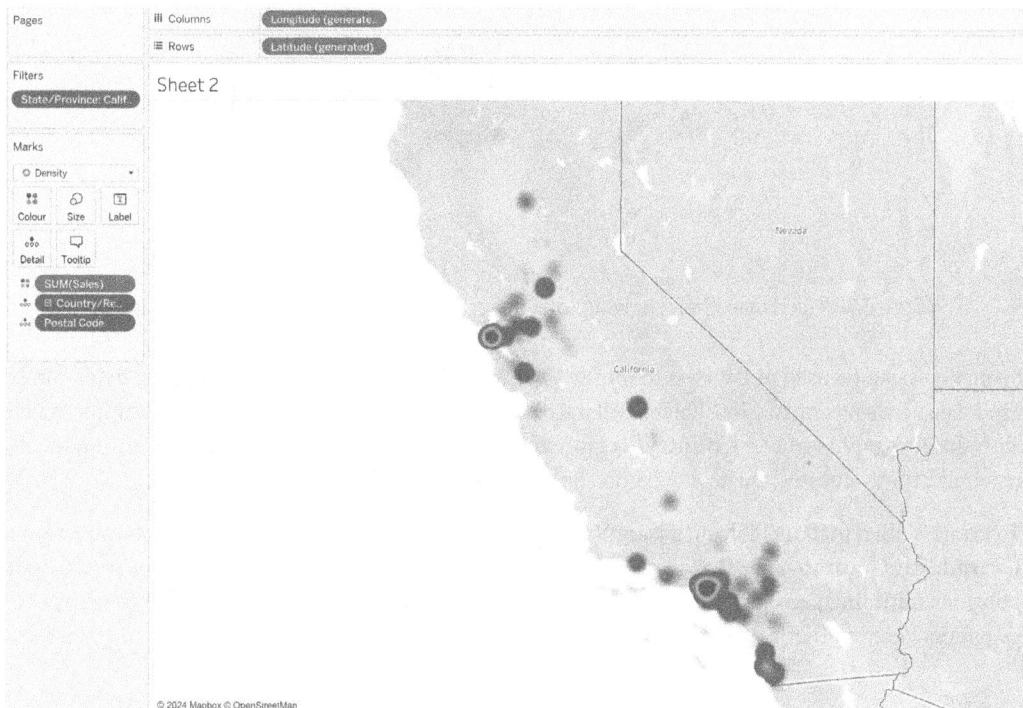

Figure 5.31: Density chart created showing the clustering of sales in specific areas of California

In this exercise, you plotted the density of total sales value for postal codes in California. You now know how to create a density map chart, select a color scheme for it, and modify the intensity of the density coloring.

Filled Maps

Filled maps in Tableau (sometimes called choropleth maps) differ from symbol and density maps. While the latter plots individual points using longitude and latitude, the former maps polygons around specified geographic areas. The polygons are colored by the amount of data points within or by the value of a specified metric for all the data points within. They are great for showing comparisons between specific areas – for example, the **Gross Domestic Products (GDPs)** of countries, with countries with higher GDPs being colored darker than countries with low GDPs.

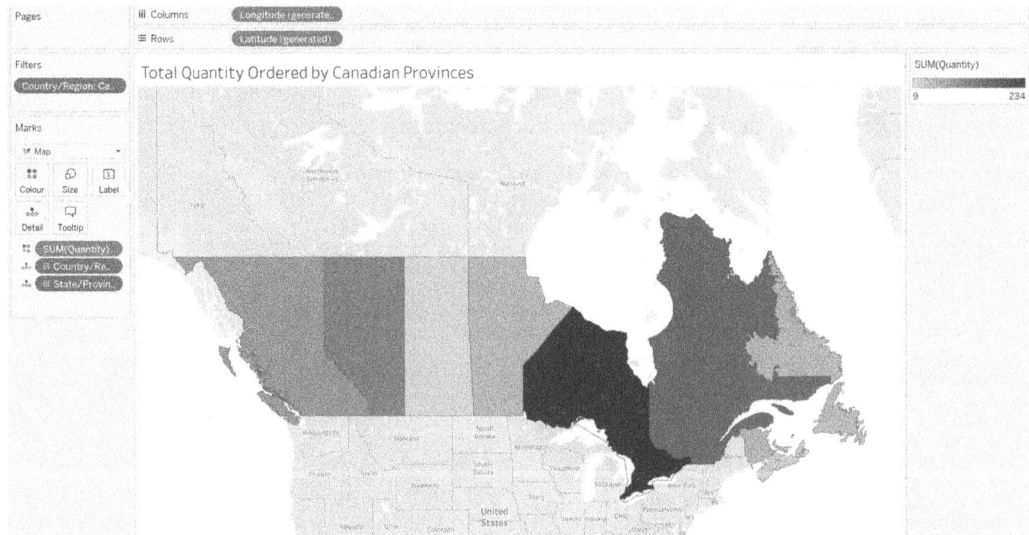

Figure 5.32: Filled map chart showing total quantity ordered by State/Province in Canada

Spatial files can be used as datasets to import custom polygons or Tableau's geographic fields can be used. Locations recognized by Tableau will have polygons predefined for that area; for example, a field with the geographic role of **Country/Region** and the value of **United Kingdom** will automatically recognize that value and create a polygon shape of the UK overlaid on the map visual.

To create a filled map in Tableau, a geographic field should be placed on **Detail** with **Longitude** and **Latitude** on **Columns** and **Rows**, respectively. The mark type should be set to **Map**, resulting in polygons outlining each geographic area. A measure should then be placed on **Color** to color each geographic area so that they can be compared.

Creating a Filled Map in Tableau

In this section, you will create a filled map chart in Tableau to allow for a state-by-state comparison of profit. Filled maps are a common chart type in Tableau and are great for comparisons between geographic regions. Follow these steps:

1. On a new sheet, double-click the **State/Province** field to automatically create a symbol map with **Longitude** on **Columns**, **Latitude** on **Rows**, and **State/Province** on **Detail** on the **Marks** card.

2. Change the mark type to **Map** to create polygon shapes for each state in the USA and Canada.

3. Drag **Profit** onto **Color** on the **Marks** card to color each state by total profit. States with a negative total profit are colored orange while states with a positive profit are colored blue.

4. The color scheme and opacity of the colors can be updated via the **Color** configuration on the **Marks** card.

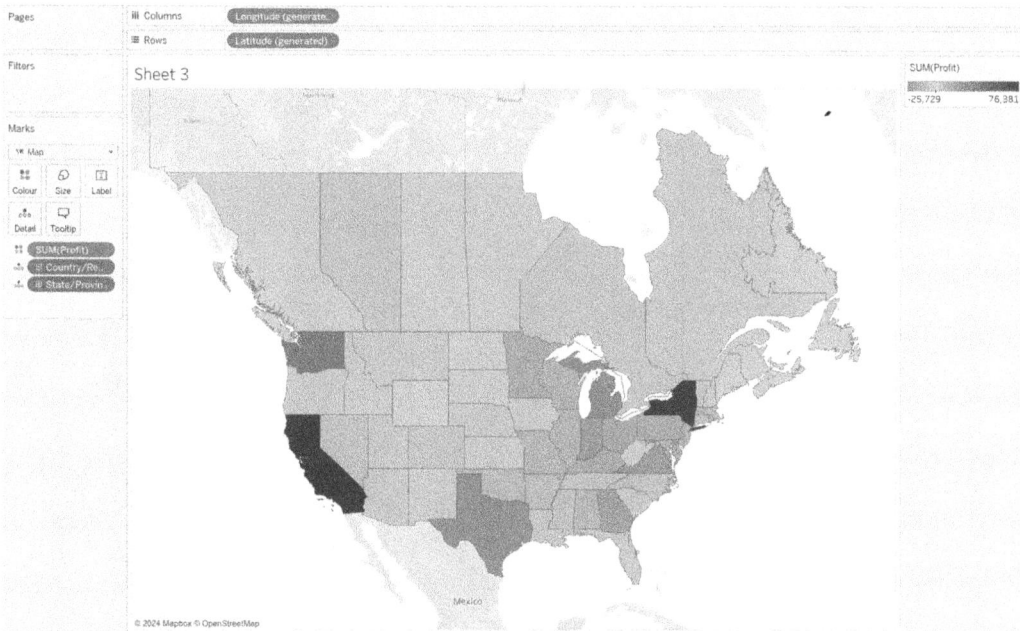

Figure 5.33: A filled map showing the difference in total profit between states of the USA and Canada

In this exercise, you have created a filled map chart that compares profit across states/regions in North America. You now know how to create filled map charts at whatever geographic region level is required, how to color the regions, and also how to modify the color scheme and opacity applied.

Analytics

Tableau charts can be enhanced with the addition of analytical objects that add further context to the visual. Most analytical objects can be found in the **Analytics** pane in the interface found by switching from the **Data** pane tab on the left-hand side to the **Analytics** tab. Objects that can be used on the current visual will be shown in black text and can be dragged onto a chart. Unavailable analytical objects are grayed out.

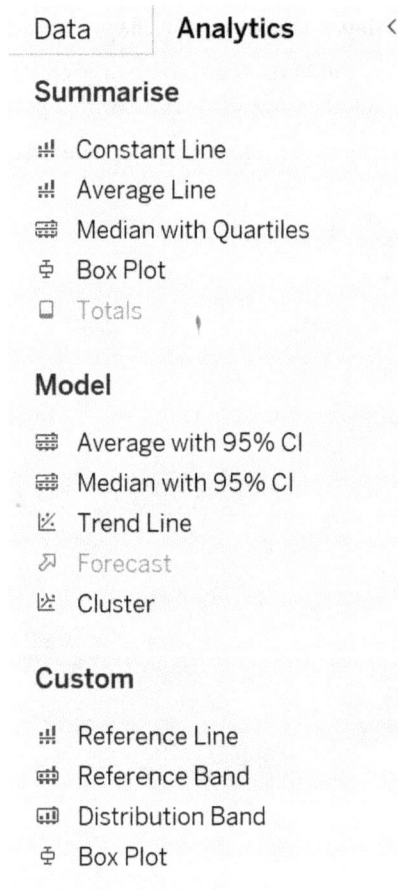

Data   **Analytics**   <

Summarise

- Constant Line
- Average Line
- Median with Quartiles
- Box Plot
- Totals

Model

- Average with 95% CI
- Median with 95% CI
- Trend Line
- Forecast
- Cluster

Custom

- Reference Line
- Reference Band
- Distribution Band
- Box Plot

Figure 5.34: The Analytics pane in Tableau's user interface

Dragging an analytical object over a chart will result in a popup that allows the user to specify how the analytical object will be placed in the view. The options differ based on the object type and view configuration. The options are to create the object across the whole chart/view with **Table**, create the object within each pane using **Pane**, or use each individual data point via **Cell**. The object being dragged onto the view can be dropped onto any of the options in the popup to select that option.

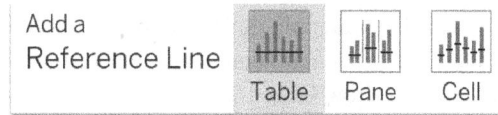

Figure 5.35: A reference line analytics object being dragged onto the view across the whole table

Analytics objects are specific to a single measure/axis on the view. If there are multiple measures on the view, when dragging the object from the **Analytics** pane, the configuration popup will require the user to specify which measure to create the object for as well as at which level.

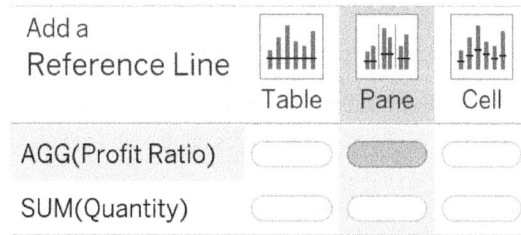

Figure 5.36: A view with both Profit Ratio and total Quantity measures included having a reference line added in each Profit Ratio pane

Once an analytics object has been dropped on the view, it can be configured or removed by right-clicking the object and selecting the relevant options. There are also some configuration options for analytics objects available in the **Analysis** menu in the top toolbar.

Totals and Subtotals

Totals and subtotals can be added to charts in Tableau from both the **Analytics** pane and via selection in the **Analysis** menu. Totals and subtotals, by default, use the same level of aggregation as the measure in the view. If the view uses the sum of sales, then a total will be the total sales overall, whereas if the measure used is the average of sales, then the total value will be the average across the data points in the view.

Totals are calculated across the whole table and can be positioned either at the top or bottom of the view. For a grand total to be added, the view must have at least one header created by placing a dimension on either the **Rows** or **Columns** shelf. Subtotals are calculated by **Pane** and, if there are multiple dimensions, the level the subtotal is calculated and displayed at can be specified.

Totals and subtotals can be created by dragging the **Total** object from the **Analytics** pane onto the view. When dragging the **Total** object, a configuration popup appears allowing the user to select either **Subtotals, Column Grand Totals,** or **Row Grand Totals.** Choosing **Subtotals** will create subtotals in each pane. **Column Grand Totals** will create a grand total across the whole view/table calculated from the top to the bottom for each column. **Row Grand Totals** will create totals across the whole view/table from left to right for each row.

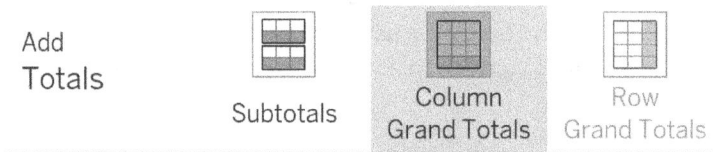

Figure 5.37: Total and subtotal pop-up options when dragging the
Total object from the Analytics pane onto the view

When a grand total is created by default, it is created across all the relevant underlying data. To have the total calculated based on the values displayed in the view, select **Totals** from the **Analysis** menu in the top toolbar, followed by **Total All Using**, then select the level of aggregation. **Total All Using** can also be used to specify how to aggregate grand totals. If, for example, the measure on the view is aggregated as a sum but you want to display the average as a total, then selecting **Average** in the **Total All Using** menu can achieve this.

Totals, by default, appears at the bottom or right-hand side of the view. To move **Totals** to the top or left-hand side, select **Analysis** followed by **Totals** and then **Column Totals to Top** or **Row Totals to Left**.

Creating Totals and Subtotals

In this exercise, you will learn how to add grand totals and subtotals to visuals in Tableau. Totals can be used in many charts but are most often displayed in table visuals and can provide a useful additional layer of context. Follow these steps:

1. On a new sheet, drag **Category** onto the **Rows** shelf followed by **Sub-Category**.

2. Place **Segment** on the **Columns** shelf and then drag **Sales** onto **Text** on the **Marks** card. The view now displays a table breaking each category out by subcategory and then showing the sales for each by segment.

3. To show the total sales by column (**Segment**), select the **Analytics** pane, then drag the **Totals** object onto the view and drop it on the **Column Grand Totals** option of the popup. Now, each column has a grand total that shows the total sales for each segment.

4. To add subtotals for each category, drag the **Totals** object onto the view and drop it on **Subtotals**. The table now has totals and subtotals.

5. To show the totals for each subcategory, the **Totals** object can be dragged onto the view again but, to try a different method, go to **Analysis** followed by **Totals** and then select **Show Row Grand Totals**. The view now has totals for each subcategory.

6. To move these to the left-hand side, go back to **Analysis** followed by the **Totals** menu and select **Row Totals to Left**.

7. Finally, to switch the totals from showing the summed values for each row/column to the average, go to **Analysis** followed by **Totals** and then select **Total All Using** and **Average**. Each total and subtotal now shows the average as opposed to the summed total.

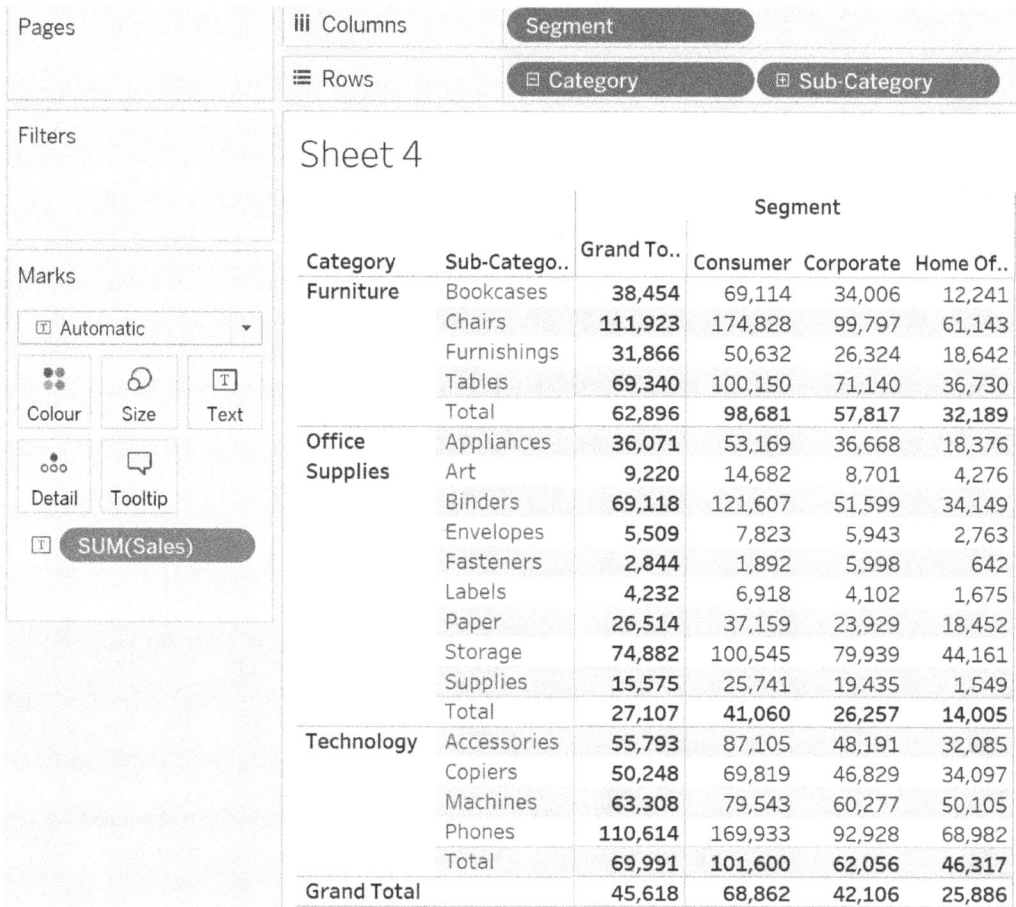

Pages			iii Columns		Segment		
			☰ Rows		⊟ Category		⊞ Sub-Category

Filters

Sheet 4

Marks

				Segment			
Category	Sub-Catego..	Grand To..	Consumer	Corporate	Home Of..
Furniture	Bookcases	38,454	69,114	34,006	12,241
	Chairs	111,923	174,828	99,797	61,143
	Furnishings	31,866	50,632	26,324	18,642
	Tables	69,340	100,150	71,140	36,730
	Total	62,896	98,681	57,817	32,189
Office Supplies	Appliances	36,071	53,169	36,668	18,376
	Art	9,220	14,682	8,701	4,276
	Binders	69,118	121,607	51,599	34,149
	Envelopes	5,509	7,823	5,943	2,763
	Fasteners	2,844	1,892	5,998	642
	Labels	4,232	6,918	4,102	1,675
	Paper	26,514	37,159	23,929	18,452
	Storage	74,882	100,545	79,939	44,161
	Supplies	15,575	25,741	19,435	1,549
	Total	27,107	41,060	26,257	14,005
Technology	Accessories	55,793	87,105	48,191	32,085
	Copiers	50,248	69,819	46,829	34,097
	Machines	63,308	79,543	60,277	50,105
	Phones	110,614	169,933	92,928	68,982
	Total	69,991	101,600	62,056	46,317
Grand Total		45,618	68,862	42,106	25,886

- Automatic
- Colour
- Size
- Text
- Detail
- Tooltip
- SUM(Sales)

Figure 5.38: Totals and subtotals added to the view in Tableau and configured positionally in terms of aggregation

In this exercise, you have used various methods to add totals and subtotals to a table visual in Tableau. You have also configured the totals positionally. You can also apply these methods to other chart types such as bar charts.

Reference Lines and Average Lines

Reference lines in Tableau are visual points of reference on an axis that can be used to provide context to the data points displayed. Reference lines can take the form of a constant line that stays fixed to a point on an axis regardless of how the data changes around it. This could be fixed to a specific date on a date axis as a point-in-time indicator. It can also be a reference point on a measure axis used as an indicator for things such as targets and thresholds.

Dynamic reference lines that update as the data is updated or the view is filtered can also be set up. **Min** and **Max** values on the axis can be identified with a reference line and lines can also be added for the average and the median. Average and median reference lines are a great way to indicate the distance of data points from the average. Reference lines can also be created for the total or sum of all data points on an axis but this is rarely useful.

Reference lines can be created by dragging either **Constant Line**, **Average Line**, or **Reference Line** from the **Analytics** pane onto a view with an axis. For views with more than one measure, the popup will allow the user to select which measure/axis to create the reference line on and also whether to create it across the whole table, pane by pane, or for each cell.

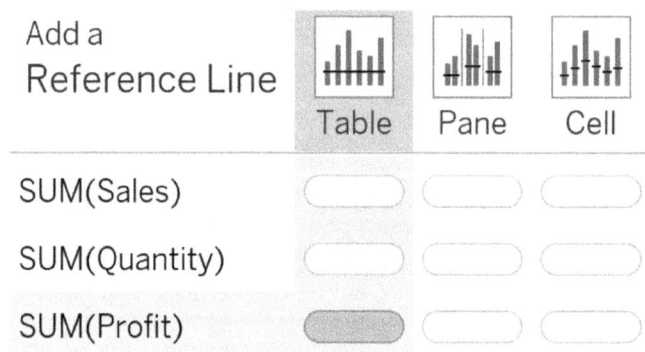

Figure 5.39: Options when dragging a reference line onto a view with multiple measures; here, a reference line is being created on the Profit axis across the whole view/table

Whether to create a reference line across the whole table, pane by pane, or for each cell is referred to as the scope of the reference line. This option is relevant to most analytical objects that can be added to a Tableau chart. **Table** refers to the creation of a line that runs from the axis across the whole chart. **Pane** refers to a new line being calculated for each individual pane. **Cell** refers to a line being calculated for each individual cell in the view.

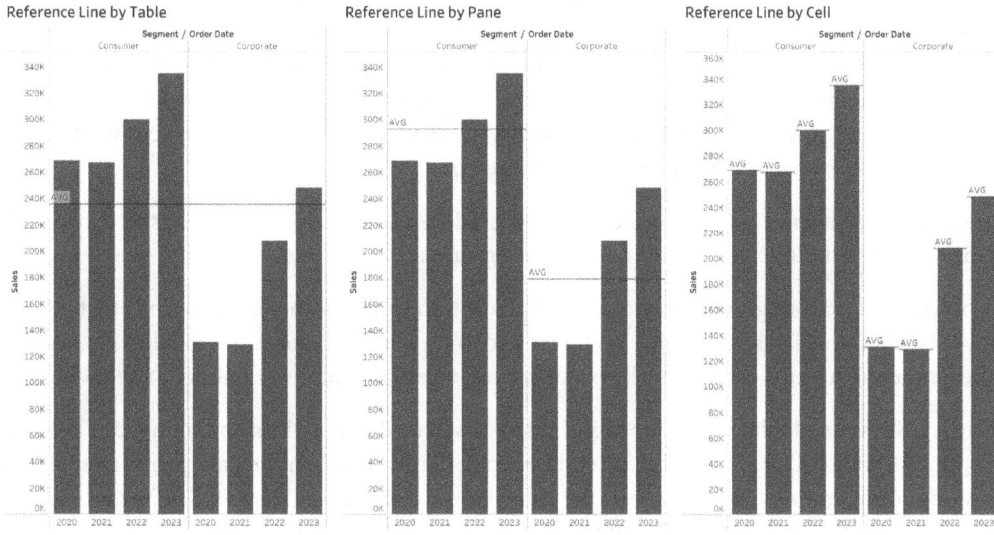

Figure 5.40: The same bar chart has reference lines with different scope settings; the left chart shows an average line across the whole table, the middle chart shows an average line per pane, and the rightmost chart shows an average line per cell

When dragging **Constant Line** onto the view, there will be a popup that asks the user to specify a value to set as the constant for the line on the axis. After inputting the desired value and pressing **Enter**, the constant line will be created. When dragging **Average Line** onto the view, no additional configuration is required. The average line is automatically added to the selected measure/axis.

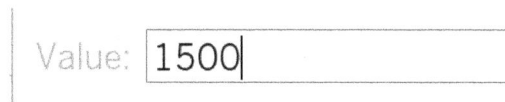

Figure 5.41: Popup for constant line value configuration

If **Reference Line** is dragged onto the view, then the reference line configuration popup opens. A customizable reference line can also be created on any axis by right-clicking and selecting **Add Reference Line**. The same reference line configuration popup appears for both methods of creation and can be opened for any reference line (including constant and average lines) by right-clicking a reference line and selecting **Edit**.

The reference line configuration popup allows the user to change the scope of the line from **Table** to **Pane** or **Cell**. The value the line is plotted against can also be modified. By default, the value will be the aggregated measure or the date used to create the axis. If the axis is a date, then the line can be set to the minimum, maximum, or constant date. If the value is a measure, then the value can be set to the total, sum, minimum, maximum, average, median, or constant value. If a constant value is selected, then the value can be typed into the **Value** box.

A parameter can also be used instead of the field used for the axis; this allows users to dynamically change the reference line positioning based on user input.

The reference line configuration pane also allows customization of the line in terms of whether to show a label on the line and a tooltip on hover and, if so, what to display. It also allows for the line to be formatted in terms of style, thickness, color, and whether to fill a color above or below the line.

There is a **Show recalculated line for highlighted or selected data points** option that can be ticked or unticked. If ticked, the reference line will be repositioned based on the data points selected and the value configuration. For example, if the value type is **Average** and five data points are selected on the edge of the chart, the reference line will move to the average value of the five selected points.

Reference lines can be removed from a chart by right-clicking and selecting **Remove**.

Creating a Constant and Average Reference Line in Tableau

In this exercise, you will add reference lines to a chart in Tableau. Reference lines can be a great visual aid in charts to signify the extremities of the data or points of significance. In this exercise, a reference line will be created on a constant value and a dynamic reference line will be created on the average value. The same techniques can be applied to create other dynamic reference lines such as minimum and maximum values. Follow these steps:

1. On a new sheet, drag **Sales** onto **Columns** and **Profit** onto **Rows**, then place **Order ID** onto **Detail**. The chart is now a scatter plot showing the **Sales** and **Profit** totals for each order.

2. To create a reference point at the **5000** sales mark for an order, open the **Analytics** pane and drag on the **Constant Line** object.

3. Make sure to drop **Constant Line** on the **SUM(Sales)** option in the popup to create the reference line on the **Sales** axis.

4. Set the value for the constant line to **5000** and press **Enter**. There is now a reference line at the **5000** sales mark that allows users to more easily see orders with sales above and below a total of 5,000.

5. To create a reference line showing the average sales, right-click the **Profit** axis and select **Add Reference Line**.

6. In the reference line configuration, make sure the value is **SUM(Profit)** and set the type to **Average**, then press **OK**.

7. This has created a reference line showing the average profit across all orders. If a point is selected, the reference line is recalculated for that point.

8. The label for the line also just says **Average**. It would be more useful if the reference line was fixed and showed the value of the average order. Right-click the line and select **Edit** to open up the configuration options again.

9.   Deselect **Show recalculated line for highlighted or selected data points**, change the label from **Computation** to **Value**, and press **OK**. The average line now shows the average order profit and the line is not recalculated when selecting any points.

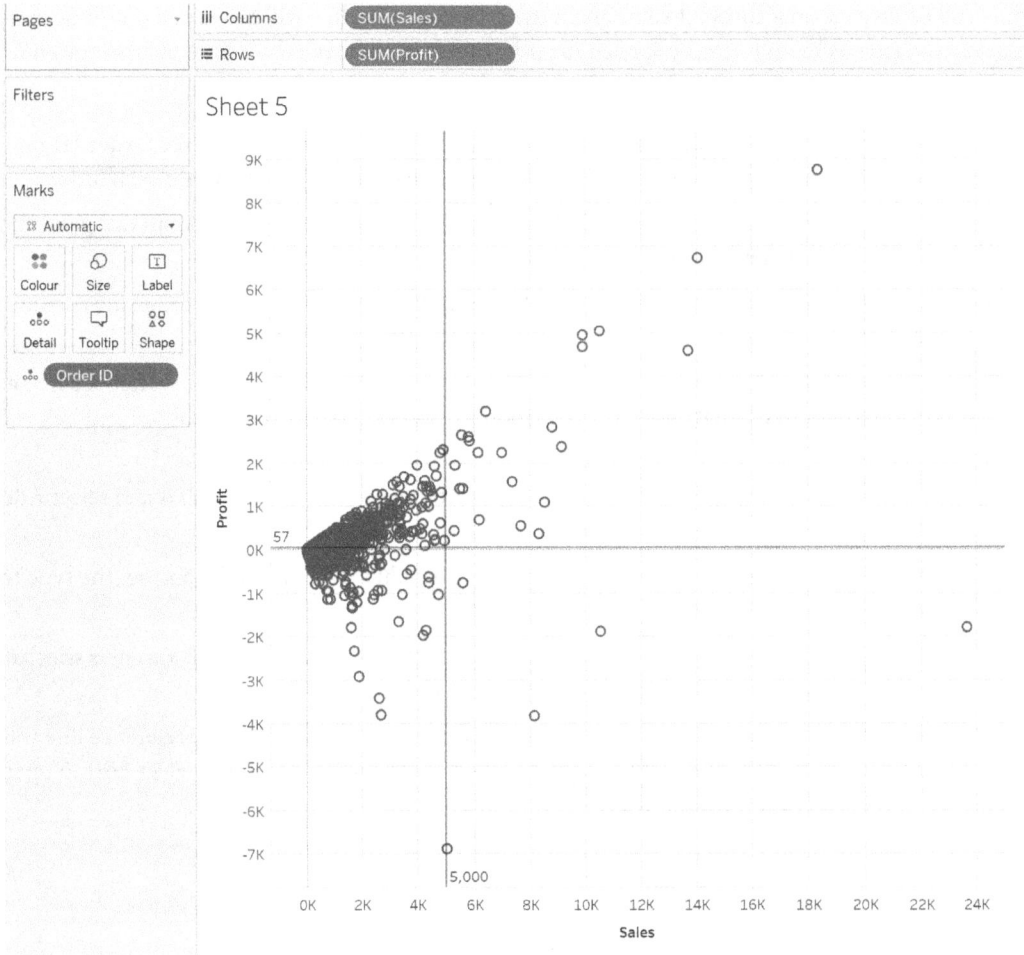

Figure 5.42: Constant line and average line added to a Sales versus Profit
scatter plot and the average line further customized

In this exercise, you have added two reference lines to a scatter plot visual. You now know two different methods for creating reference lines and you know how to set the value for a reference line, both for dynamic and static/constant reference lines. You also know how to configure reference lines to include a custom label and to be repositioned based on the user selection.

Reference Bands and Distribution Bands

Reference bands and distribution bands in Tableau are similar to reference lines, but instead of referencing a single point on an axis, the area from one reference point to another reference point is shaded. The shading is behind the marks on the view so it will not cover any data points. Therefore, both reference bands and distribution bands require a continuous axis to be placed along.

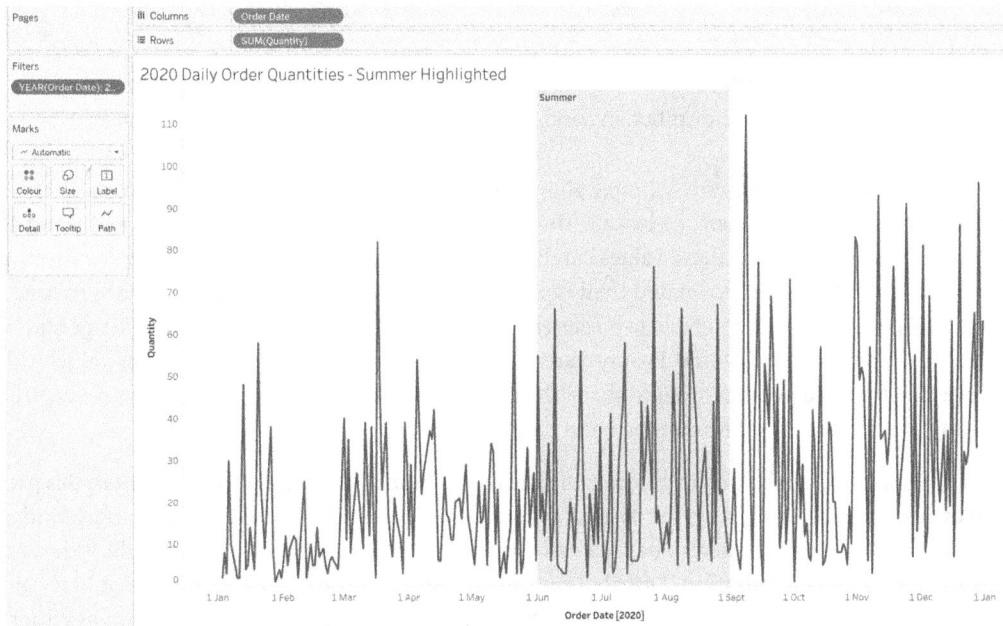

Figure 5.43: Reference band highlighting data points that fall during summer

Reference bands shade the area between either two constant or computed values in a solid color that can be specified by the user. Reference bands can be created on measure axes and date axes. To create a reference band, drag on the **Reference Band** object from the **Analytics** pane and drop it on the desired axis at the desired scope for the range (**Table**, **Pane**, or **Cell**). Alternatively, right-click the desired axis and select **Add Reference Line**. Creating a reference line from the axis will open up the reference line configuration window, but the options along the top allow the user to select **Line** for a reference line, **Band** for a reference band, **Distribution** for a distribution band, or **Box Plot** to create a box plot.

Dragging the **Reference Band** object from the **Analytics** pane onto the view will open the same configuration window but already set to **Band**.

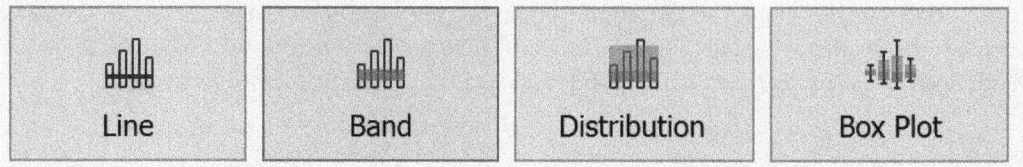

Figure 5.44: The configuration window allows users to select whether to add a reference line, reference band, distribution band, or box plot; here, a reference band is selected

The reference band configuration window allows the user to set the scope of the band to occur either across the whole table, by pane, or by cell. The user is then asked to set a **Band From** and **Band To** reference point on the axis. These values can be typed in as hardcoded constant values by selecting **Constant** in the **Type** dropdown and then typing the desired value in the **Value** box. Alternatively, the same dynamic options available to a reference line can be used; for example, the average sum of sales value can be set as the **Band From** reference point and the maximum sum of sales value could be set as the **Band To** point in order to shade the top half of sales values. Parameters can also be used, allowing the end user to customize the band range.

A distribution band can be created by dragging the **Distribution Band** object from the **Analytics** pane onto the view and dropping it on the desired axis and scope. This will open the configuration window for a distribution band. It can also be created by right-clicking an axis and selecting **Add Reference Line**, then selecting **Distribution**. Distribution bands are similar to reference bands in that they color the area behind the marks on a chart but are different in that the reference points are set as either a percent of a value, percentiles, quantiles, or a range of standard deviation.

A single reference point can be specified when customizing a distribution band, which will result in a reference line at that point. Adding another reference point will result in the area between those points being shaded. More points can be added to the distribution band resulting in differing shades of gray between each reference point.

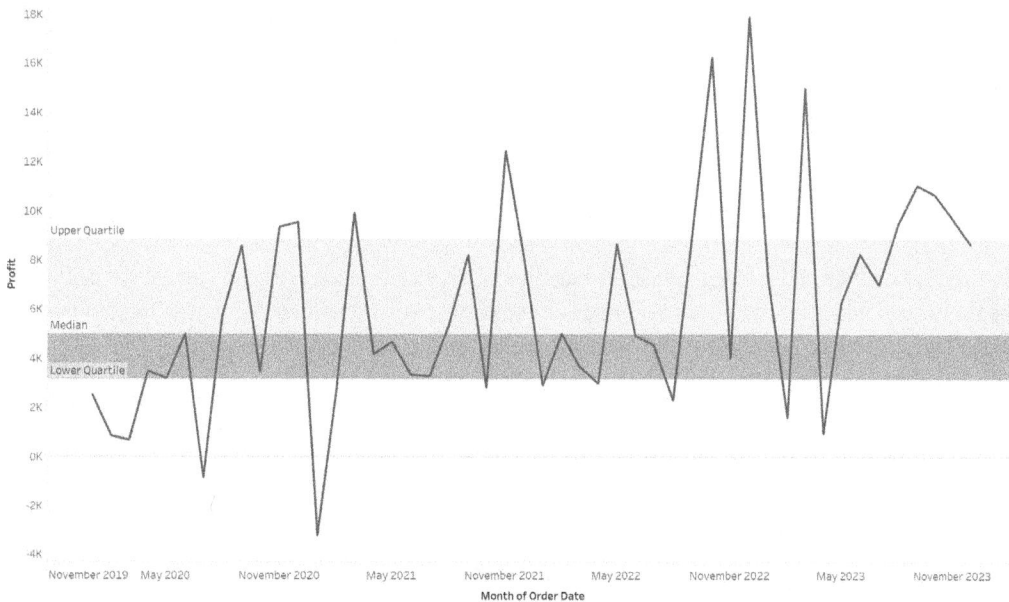

Figure 5.45: Distribution band calculated with four quartiles; the lower quartile
to the median is shaded darker than the median to the upper quartile

The **Percentages** option for distribution bands allows the user to set two percentage values and a value to calculate the percentage from. The value to calculate the percentage from can be a constant, a parameter, or any of the aggregation options available to a reference line on the axis measure. The percentage values are set by typing into the **Percentages** box, and multiple values can be set delimited by commas.

The **Percentiles** option for distribution bands allows the user to select **80**, **90**, **95**, or **99** to create a reference line (which then needs to be formatted) at that percentile point for the data. Alternatively, **Enter one or more values** can be selected, which allows the user to type in as many percentile values as desired delimited by a comma.

The **Quantiles** option allows the user to break the distribution band up by between 3 and 10 tiles of differing shades of gray. Reference points are calculated by Tableau for both quantiles and percentiles using estimation type 7 in the R standard.

The **Standard Deviation** option sets reference points at the specified number of standard deviations above and below the mean. Options to calculate the standard deviation by a sample or on the population can also be selected.

To edit a reference or distribution band once it has been created, right-click either end of the band and select **Edit**. By default, the reference points for reference bands and distribution bands do not have a line but one can be set to emphasize the limits of the band. The fill color or shading of the reference band or distribution band can also be formatted in the configuration pane. To remove either band type, right-click either end of the band and select **Remove**.

Creating a Reference Band and a Distribution Band

In this section, you will create a reference band and a distribution band in Tableau Desktop. This will be useful as it will allow you to differentiate between the two and will enable you to implement both band types to enhance your visuals with additional context. Follow these steps:

1. On a new sheet, drag **Segment** onto **Columns** and right-click and hold **Order Date** before also dragging it onto **Columns**.

2. Dropping the **Order Date** field onto **Columns** will open up the aggregation selection pane; select the continuous **MONTH** option.

3. Now, drag **Sales** and **Profit** onto **Rows** to create a view displaying six line charts. Monthly sales and profit are shown for each segment.

4. To add a distribution band on **Sales** specific to each segment, drag **Distribution Band** from the **Analytics** pane onto the view and drop it on the **Sales** measure and the **Pane** scope option.

5. The distribution band configuration window will now pop up correctly set to **Pane**. Switch from the default **Percentages** computation to **Quantiles** to show the quantile distribution of sales for each segment.

6. Ensure the value is set to **4**, then press **OK**.

7. To add a reference band that will show when profit is within a specified target range, right-click the **Profit** axis and select **Add Reference Line**.

8. In the configuration window, select **Band** and ensure the scope is **Table** as we want to add a shaded target area that is consistent across the whole chart.

9. Set the **Band From** type to **Constant** and type **10000** as the value. This sets the minimum target value for **Profit** to 10,000.

10. For the **Band To** configuration, set the sum of **Profit** as the value and set the type to **Maximum**. The reference band now colors from the **10000** target point to the maximum profit value across the chart.

11. The target area can be formatted by making the line more pronounced with less opacity and setting a fill color.

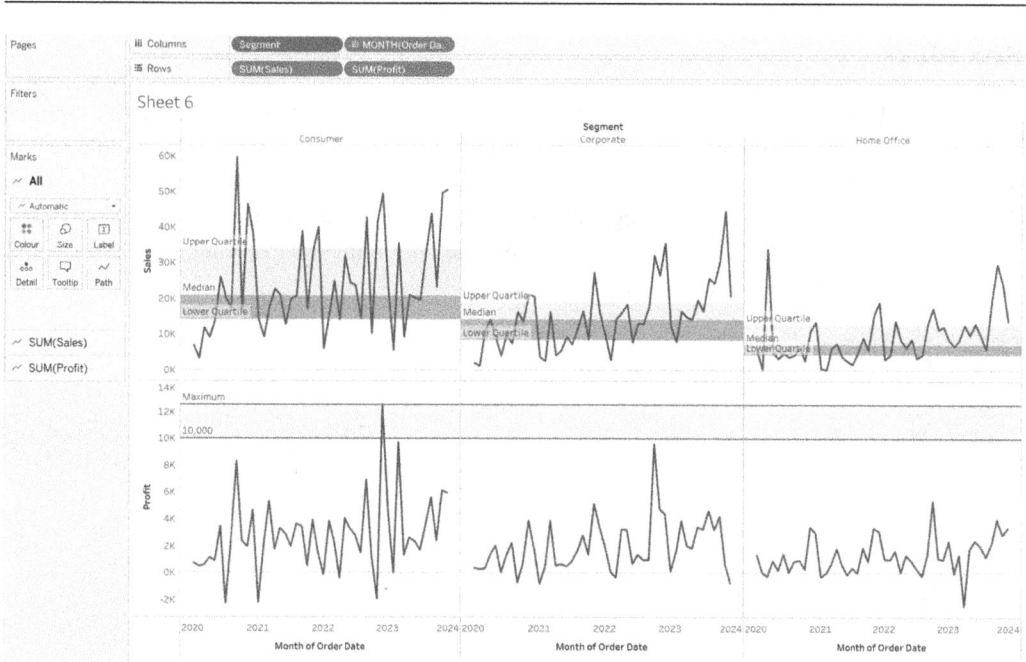

Figure 5.46: Distribution band created on the Sales axis by pane showing
the quantile breakdown for each segment; reference band created for Profit
highlighting the areas between 10k and the most highly profitable month

In this exercise, you created a distribution band and a reference band and observed the differences between the two. You also added custom formatting and included both constant and dynamic reference points in the reference band.

Box Plots

Box plots, or box-and-whisker plots, can be added in Tableau to show the distribution of the values on the selected axis. The middle section of the box plot is shaded and represents the middle 50% or the middle two quartiles of the data. The shaded area has two colors, the intersection of which shows the median. There are then lines coming out of each end of the shaded area, which are referred to as the whiskers. These can be configured to show either the minimum and maximum data points or all points within 1.5 times the interquartile range.

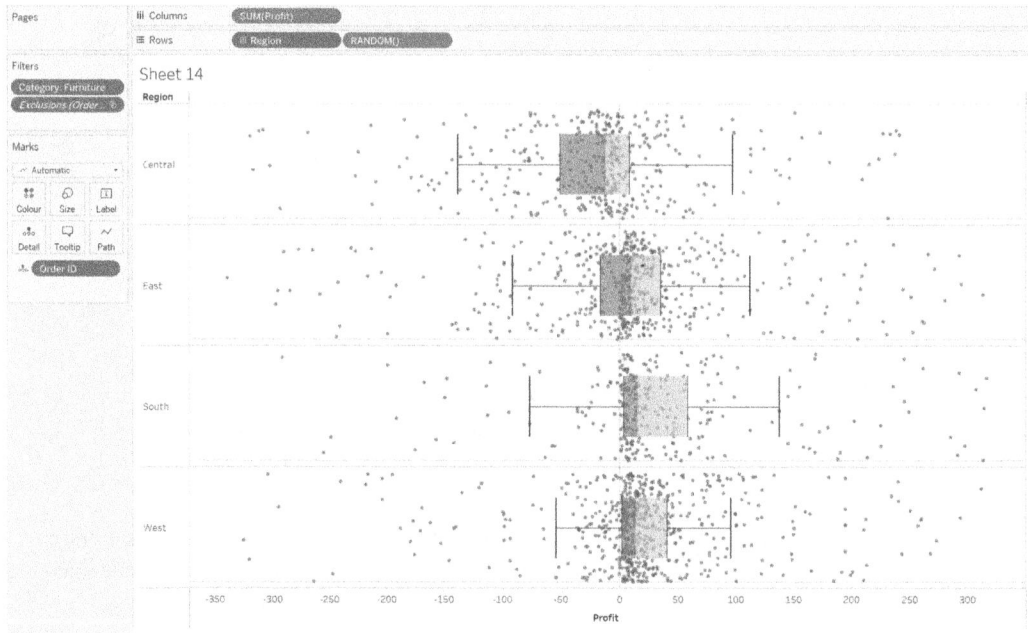

Figure 5.47: Box plots showing the distribution of orders by Profit for each region

To create a box plot, drag the **Box Plot** object from the **Analytics** pane onto the view and drop it on the desired measure axis. Box plots can be created on measure axes only (not dates). The scope for box plots is **Cell** only, and if multiple measures are on the view, then the correct axis can be chosen by dropping the object onto that measure. Dropping the **Box Plot** object onto **Cell** will add box plots to all measures. A box plot can also be created by right-clicking the desired axis and selecting **Add Reference Line**, then, in the configuration window, selecting **Box Plot**.

The configuration window for box plots allows the user to select whether the whiskers should extend to data within 1.5 times the interquartile range or whether they should span the maximum extent of the data. There is also a checkbox for hiding the underlying marks in the view (except outliers). This results in a chart that is just a box plot. The style, box shading (or fill), and whiskers can all be formatted to specific colors, thickness, and so on.

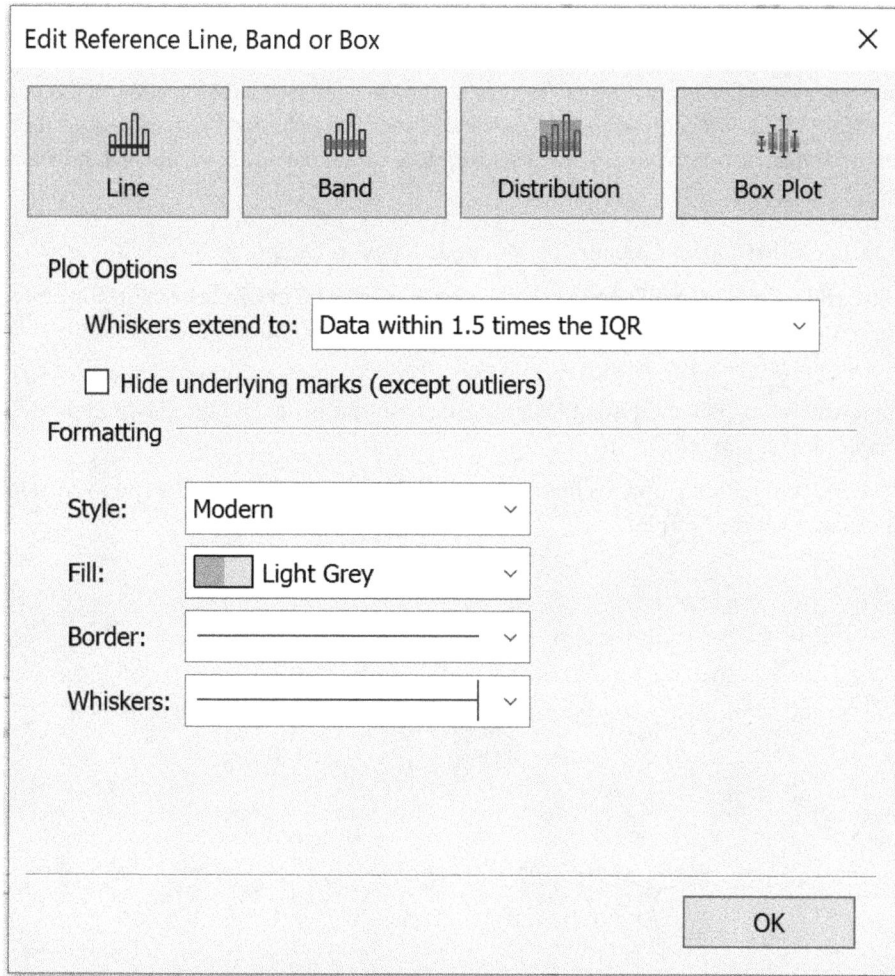

Figure 5.48: Box plot configuration window

Box plots have their own tooltips when hovering over them that display the upper and lower whisker values (which refer to the tips of the whiskers), the upper and lower hinges (which refer to the edge of the boxes), and the median.

To edit a box plot, right-click it and select **Edit**. To remove a box plot, right-click it and select **Remove**.

Creating a Box Plot

In this exercise, you will add a box plot overlay to your data in Tableau Desktop and you will customize the formatting to make it more appealing to the end user. Box plots are a great analytical object to learn how to implement as they show the distribution of values in an easy-to-understand way. Follow these steps:

1. On a new sheet, place **Category** on **Rows** and **Sales** on **Columns**.

2. Place **Sub-Category** on **Detail**, then change the mark type to **Circle** because box plots cannot be added to bar charts, which is the default selection. The chart now shows the distribution of total sales for each subcategory within each category.

3. To create a box plot, drag **Box Plot** from the **Analytics** pane onto the view and drop it on the **Sales** measure option.

4. Use the configuration options that pop up to hide the underlying marks and to customize the formatting of the box plot.

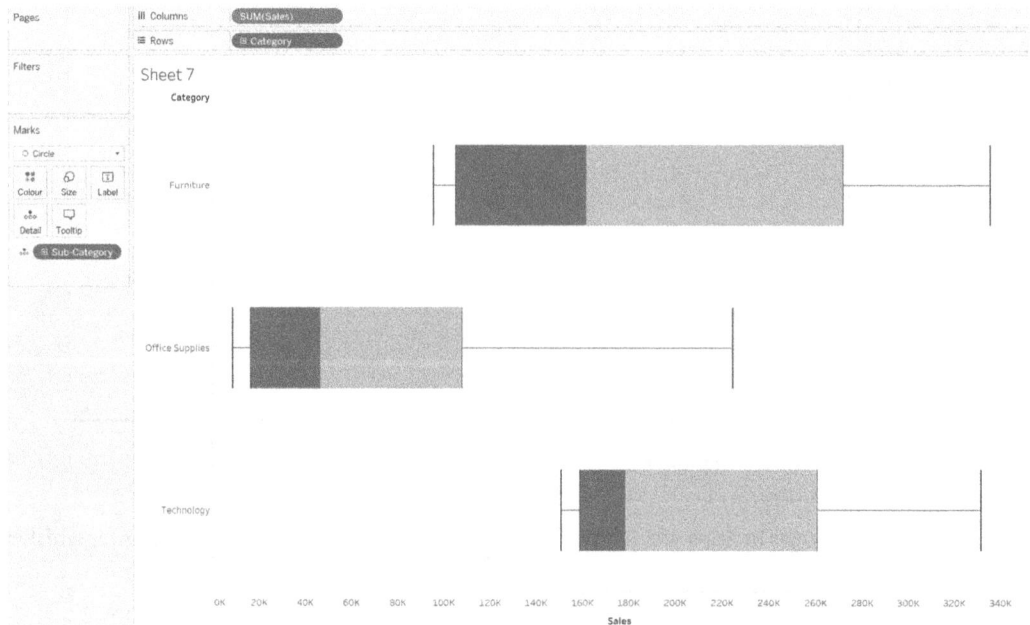

Figure 5.49: Box plots created showing the distribution of total
sales for subcategories within each category

In this exercise, you have added a box chart to the top of your data and have hidden the points below. This results in a pure box plot chart, but sometimes it can also be good to leave the data points on the chart for additional context. You have also learned how to apply formatting to box plots.

Trend Lines

Trend lines can be added to charts in Tableau to help visually identify trends in the data. To create a trend line, drag the **Trend Line** object from the **Analytics** pane onto the view and drop it on the desired model type. The available trend line model types are **Linear**, **Logarithmic**, **Exponential**, **Polynomial**, and **Power**. The scope within which the trend lines are created is within each pane of the view.

Figure 5.50: Trend line model options with Linear being selected

For a trend line to be added, the chart needs to have two axes. This can be a scatter plot with two measures as axes and can also be a chart with a measure and a continuous date axis.

There is no configuration window popup when creating trend lines, but trend lines can be configured by right-clicking the line and then selecting **Edit All Trend Lines**. The configuration popup when modifying trend lines allows the user to switch between the available models. If there are dimensions breaking up the view, then these can be selected and deselected to allow for trend lines per breakdown. If no dimensions are created, then each pane in the view will show the same trend line. There are also additional options to show or hide tooltips on trend lines, show or hide confidence bands around the trend line, and allow or disallow a different trend line per color (per dimension breakdown that is placed on **Color**). The y intercept can also be forced to be **0**, and the option to recalculate the line when data points are selected can be toggled on or off.

Hovering over a trend line will display the summary statistics for the line as a tooltip. The tooltip includes the equation that the line represents, the R-squared value for the trend line, as well as the P-value. For a more detailed statistical breakdown, right-click the line and select **Describe Trend Model**.

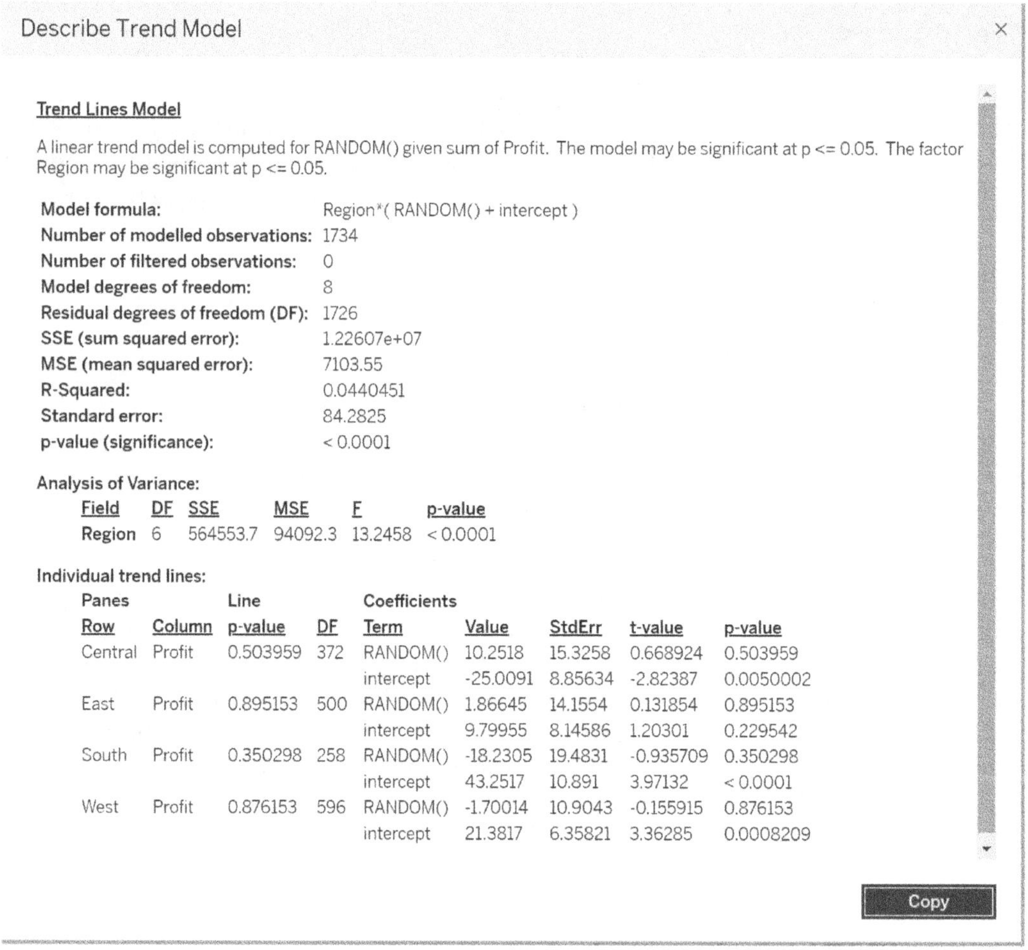

Describe Trend Model   ×

Trend Lines Model

A linear trend model is computed for RANDOM() given sum of Profit. The model may be significant at p <= 0.05. The factor Region may be significant at p <= 0.05.

Model formula: Region*(RANDOM() + intercept)
Number of modelled observations: 1734
Number of filtered observations: 0
Model degrees of freedom: 8
Residual degrees of freedom (DF): 1726
SSE (sum squared error):   1.22607e+07
MSE (mean squared error):   7103.55
R-Squared: 0.0440451
Standard error: 84.2825
p-value (significance):   < 0.0001

Analysis of Variance:

Field	DF	SSE	MSE	F	p-value
Region	6	564553.7	94092.3	13.2458	< 0.0001

Individual trend lines:

Panes		Line		Coefficients				
Row	Column	p-value	DF	Term	Value	StdErr	t-value	p-value
Central	Profit	0.503959	372	RANDOM()	10.2518	15.3258	0.668924	0.503959
				intercept	-25.0091	8.85634	-2.82387	0.0050002
East	Profit	0.895153	500	RANDOM()	1.86645	14.1554	0.131854	0.895153
				intercept	9.79955	8.14586	1.20301	0.229542
South	Profit	0.350298	258	RANDOM()	-18.2305	19.4831	-0.935709	0.350298
				intercept	43.2517	10.891	3.97132	< 0.0001
West	Profit	0.876153	596	RANDOM()	-1.70014	10.9043	-0.155915	0.876153
				intercept	21.3817	6.35821	3.36285	0.0008209

Copy

Figure 5.51: The Describe Trend Model popup for a Tableau trend line

Adding a Trend Line to a Chart in Tableau

In this exercise, you will add a trend line to a Tableau scatter plot and customize the trend line. You will add a linear trend line on a scatter plot, but the skills you will learn will enable you to apply trend lines to multiple charts, including line charts, and also to add multiple types of trend lines, such as logarithmic. Follow these steps:

1. On a new sheet, drag **Segment** onto **Columns** followed by **Sales**.

2. Drag **Profit** onto **Rows** and **Order ID** onto **Detail** to create a Sales versus Profit scatter plot for each order scatter plot by segment.

3. Drag the **Trend Line** object from the Analytics pane onto the view and drop it on the **Linear** model type to create a linear trend line for each scatter plot.

4. Hover over the lines to see the summary and right-click and select **Describe Trend Model** to see more details on the model.

5. Drag **Category** onto **Color** on the **Marks** card to create a line in each scatter plot for each category.

6. Right-click one of the trend lines and select **Edit all Trend Lines**.

7. Toggle the **Allow a trend line per color** option to switch between a single line per scatter plot and a line per category.

8. Uncheck **Segment** in the **Factors** section to standardize the model across all scatter plots.

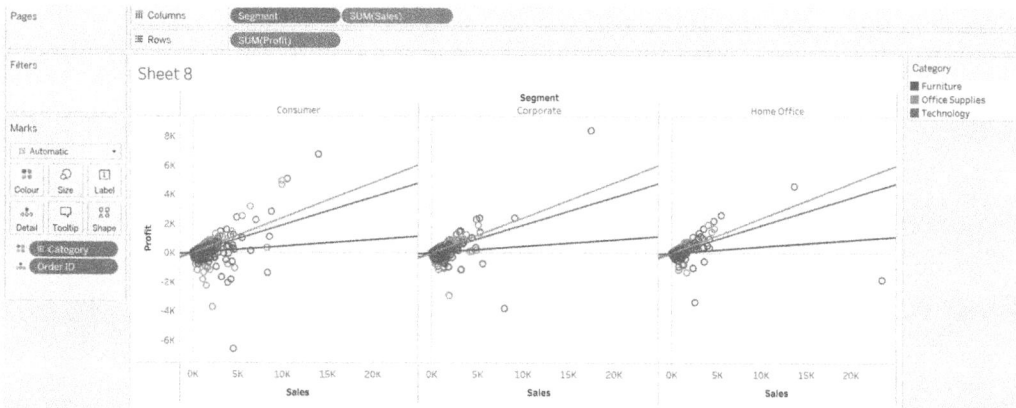

Figure 5.52: Linear trend line added to the scatter plots with a line
for each category but standardized across segments

In this exercise, you added a trend line to a scatter plot chart that is broken down by segment. By default, there was a single trend line per scatter plot, but you customized the trend line to include a line per category within the scatter plot and for the trend line to be calculated across all data points, as opposed to per segment. You now know how to add and customize trend lines in Tableau.

Forecasting

Forecasts in Tableau can be created to predict and visualize future trends in the data. Forecasts require a measure and a **Date Field** in the view (date fields can also be replaced by dimensions made up of integer values). Tableau automatically forecasts the data using exponential smoothing models that give more recent observations greater weight than older observations. The models are good for picking up seasonality and developing trends and visualizing the continuation of these into the future.

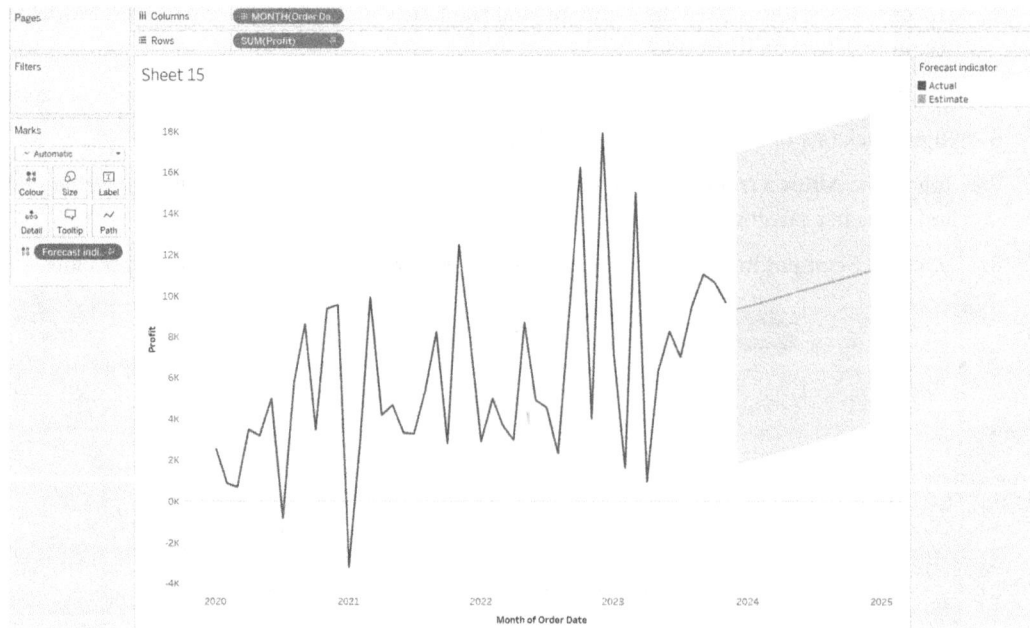

Figure 5.53: Forecast added to a monthly profit analysis chart

To create a forecast in Tableau, either right-click the visual and select **Forecast** followed by **Show Forecast** or drag the **Forecast** object from the **Analytics** pane onto the view. This will create an automatic forecast that continues the trend axis with estimated values shown in a lighter color. The legend is created automatically with a **Forecast Indicator** field being created and placed on **Color** on the **Marks** card.

The **Forecast Estimate** value can be hovered over to reveal the estimated values in the tooltip. The tooltips can be further enhanced by dragging the measure being forecasted from the **Data** pane onto the **Marks** card, right-clicking it, and then selecting **Forecast Result**. That field can then be set to represent the trend, precision, precision percentage, quality, and upper or lower prediction interval for the estimated value for the forecast in the tooltip.

The upper and lower prediction intervals are represented visually by default when creating a forecast. If the forecast is created on a line chart, then a shaded area is created above and below the estimated line representing the 95% prediction interval for the forecast. This means that the model predicts there is a 95% chance that the value for the future time period will fall within the shaded range, with the estimate line in the middle being the best prediction. If the chart mark type is **Shape**, **Square**, **Circle**, **Bar**, or **Pie**, then the forecasted marks will have the prediction intervals displayed as whiskers from the predicted point.

Forecasts can be configured by right-clicking the chart and selecting **Forecast** followed by **Forecast Options**. The forecast length can be updated from **Automatic** to an exact time period or until a specific point in the future. The **Aggregate by** option allows users to select a lower level of detail than what is represented on the view for the creation of the forecast. The **Ignore last** option allows users to ignore a specified number of recent periods when creating the forecast, and this is by default set to **1**. This could be useful when data for the current period is known to be incomplete. If the measure being forecasted has missing values, then these can be set to be filled in with zeros. The prediction intervals can also be toggled on or off for a forecast and can be customized to a value different from the default 95%.

Forecast Options customization also allows users to update the forecasting model. By default, Tableau will use whatever model it thinks is best. Tableau's selection (**Automatic**) can be switched to **Automatic without seasonality** to ignore seasonality, or to **Custom**. Custom models allow users to customize both the trend and the season by setting either to **None**, **Additive**, or **Multiplicative**. **None** means that the model will not assess the data for trend and/or seasonality, respectively. Otherwise, the effects of each isolated factor in the model can be summed or multiplied to create a combined forecasted effect via the **Additive** and **Multiplicative** options, respectively.

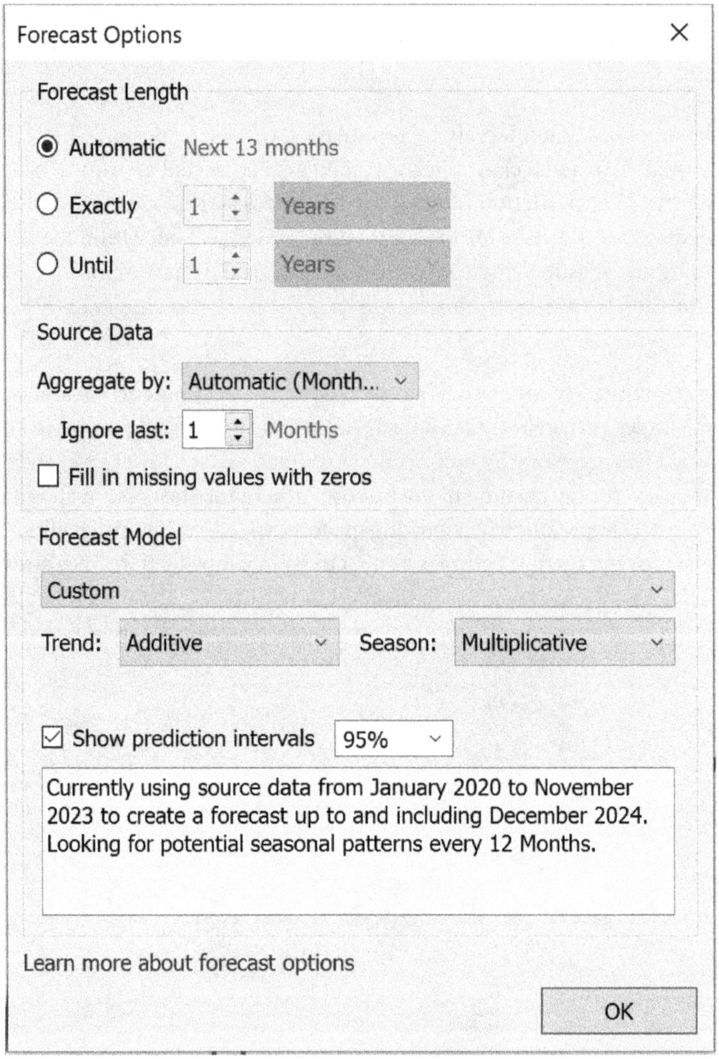

Figure 5.54: Forecast Options configuration for a Custom forecast model

To get a summary of a forecast, right-click the chart and select **Forecast** followed by **Describe Forecast**. A popup will appear that provides a description of the options used to create a forecast, including the measure being forecasted and along what axis/time series. The length of time of the forecast is shown as well as the time period the forecast was based on. Configuration options such as ignored time periods and seasonal patterns are also called out.

Below the description of the options used in the creation of the forecast is a summary table that describes the forecast including the initial value in the forecast and the change from this initial value to the final value. The seasonal effect calls out a high and a low period to represent the seasonal trend and the contribution provides a percentage of both the trend and seasonality components contributing to the forecast. Finally, a quality indicator tells the user whether the forecast fits the data well or not with possible results being **GOOD**, **OK**, or **POOR**.

The forecast description and the summary table are displayed on a **Summary** tab, but there is also a **Models** tab that provides more detailed statistics about the model.

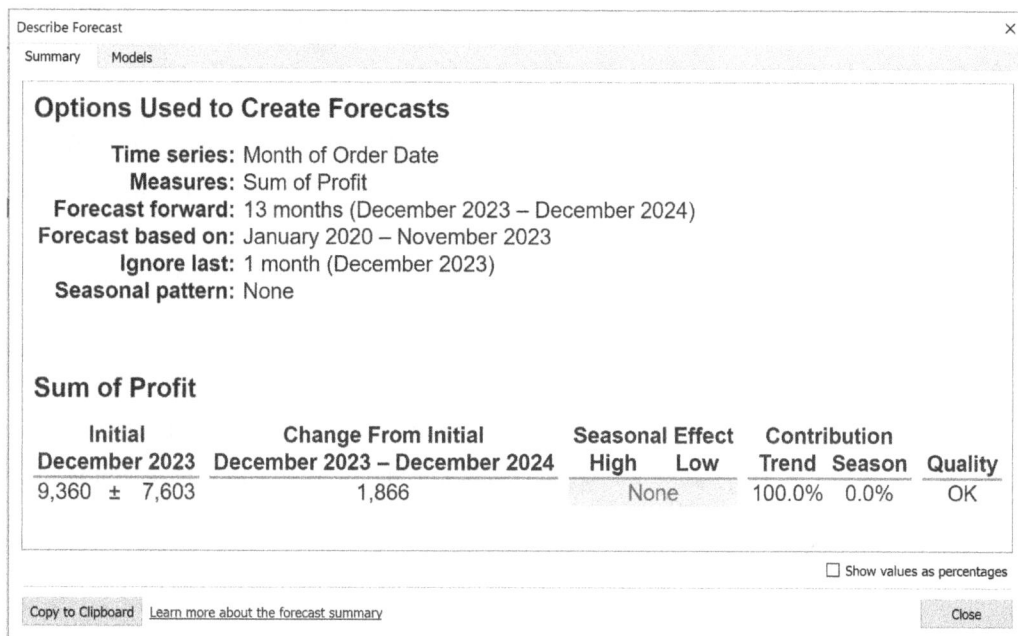

Figure 5.55: Tableau forecast model description

To remove a forecast from a chart, right-click the chart, select **Forecast**, and then uncheck the **Show Forecast** option.

Creating and Customizing a Forecast in Tableau

In this exercise, you will create a line chart and add a forecast to the line chart. You will also learn how to customize the forecast. Forecasting is a useful skill to learn in Tableau as it can provide end users with some indication of how their data might look in the future. Follow these steps:

1. On a new sheet, place **Order Date** onto **Columns** as a continuous month, and place **Sales** on **Rows**. There is now a line chart showing total sales by month.

2. To add a forecast, right-click the chart and select **Forecast** followed by **Show Forecast**. An additional 13 months have been added to the axis with the forecasted values and a legend has been added differentiating the forecast values from the original actual data points.

3. Hover over the forecast to see the forecasted values.

4. To add the precision percentage to the tooltip, drag the **Sales** field from the **Data** pane onto the **Marks** card.

5. Right-click the **Sales** field on the **Marks** card and select **Forecast result** followed by **Precision %**.

6. Hovering over the estimated data points now results in the precision percentage being shown in the tooltip.

7. To customize the forecast, right-click the chart and select **Forecast** followed by **Forecast Options**.

8. In the configuration popup, increase the forecast length by selecting **Exactly** and setting the values to two years.

9. To create a custom forecast model, use the dropdown in the **Forecast Model** section and select **Custom**.

10. **Trend** and **Season** dropdowns are now visible and these can be customized to create a new model, for example, by setting both to **Multiplicative**.

11. Untick **Show prediction intervals** to show the forecast as a single line, then press **OK**.

12. To see the summary of the model, right-click the chart and select **Forecast** followed by **Describe Forecast**.

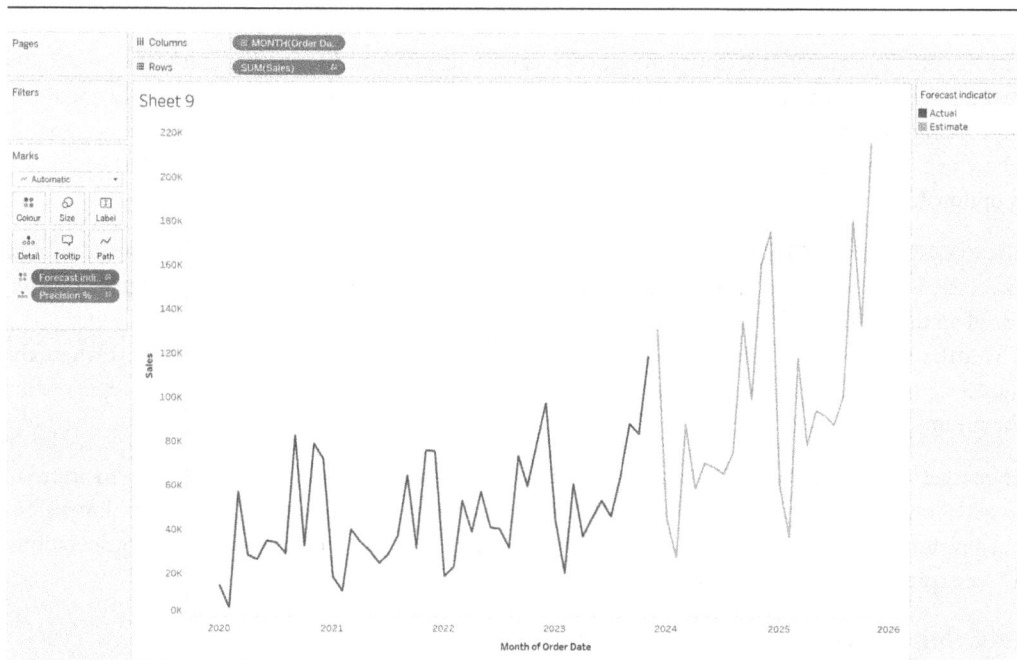

Figure 5.56: Two-year forecast created using multiplicative trend and seasonality

In this exercise, you added a forecast to a line chart that predicts sales values going into the future. You customized the forecast model to make further predictions with the model and also updated the logic the model uses when creating the forecast. You also learned how to add additional information to the forecast tooltip and observed how to inspect the forecast model summary.

Predictive Models

Predictive modeling can be done in Tableau using the MODEL_PERCENTILE and MODEL_QUANTILE calculated field functions. The resulting calculated fields are table calculations that need to have the direction/level of detail set in the view. Table calculations are the last thing to be calculated in Tableau, which means that updating the view in any way (for example, by filtering the data) will also impact the results of the model.

The models used for prediction are, by default, linear regressions but can be specified as a regularized linear regression or a Gaussian process regression. The function's first parameter is an optional model specification, which can be updated by typing, surrounded by quotation marks, either model=linear for a linear regression, model=rl for a regularized linear regression, or model=gp for a Gaussian process regression.

The MODEL_PERCENTILE function returns a value between 0 and 1 that represents the probability that a target expression (aggregated measure) is predicted by a predictor expression or predictor expressions. The function is written as MODEL_PERCENTILE([model_specification], [target expression], [predictor expression(s)]), where the model specification is optional.

The MODEL_QUANTILE function takes a quantile, a target expression, and at least one predictive expression and returns a target value in the given quantile within the probable range calculated based on the expressions. It is written as MODEL_PERCENTILE([model_specification], [quantile], [target expression], [predictor expression(s)]), where the model expression is optional. The quantile is a decimal value between 0 and 1, with 0.5 representing the mean.

The target expressions in both functions must be aggregated measures and identify the measure to target or predict. The predictor expressions are comma delimited and can be measures or dimensions but any dimensions must be wrapped in ATTR as all fields need to be aggregated for table calculations. These expressions are used to calculate the prediction.

Creating a Predictive Model in Tableau

In this exercise, you will create a predictive model in Tableau that uses one measure to predict another. Predictive models can provide a lot of insight to end users if calibrated correctly using relevant fields.

To complete this exercise in Tableau Desktop, connect to Tableau's readily available training data source, **Superstore Data**, which can be found in the **Saved Data Sources** section of the home screen:

1. On a new sheet, create a scatter plot by placing **Sales** on **Columns**, **Profit** on **Rows**, and **Order ID** on **Detail**, then set the mark type to **Circle**.

2. To create a predictive model to see how well profit is predicted by sales, create a new calculated field and call it MODEL_PERCENTILE.

3. Populate the calculated field with the following expression: MODEL_PERCENTILE(SUM([Profit]), SUM([Sales])). Press **OK**.

4. The MODEL_PERCENTILE calculated field has no model defined so it uses the default linear regression and uses the sum of **Sales** to predict the sum of **Profit**.

5. Place the MODEL_PERCENTILE field onto **Color**.

6. By default, the color of the circles is unchanged because the table calculation has not been configured to compute using **Order ID**. Edit the table calculation, select **Specific Dimensions**, tick **Order ID**, and close the configuration window.

7. The color legend now colors each circle from light blue to dark blue for a value between 0 and 1 for how well **Sales** predicts **Profit** in that order.

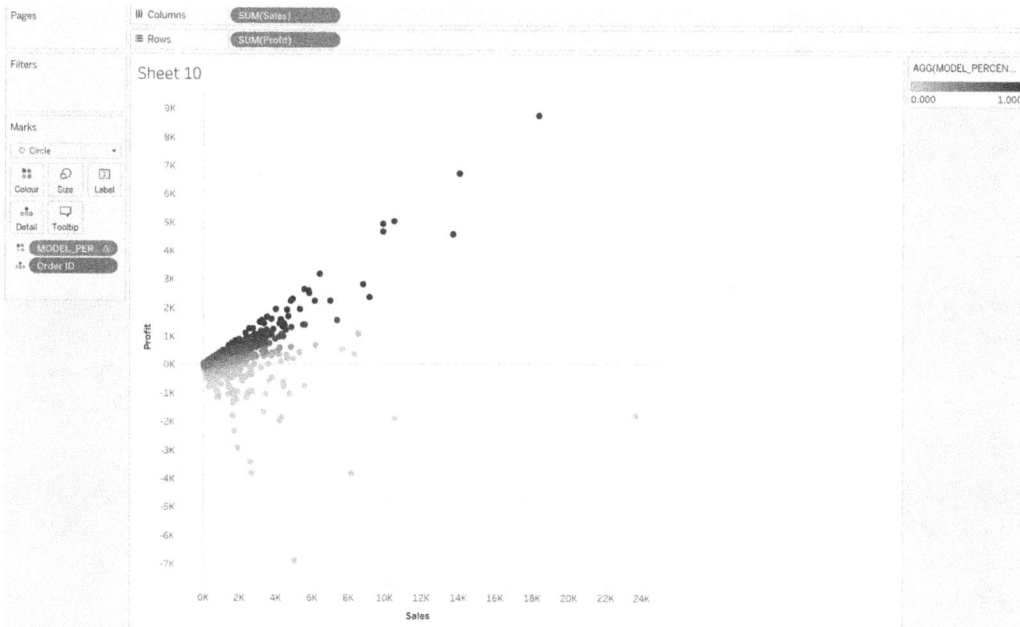

Figure 5.57: Predictive model created that predicts profit total based on sales total, placed on a view to clearly demonstrate the prediction works correctly with higher sales orders that have higher profit being colored darker

In this exercise, you created a predictive model that predicts profit based on sales. You also learned how to specify the way the model is calculated in terms of which fields are relevant.

Summary

The aim of this chapter was to look into each chart type that can be built in Tableau. Each section showed the requirements to build a particular chart, with detailed steps, helping build confidence among candidates in the use of these charts. Take the time to try building these charts with **Superstore Data** to get familiar.

Tableau allows for the creation of a wide range of chart types and these are configured based on the field types used, where they are placed in the visual, and the mark type selected. Geographic charts can be created using geographic fields and analytical objects can be added to charts to provide a greater level of insight.

Geographic charts require a **Longitude** and **Latitude** field placed on **Rows** and **Columns**. When a field is given a geographic role in Tableau, longitude and latitude are automatically generated and these fields can be used to create a map visual. Symbol maps can be created to plot specific points on a map, density maps can be created to highlight clusters on a map, and filled maps allow users to compare geographic polygons overlaid on a map.

The analytical features available in Tableau allow users to add additional pieces of insight to their visuals such as totals and subtotals for column and row comparisons. Reference lines can be added to emphasize a specific point on an axis. The specific point can be a constant value such as an important date or a target point to reach or can be more dynamic such as an average across the data. Reference bands and distribution bands color the area between two reference points and, again, these reference points can be constant or dynamic values and can be a single shaded area across the whole chart/table, within each pane, or cell by cell. Box plots are an additional analytics feature that shows the distribution of data with the middle 50% being contained within the box.

Analytical features in Tableau can also involve the creation of models to plot the trend of existing data points, forecast future data points based on trend and/or seasonality, and predict a target measure based on predictor variables.

In the next chapter, you will look at dashboarding and stories in Tableau, starting with dashboard creation and the objects available that can be included in dashboards.

Exam Readiness Drill – Chapter Review Questions

Apart from a solid understanding of key concepts, being able to think quickly under time pressure is a skill that will help you ace your certification exam. That is why working on these skills early on in your learning journey is key.

Chapter review questions are designed to improve your test-taking skills progressively with each chapter you learn and review your understanding of key concepts in the chapter at the same time. You'll find these at the end of each chapter.

> **How to Access these Resources**
>
> To learn how to access these resources, head over to the chapter titled *Chapter 9, Accessing the Online Practice Resources*.

To open the Chapter Review Questions for this chapter, perform the following steps:

1. Click the link – `https://packt.link/TDA_CH05`.

    Alternatively, you can scan the following **QR code** (*Figure 5.58*):

Figure 5.58: QR code that opens Chapter Review Questions for logged-in users

2.   Once you log in, you'll see a page similar to the one shown in *Figure 5.59*:

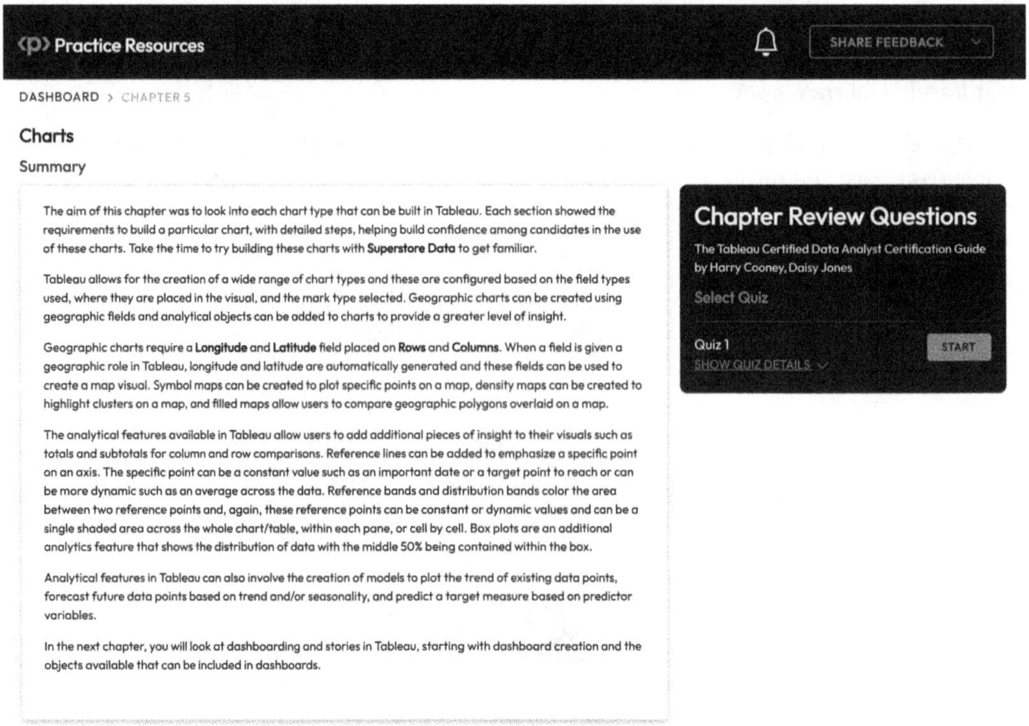

Figure 5.59: Chapter Review Questions for Chapter 5

3.   Once ready, start the following practice drills, re-attempting the quiz multiple times.

Exam Readiness Drill

For the first three attempts, don't worry about the time limit.

ATTEMPT 1

The first time, aim for at least **40%**. Look at the answers you got wrong and read the relevant sections in the chapter again to fix your learning gaps.

ATTEMPT 2

The second time, aim for at least **60%**. Look at the answers you got wrong and read the relevant sections in the chapter again to fix any remaining learning gaps.

ATTEMPT 3

The third time, aim for at least **75%**. Once you score 75% or more, you start working on your timing.

> **Tip**
> You may take more than **three** attempts to reach 75%. That's okay. Just review the relevant sections in the chapter till you get there.

Working On Timing

Target: Your aim is to keep the score the same while trying to answer these questions as quickly as possible. Here's an example of how your next attempts should look like:

Attempt	Score	Time Taken
Attempt 5	77%	21 mins 30 seconds
Attempt 6	78%	18 mins 34 seconds
Attempt 7	76%	14 mins 44 seconds

Table 5.17: Sample timing practice drills on the online platform

> **Note**
> The time limits shown in the above table are just examples. Set your own time limits with each attempt based on the time limit of the quiz on the website.

With each new attempt, your score should stay above **75%** while your "time taken" to complete should "decrease". Repeat as many attempts as you want till you feel confident dealing with the time pressure.

6

Dashboards

Introduction

This chapter will cover dashboarding and stories in Tableau. Dashboards are collections of charts that can be made to interact with each other to facilitate more advanced analysis. A dashboard is the most common way to share analysis with end users. Dashboard creation and the available objects that can be included in dashboards will be discussed first. Making dashboards interactive will be covered second. Some general tips for best practices when designing dashboards will also be discussed. Finally, how to combine dashboards into Tableau's stories will be looked at.

There are three types of tabs along the bottom of Tableau's interface – **Sheets**, **Dashboards**, and **Stories**. Sheets are where individual charts are created, dashboards allow for the collection of multiple sheets in a single view, and stories provide functionality for sequences of dashboards and/or sheets to be created.

Combining sheets into dashboards is the primary recommended methodology for sharing analysis with end users in Tableau. Knowledge of dashboard functionality is therefore important for any Tableau user and will be a key component of the Tableau Certified Data Analyst exam. Stories are great for crafting a data narrative and are also a topic an exam taker may be asked about.

In this chapter, the following topics will be covered:

- Dashboard creation
  - Dashboard objects
  - Layout options

- Dashboard interactivity

  - Dashboard actions

  - Filtering multiple sheets

  - Sheet swapping

- Stories

Dashboard Creation

Dashboards can be created by clicking the **New Dashboard** tab button along the bottom toolbar or by selecting **Dashboard** in the top toolbar followed by **New Dashboard**. Either one of these methods will create a new tab along the bottom toolbar called **Dashboard N** with the N being replaced by however many dashboards have been created in the workbook.

The dashboard creation interface differs from that of sheet creation. There is no data pane that can be toggled to analytics; instead, there is a dashboard pane that can be toggled to layout. The dashboard pane contains options to select either a default dashboard view or a device-specific view for the user to configure and updating this will transform the size of the canvas in the middle of the page. The size of the canvas can also be configured directly using the **Size** options in the dashboard pane.

Below the **Size** options on the dashboard pane is the **Sheets** section, which lists all of the sheets in the workbook. These can be dragged from the dashboard pane onto the canvas to create a collage of multiple sheets. In addition to sheets, there are a variety of dashboard objects that can be dragged onto the canvas to create the dashboard. How the sheets and objects are placed onto the canvas is configurable – they can be placed in either tiled format or floating on the canvas. A dashboard title can also be toggled on or off, resulting in a textbox appearing or disappearing at the top of the dashboard canvas.

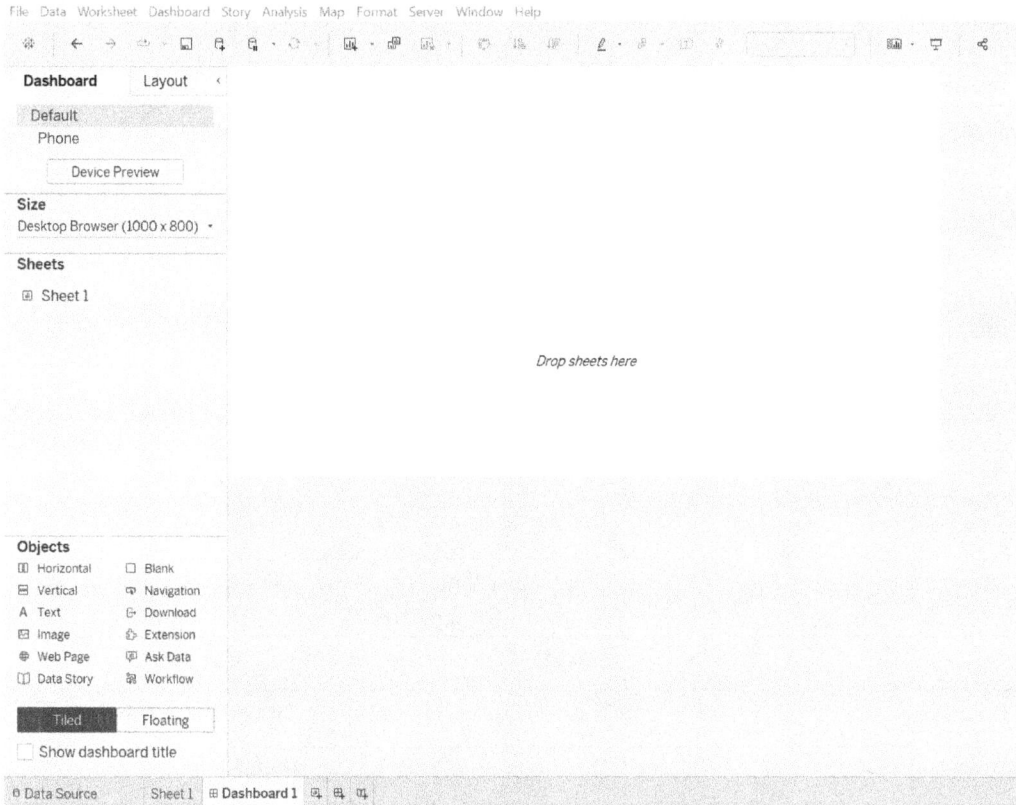

Figure 6.1: New dashboard created, ready to be configured

Dashboard Objects

Dashboard objects are items that can be placed on a dashboard. The available objects can be found in the **Dashboard** pane and can be dragged from there onto the dashboard canvas. Whether the object is dragged and dropped relative to other objects in a tiled fashion or whether the object is placed floating exactly where it is dropped depends on whether **Tiled** or **Floating** is selected in the dashboard pane when the object is dragged to the canvas.

The following table shows some dashboard objects and explains their functions:

Dashboard Object	Icon	Function
Sheet		This charts visuals that were created in worksheets
Container		This groups objects that hold multiple other dashboard objects within
Text	A	This displays textual information
Image		This displays a static image
Blank		A filler object to take up space
Button		Workbook navigation, showing and hiding dashboard objects, and downloading the dashboard
Web Page		This displays a web page on the dashboard
Extension		Objects that contain non-native Tableau functionality often created by third parties
Ask Data		This allows users to ask questions about data
Data Story		This provides a description of data on the dashboard
Workflow		This connects to and runs a Salesforce flow

Table 6.1: Dashboard object icons and functions

Sheets

Sheets are the charts created throughout a Tableau workbook. These can be dragged from the **Sheets** section of the dashboard pane directly onto the canvas. If the sheet has any legends, filters, or parameters on it, then these will automatically be added to the dashboard as well as the chart itself. These items will be dynamically grouped into a specific Tableau-created container. Updating anything on a sheet will cause the sheet to update on the dashboard as well; this includes adding or removing fields and filtering. Tooltips created on a sheet also function on the dashboard. Only one instance of a sheet can exist on the same dashboard, but the same sheet can be placed on multiple dashboards.

A sheet can be selected on a dashboard by selecting it on the list of sheets in the dashboard pane or by selecting the sheet directly on the dashboard. Selecting a sheet (or any dashboard object) on a dashboard will result in a gray outline around the edge of the sheet. There is a small selection area in the top center of this outline that has two white lines going through it. Hovering over and then holding select on this section will allow the user to move the sheet or any dashboard object around the dashboard.

To the top right of the grayed-out sheet outline is an x for removing the sheet from the dashboard. There is also a go-to-sheet button that allows the user to quickly navigate to the sheet to configure it further. Below the go-to-sheet button is a carrot that opens some additional options for configuring the sheet itself as well as the sheet's behavior on the dashboard.

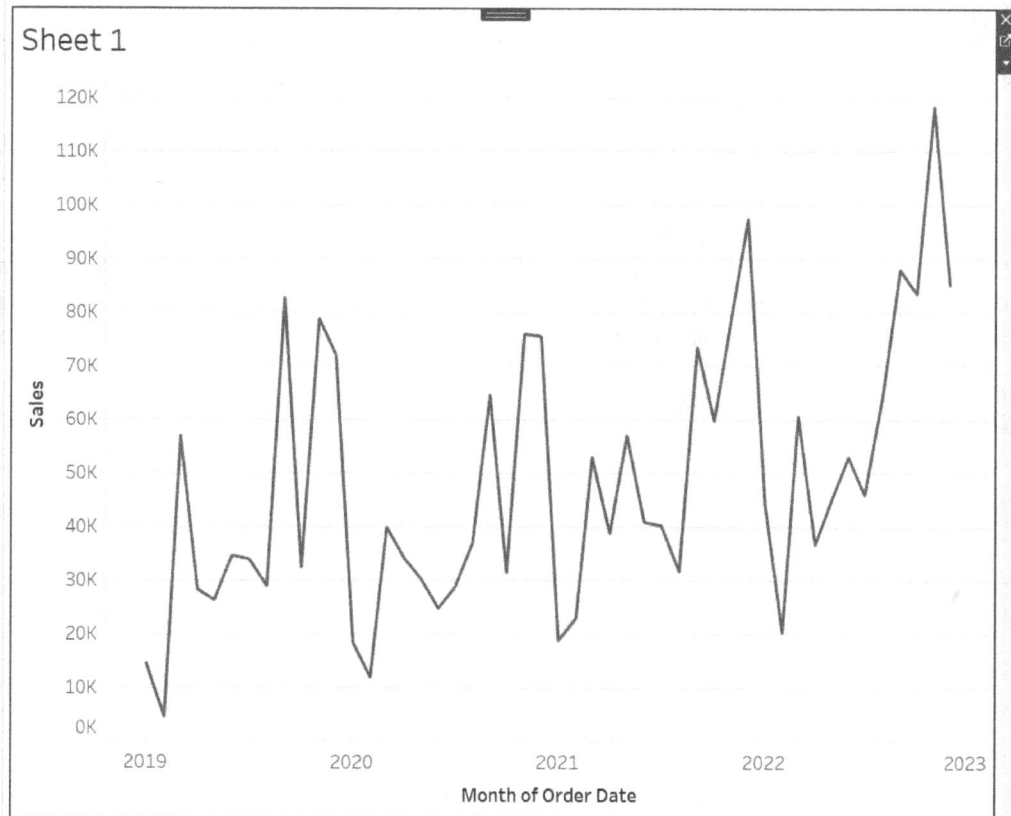

Figure 6.2: A single sheet on a dashboard is selected, resulting in a gray outline

Containers

Containers are dashboard objects that hold groupings of other dashboard objects, and that can include other containers. Container objects are called **Horizontal** and **Vertical** in the dashboard pane because objects are grouped within a container either side by side horizontally or are stacked on top of each other vertically. A container is always either horizontal or vertical, meaning objects must be grouped in one direction only. Whether the **Horizontal** or **Vertical** object is dragged onto the canvas is irrelevant. The behavior of the container is actually defined when a second object is placed into the container either to the side of or above/below the first object.

Containers are unique among dashboard objects in that when they are selected, their outline is blue as opposed to gray. This enables the user to easily see the boundaries of the container and the objects within. Containers have a selection section in the center of the top outline that allows users to drag the container to another location on the dashboard canvas.

To place a sheet or a dashboard object into a container, simply drag it over the container on the canvas until the container is fully shaded. Once the container is shaded and the blue outline is visible, let go of the sheet or object to drop it into the container. Trying to select the container will now result in selecting the object within the container only, identified by the gray outline. To select the container, double-click the shaded area with the two white lines used for object movement; this will turn the outline blue, indicating that the container is selected.

To add another object to the container, repeat the process but observe that the whole container is not shaded when dragging the object over it. Instead, one of the edges of the container is shaded. Which edge is shaded depends on where positionally the user is dragging the object in the container, and where the edge is shaded indicates where the object will be placed in the container.

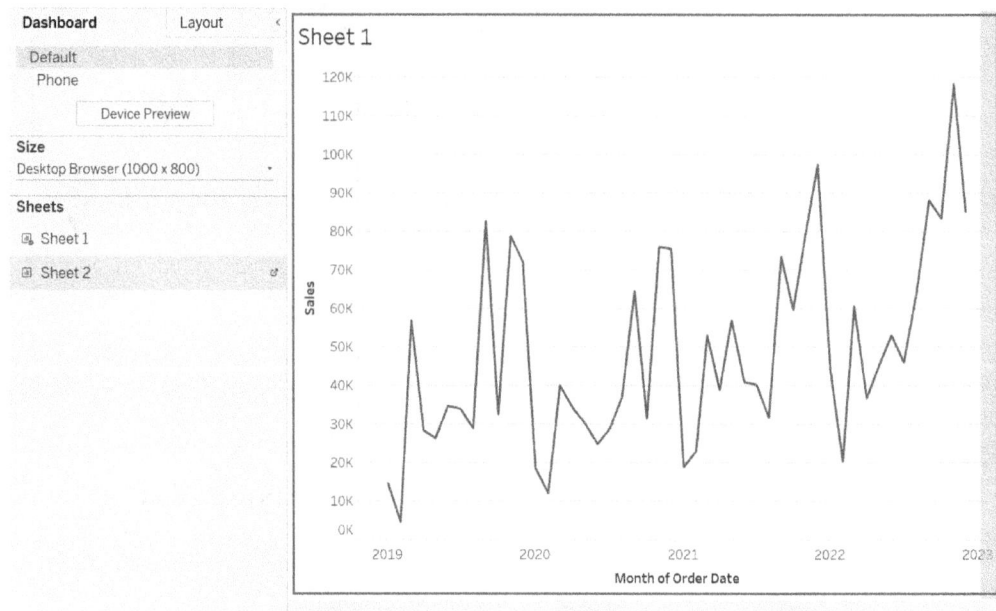

Figure 6.3: Sheet 2 is being added to a container along with Sheet 1 – it is being placed to the right, resulting in a horizontal container

To place an object to the right of an existing object, drag it to the right-hand side of the container until the desired edge is shaded. Do the opposite for the left and drag the object to the center, and then to the top or bottom to place it above or below. How the second object is placed in a container determines whether the container is a horizontal (left or right) or vertical (above or below) container.

After the second object has been placed in a container, more objects can be added, but these can now only be added horizontally or vertically to existing objects based on the container type. Objects can also be placed between existing objects, in which case the area between will be shaded.

Objects within a container can be manually sized by dragging the edges between any two objects or by fixing the size of an object. Objects can also be set to be distributed evenly within the container by using the carrot dropdown in the top right of the container followed by **Distribute Contents Evenly**.

Containers can be placed within containers and, as a result, it can become difficult to directly select an outer container to move all the elements within. On the dashboard pane, the **Layout** tab provides an item hierarchy at the bottom. From here, any dashboard object can be selected and the items within each container can be identified. All dashboard objects other than sheets can also be renamed here or by selecting the object options menu in the top right of the object followed by **Rename Dashboard Item**.

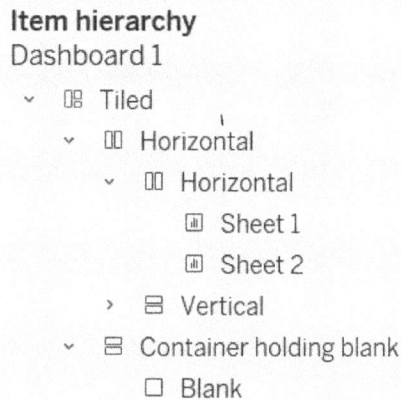

Item hierarchy
Dashboard 1
⌄   ▣ Tiled
    ⌄   ◻◻ Horizontal
        ⌄   ◻◻ Horizontal
                ▣ Sheet 1
                ▣ Sheet 2
            ›   ▤ Vertical
    ⌄   ▤ Container holding blank
            ◻ Blank

Figure 6.4: Item hierarchy showing multiple containers and objects within
– there is a horizontal and vertical container within a horizontal container;
a container item has been renamed Container holding blank

Text

Text objects are textboxes that can be placed onto the dashboard canvas. Using the **Show dashboard title** checkbox at the bottom of the dashboard pane creates a textbox at the top of the dashboard. Textboxes can be dragged anywhere on dashboards and so can be useful as titles, sub-headers, and annotations.

After dragging a textbox onto the dashboard, the text configuration window will pop up, allowing the user to type out the text and format the font, size, color, and alignment. The text can also be made bold, italic, or underlined. Formatting is not locked to the whole text – all formatting can be applied to a single character.

There is an insert button in the configuration window that allows users to insert dynamic content, such as the sheet name or workbook name, as well as the viewer's full name or username or any parameters that exist in the workbook. The sheet name returns the name of the dashboard tab along the bottom.

A textbox can be double-clicked at any time to update the text within.

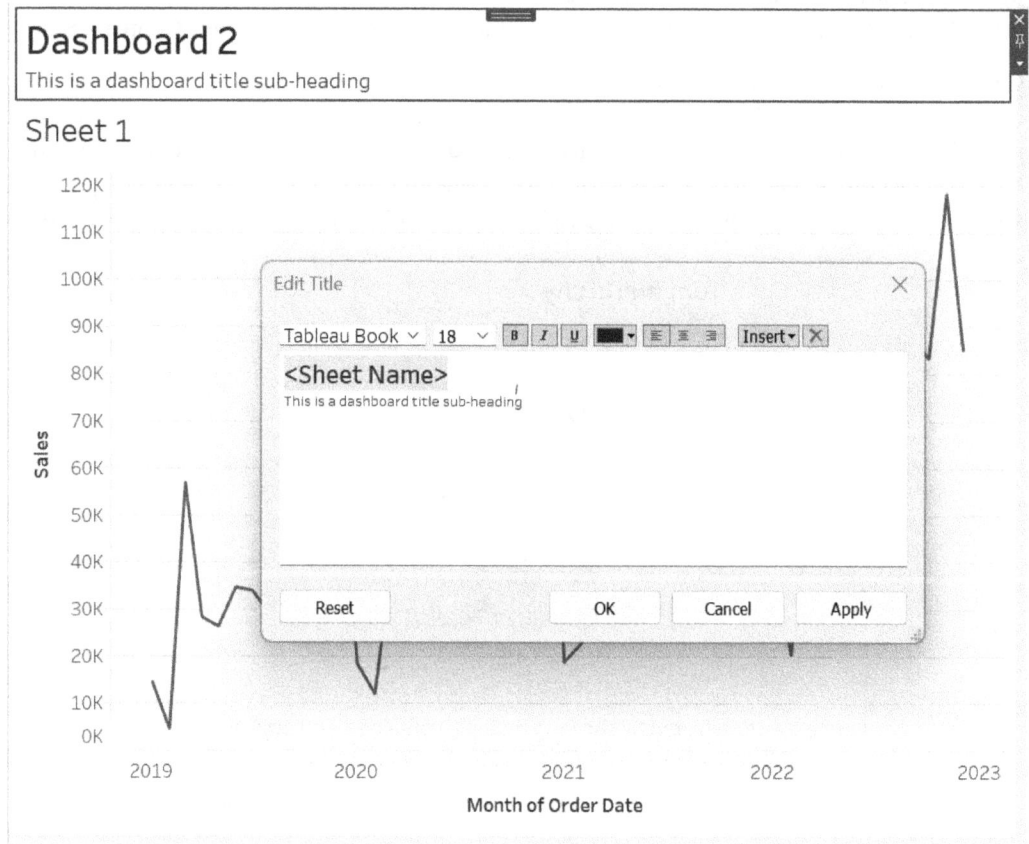

Figure 6.5: Textbox configuration with varied text formatting and
a dynamic title set using the sheet name

Image

Images can also be added to dashboards as dashboard objects. This can be useful as a background image, a company logo, an icon, and so on. When the image dashboard object is dragged onto the canvas, a configuration window pops up that asks the user to either provide an image file (.jpeg or .png) from their local machine to insert into the dashboard or provide a URL to an image that can be displayed.

A URL can be provided during configuration that opens up when the image is clicked. Additional configuration options include whether to fit the image to the area it takes up on the canvas and centering the image in that area.

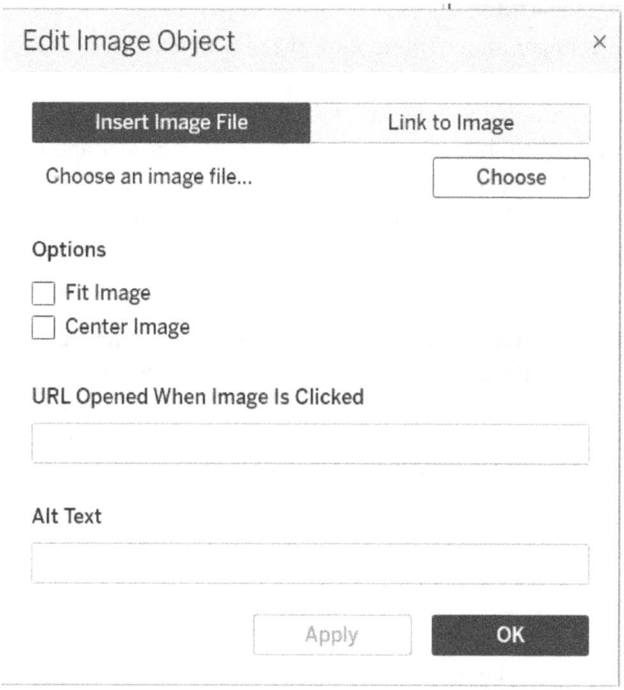

Figure 6.6: Image dashboard object configuration options

Blank

Blank objects can be dragged onto a dashboard to fill the blank space. It can be useful, for example, to provide a whitespace buffer between two other sheets or objects and this can be easier to manage using a blank object itself. This way, the size of the blank object can be modified directly. Blank objects can also be colored and are often used as lines to break apart sections in a dashboard visually; for example, a blank object that runs the width of a dashboard but has a 1-pixel height and is colored black will appear as a black line.

Buttons

There are multiple types of buttons that can be added to dashboards. Navigation buttons allow users to easily move around the workbook, download buttons facilitate downloads of the dashboard, and show/hide buttons.

Navigation buttons can be dragged onto the canvas from the objects section of the dashboard pane, which results in the button configuration window being shown. The first option in the configuration window is a drop-down selection to set where clicking the button should navigate to in the workbook. This can be any sheet or dashboard in the workbook. Below this are the standard button configuration options available for every type of button.

The standard button configuration options allow the user to first select whether the button should be a text or image button. If a text button is selected, then a title for the button can be provided and formatted and this is the text that will display. If the image button is selected, then the user can select an image on their machine to show as the button. The button border and background can also be formatted in terms of color, outline style, and opacity and this generally makes more sense for text buttons. Finally, a tooltip can be provided to give the user instructions or additional context about the button when hovering over it.

The download object can also be dragged from the dashboard pane onto the canvas. The configuration popup for the download button is the same as with navigation, other than the first option, which instead asks for a file format the download should export to. The download is PDF by default but can also be set to PowerPoint, Image, or Crosstab for Excel. Clicking on the button will open up a window, allowing the user to customize their export, for example, by selecting whether to download a PDF of just the current dashboard (sheet) or the whole workbook, the size and orientation, and so on.

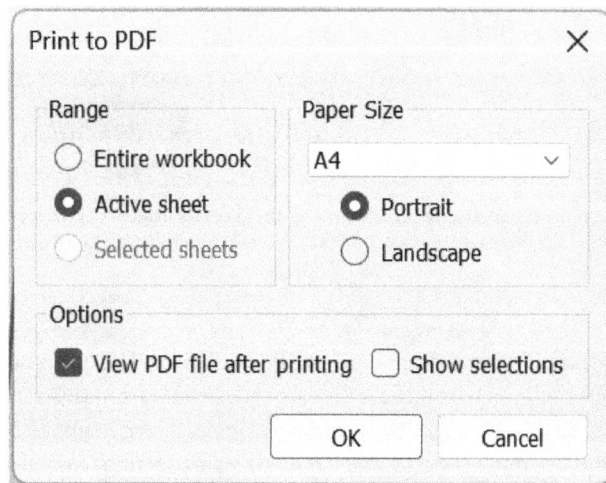

Figure 6.7: Download options after the Download to PDF button clicked

The show/hide button can be used to show and hide objects on a dashboard including containers and sheets. To add a show/hide button, go to the top-right options caret on a selected dashboard object and select the **Show/Hide** button. This will add a small **x** floating object, which, when clicked, turns into three horizontal lines and hides the object it is linked to. Clicking the three horizontal lines shows the object again and switches the button back to an **x** image. Selecting the buttons options dropdown and **Edit** button opens the configuration window. By default, the dashboard item to show and hide will be selected, but this can be changed to another dashboard object if desired. The button is an image button using the **x** and horizontal lines previously mentioned and these are switched using the **Item Shown** and **Item Hidden** tabs in the configuration. Other than the selection of which item to show and hide and the ability to have a separate configuration for when the object is shown versus hidden, the rest of the button configuration options are the same as the navigation and download button types.

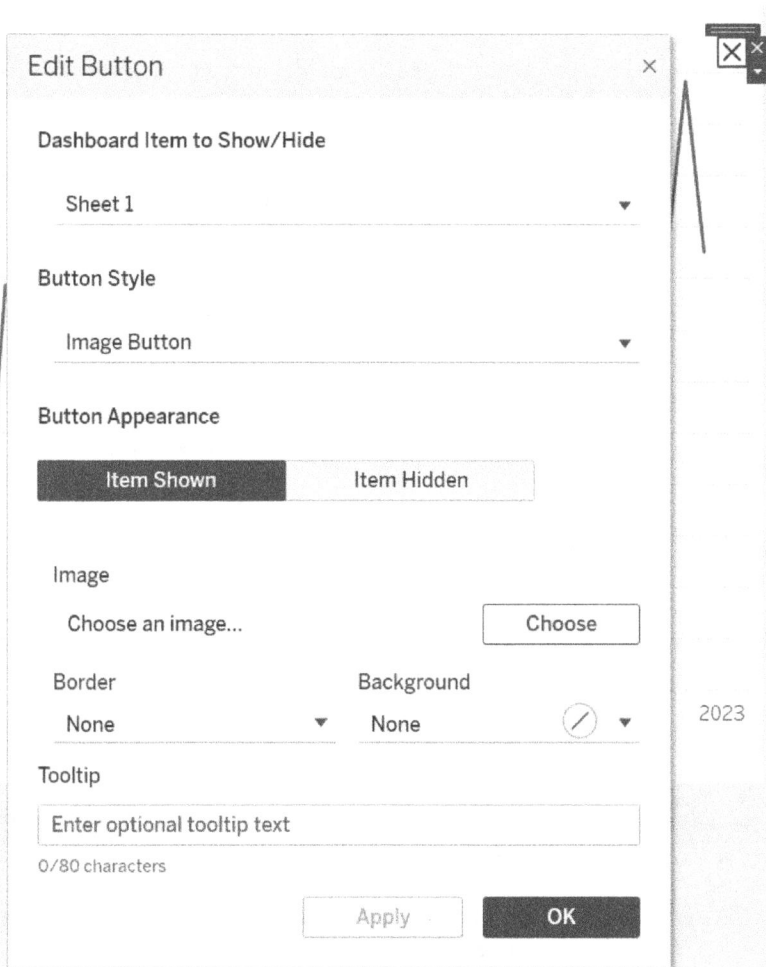

Figure 6.8: Show/hide button configuration with the default x button visible at the top right

Web Page

A web page object can be dragged onto a dashboard, and, at this point, it will ask for a URL. The URL provided can include dynamic content such as parameters allowing the user to update the web page displayed. Once the URL has been provided, the object exists as the provided web page URL.

Extensions

Extensions are dashboard objects that allow users to add functionality not native to Tableau to their dashboard. When dropping the extensions object onto the dashboard, there is a window popup that allows the user to browse through available extensions and select one for addition. Most extensions are third-party extensions made available through Tableau Exchange but there are also some Tableau-/Salesforce-created extensions, such as a tool for analysis and bookmarking previous filter settings.

Users can also download extensions to their Tableau repository or create Tableau extensions themselves, both of which can be accessed using the **Access Local Extensions** button.

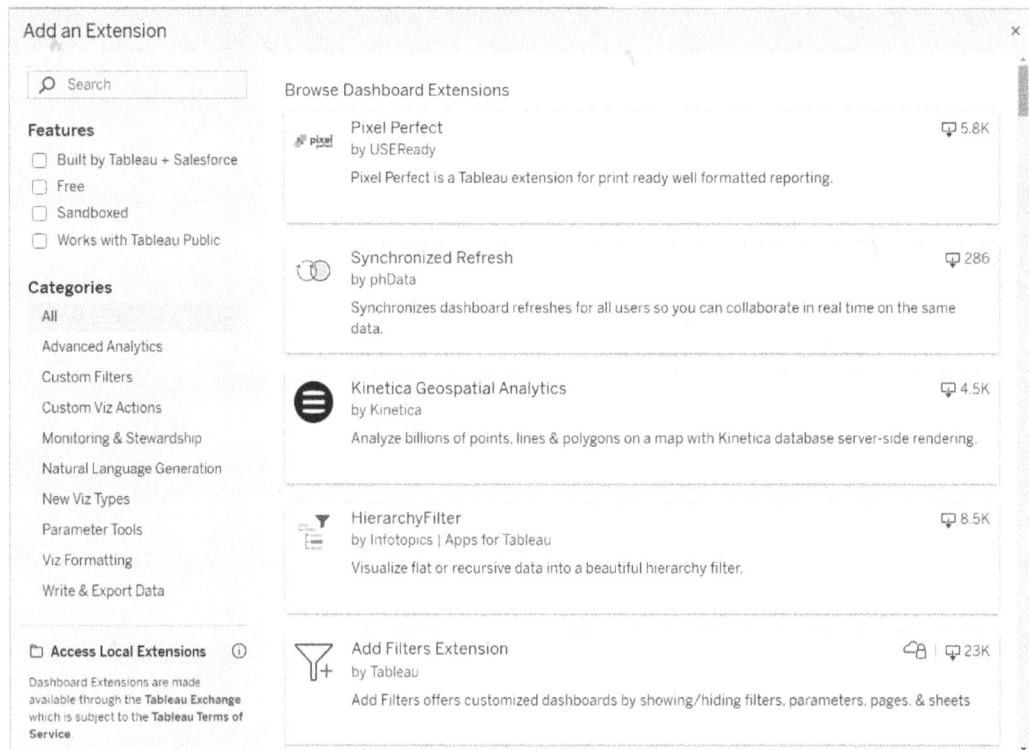

Figure 6.9: Tableau extension browsing and selection

Ask Data

The **Ask Data** dashboard object allows the user to ask questions about a specific data source on a connected **Tableau server/cloud**. A **lens** first needs to be created that looks directly into the sunset of a data source. This lens can then be selected for interrogation by the dashboard user. Placing the object on a dashboard will result in a search box for asking questions about the data and a pane that displays chats to answer the questions asked.

Data Story

The **Data Story** dashboard object is similar to Ask Data in that it can be used to provide an explanation of the data but instead of answering user input questions with charts, a descriptive sentence about one of the charts on the dashboard is provided. This can be useful as a dynamic descriptor or overview of a piece of analysis that will update as the data source updates. When placing the object onto the canvas, the configuration will ask for a sheet to analyze, as well as which dimensions and measures from that sheet to consider. Finally, it will ask for a description of the data, whether is it discrete, continuous, a scatterplot, or a percent of whole. Once the Data Story has been configured, it will be placed on the dashboard canvas as a series of insights written in natural language.

Workflow

The **Workflow** dashboard object allows users to connect to a Salesforce flow and send data points from the Tableau dashboard through that flow.

Creating a Dashboard and Adding a Sheet and a Title

In this exercise, you will create a dashboard in Tableau Desktop, and you will add a sheet as well as a title to the dashboard. Dashboards are the primary means by which analysis is shared in Tableau and this exercise is a great introduction to dashboard building. Follow these steps:

1. On a new sheet, place **Order Date** as continuous months on columns and **Sales** on rows to show total sales by month for the whole data set.

2. Rename the sheet `Monthly Sales`.

3. Create a new dashboard using the **Dashboard** option in the toolbar, followed by **New Dashboard**.

4. From the dashboard pane on the left, drag the **Monthly Sales** sheet onto the canvas. The sheet is now displayed on the dashboard.

5. To add a title to the dashboard, select **Show dashboard title** at the bottom of the dashboard pane.

6. Double-click the textbox that has appeared to modify the title.

7. Replace the dynamic sheet name text with **Sales Analysis Dashboard** and make the font bold.

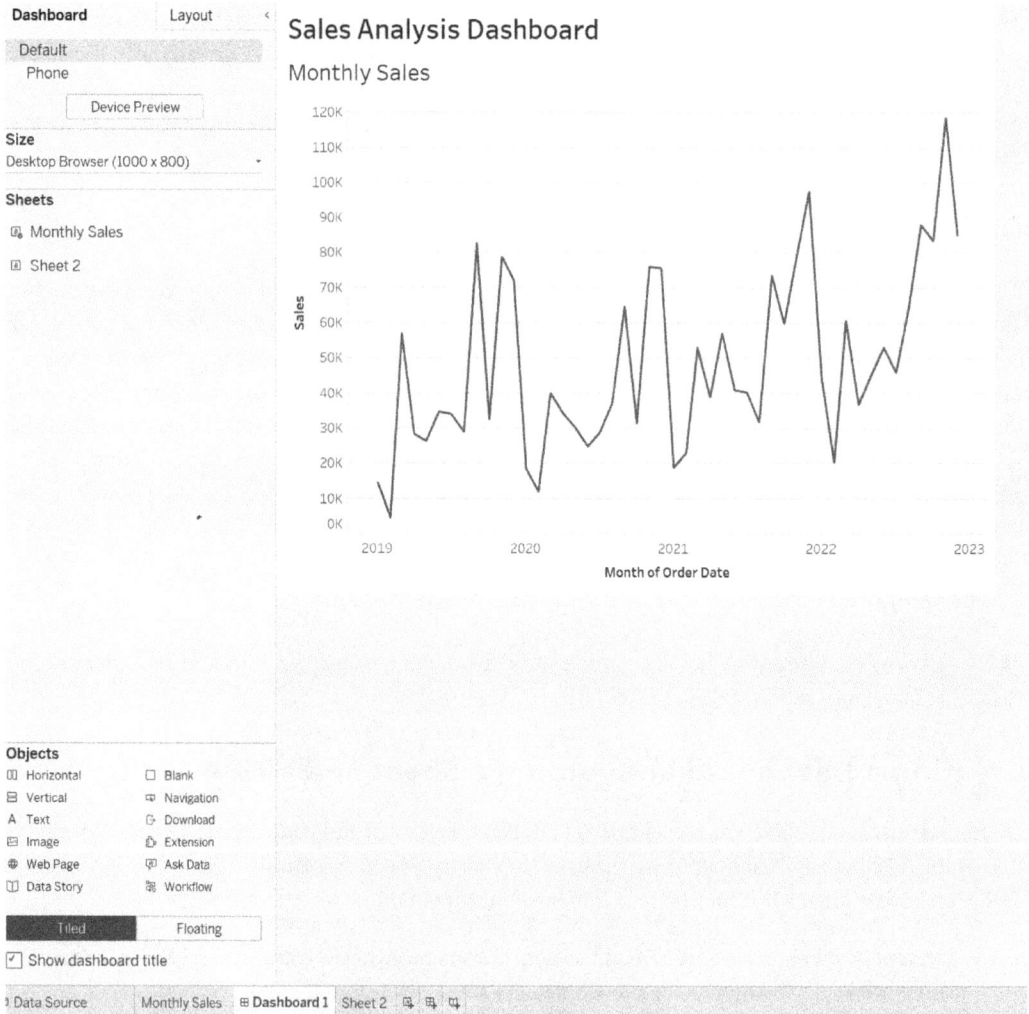

Figure 6.10: A new dashboard was created with a chart and a title added

In this exercise, you created a dashboard in a Tableau workbook and added an existing sheet and a title to the dashboard. You now know how to create dashboards and add objects to them.

Adding a Container and Another Sheet to the Dashboard

In this exercise, you will add a container object to the dashboard as well as another sheet. Containers are the best way to organize and structure dashboards, and in this exercise, you will demonstrate how to add multiple objects to a container. Follow these steps:

1. Create a new sheet for some additional sales analysis called Sub-Category Sales.

2. Place **Sub-Category** on rows and **Sales** on columns to create a bar chart showing total sales by sub-category.

3. Sort the bars from high to low.

4. Return to the **Dashboard** and drag a horizontal container **Dashboard** object onto the canvas.

5. The object placement type is set to **Tiled** by default, which means **Dashboard** objects take up a specific position on the canvas as opposed to floating on top of the canvas and potentially overlapping with each other. When dragging the container over the canvas, sections of the canvas will be shaded, indicating where the object will be placed.

6. Drop the container at the bottom of the **Dashboard**. This will cause the container to take up the bottom portion of the dashboard with the **Monthly Sales** chart being pushed upwards.

7. Select the **Monthly Sales** chart so that there is a gray outline around it.

8. Drag the selection section in the top center of the outline and use it to drag the chart into the container below.

9. The **Monthly Sales** chart now fills the majority of the **Dashboard** again. Double-clicking the selection section will show a blue outline, indicating that the chart is within the container.

10. Drag the **Sub-Category Sales** sheet from the **Dashboard** pane into the container to the right-hand side of the Monthly Sales chart. The **Dashboard** now has two charts and a title.

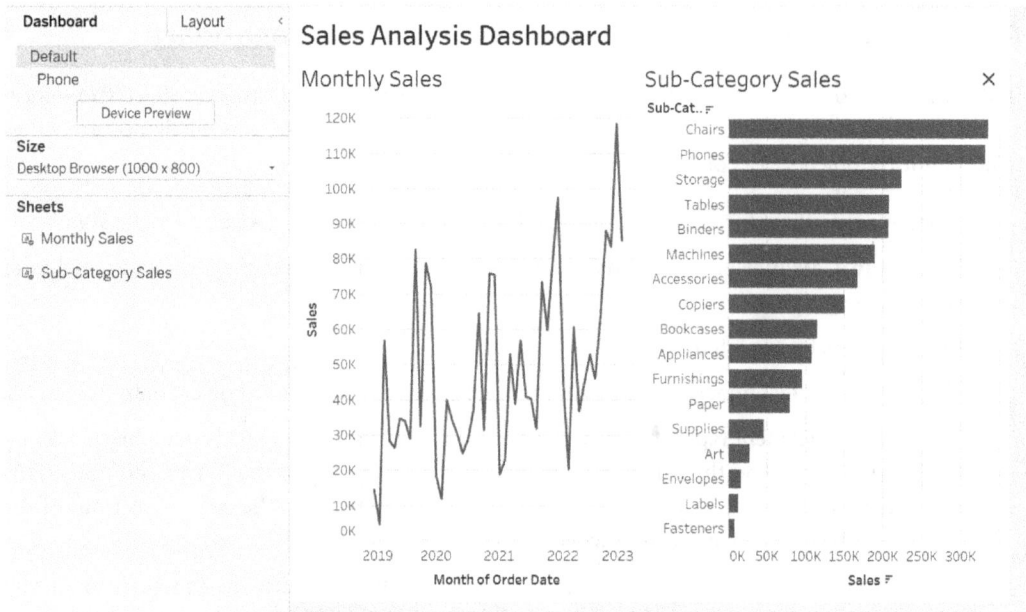

Figure 6.11: The Dashboard now has two sheets within a container

In this exercise, you added a container to a **Dashboard** and observed how tiled dashboarding works with objects being dropped into the shaded areas of the dashboard. You also added multiple sheets into a container and observed how the tiled shading is similar but different, with the container being outlined in blue, when an object is being dropped into a container.

Adding Buttons to the Dashboard

In this exercise, you will add a show/hide and a download button to the dashboard. You will also use a floating methodology when placing buttons on the dashboard. Buttons provide a lot of useful functionality to dashboards such as navigation, downloads, and showing or hiding dashboard elements. It is also important to understand how to float objects on a dashboard as overlapping elements are often preferable. Follow these steps:

1. Select the **Sub-Category Sales** dashboard and use the drop-down option in the top right. Select **Add Show/Hide Button** from the options available in the pop-up menu to create a button that will show and hide the chart.

2. The button will be floating on top of the existing charts (as opposed to tiled) and can be moved around the dashboard and dropped anywhere. Place it at the top right of the Sub-Category Sales chart.

3. To add a download button, first select **Floating** at the bottom of the dashboard pane. This means that any object dragged onto the canvas will be floating above the existing objects as opposed to taking up space positionally.

4. Drag and drop the download button in the top right of the dashboard where there is whitespace.

5. Use the **Object Options** dropdown then select **Edit Button**.

6. Use the configuration popup to allow users to download an image, give the button a title of Save Image, and set the background color for the button to a light gray.

7. Test both buttons by opening presentation mode (use the easel icon in the middle of the page on the icons toolbar or select **Window** followed by **Presentation Mode**.

8. In presentation mode, click the **x** button to hide the bar chart then the three horizontal lines to show it again.

9. Next, click the **Save Image** button and save an image of the dashboard.

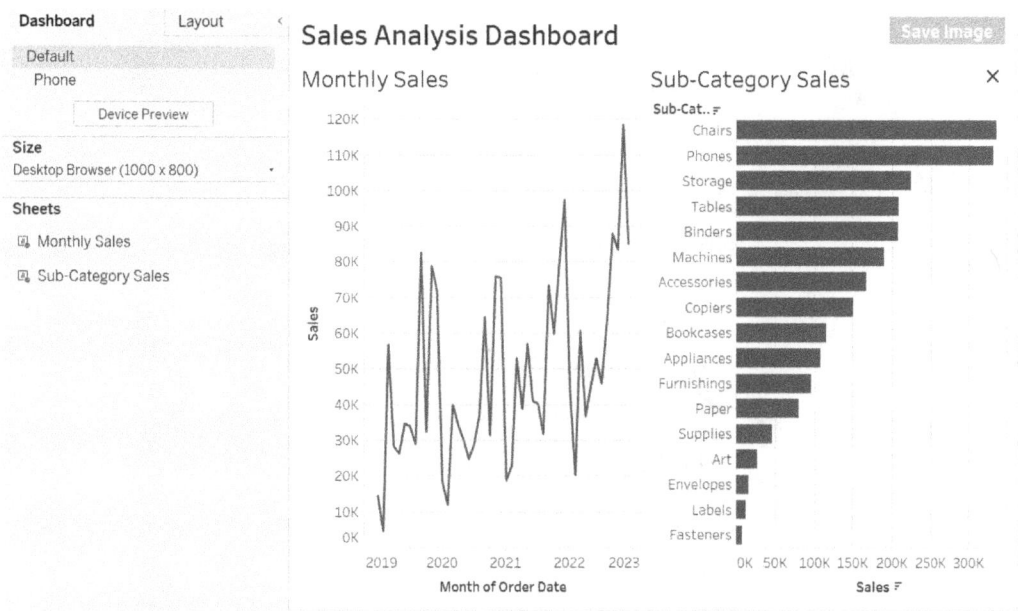

Figure 6.12: Download and show/hide buttons added to the dashboard

In this exercise, you added a show/hide button to a dashboard object, switched the dashboard placement methodology to floating, and added and configured a download button. You now know how to utilize buttons in Tableau dashboards as well as how to float objects.

Layout Options

When it comes to creating dashboards, there are a variety of options for how to lay them out. Dashboards consist of various objects such as charts and the methodology for how these are placed on the canvas can be configured. The size of the canvas can also be configured as well as the size of objects and the padding or whitespace around each object.

Dashboard Sizing/Device Layouts

When it comes to the size of the dashboard canvas, there are multiple options. The options can be accessed via the dashboard pane under the size heading. From here, the canvas size can be set to automatic, which means it will automatically fill the screen of the user's device. Alternatively, a range can be set that fixes the automatic sizing between a minimum and maximum pixel height and width. The final option is to fix a dashboard at a specific size in terms of pixel width and pixel height. There are some presets for the user to pick from if a fixed size is selected, such as a generic desktop size, PowerPoint size, or A3 size. Alternatively, the user can select **Custom** and set their own pixel height and width.

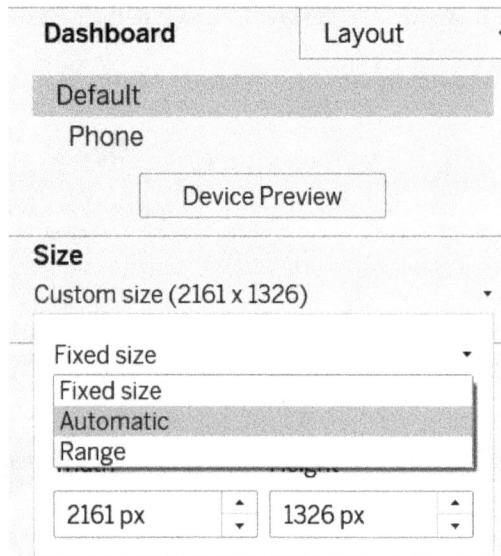

Figure 6.13: Dashboard canvas size options

In addition to the standard dashboard canvas, device-specific layouts can be created. By default, a phone-sized layout will be generated that reorganizes the dashboard content onto a phone-sized canvas with dashboard objects stacked on top of each other. An outline will be provided to show how much of the dashboard will fit onto the phone screen with the content below needing to be scrolled down to.

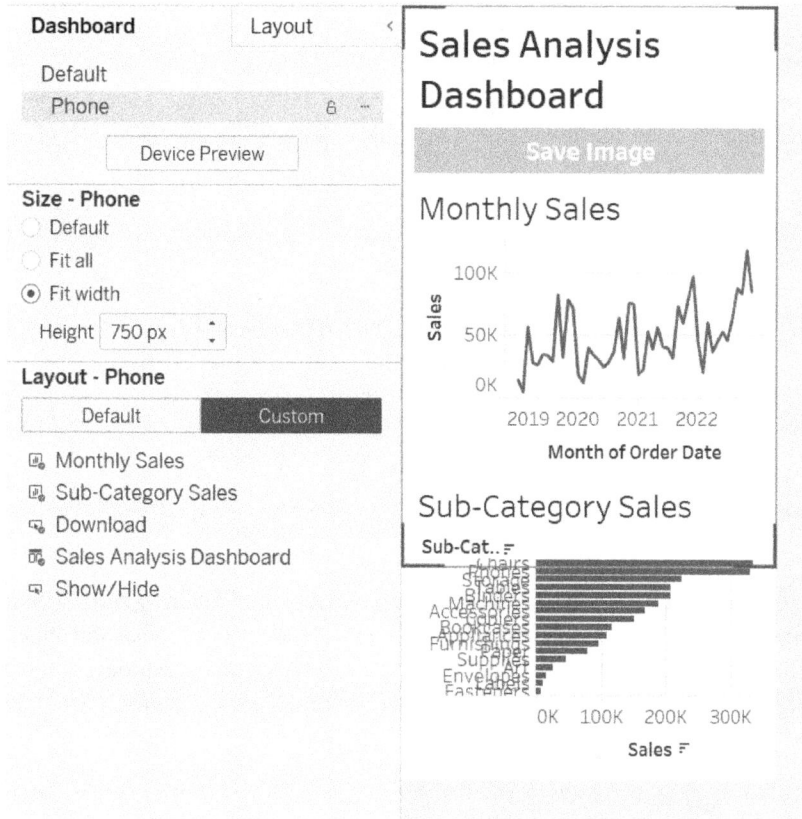

Figure 6.14: The Phone layout selected with content fit to phone width size and with the sub-category sales bar chart, which requires the phone user to scroll down to

To access and customize the **Phone** canvas layout, the user can select **Phone** below **Default** on the dashboard pane. An auto-generated phone layout will be displayed and this is locked by default. If the user selects the padlock icon, then the layout can be customized. The user can set the phone view to retain the default dashboard canvas size, meaning phone users will likely have to scroll horizontally to see the content. Alternatively, all dashboard content can be set to fit within the phone screen boundaries with **Fit all**. By default, the canvas is set to **Fit width**, which fits objects to a phone screen width but with content stacked on top of each other, with scrolling likely required to see all of the content.

Dashboard objects can also be removed from the **Phone** layout specifically so whilst they would show to non-phone users, anyone accessing the dashboard via their phone would not have visibility. The height of the canvas can also be adjusted on the dashboard pane if more space is required or if too much is provided.

In addition to the **Phone** layout, a desktop-specific layout and a tablet-specific layout can also be configured. To add these layout types, right-click near the **Default** and **Phone** layout options on the dashboard pane and select **Add Desktop Layout** or **Add Tablet Layout**. Layout types can also be deleted by selecting the three dots to the right-hand side of the layout name followed by **Delete Layout**.

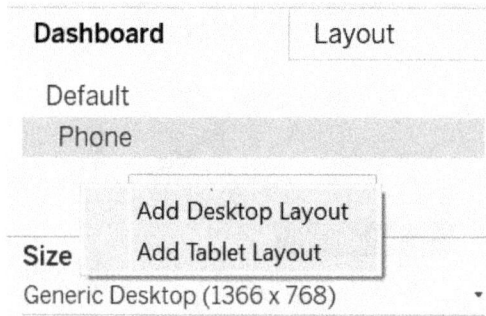

Figure 6.15: Additional layout options

To view how a dashboard will look across various device types of differing screen sizes, select the **Device Preview** button on the dashboard pane. When this is selected, a toolbar will appear above the dashboard that allows the user to select the device type to test as well as the model of the device, each with differing screen sizes that the dashboard canvas can be compared against.

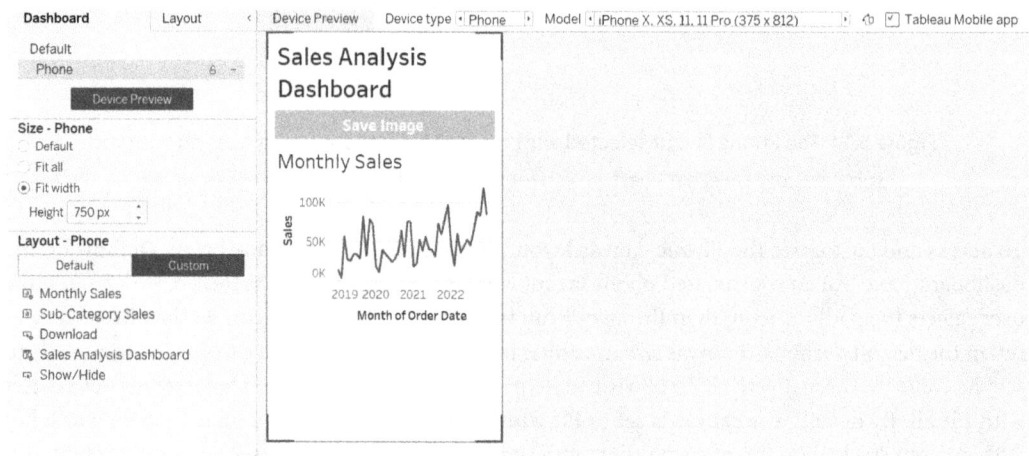

Figure 6.16: Previewing the dashboard experience of various
iPhone model users on the Tableau mobile app

Floating versus Tiled

When it comes to placing objects such as charts onto a dashboard canvas, there are two options: tiled and floating. Objects are dragged from the dashboard pane onto the canvas in either tiled or floating format and this is set via the selection at the bottom of the dashboard pane.

Tiled	Floating

Figure 6.17: The Tiled and Floating selection with Floating selected

If the setting is set to **Tiled**, then any object dragged onto the dashboard canvas will be placed onto the canvas in a grid-like fashion. This results in a single layer of objects that can be side by side or above each other but will never overlap. When dragging an object onto the canvas, an area of the canvas will be shaded. Moving the dragged object around the canvas will shade different areas. The shaded area indicates where the object will be dropped on the canvas grid.

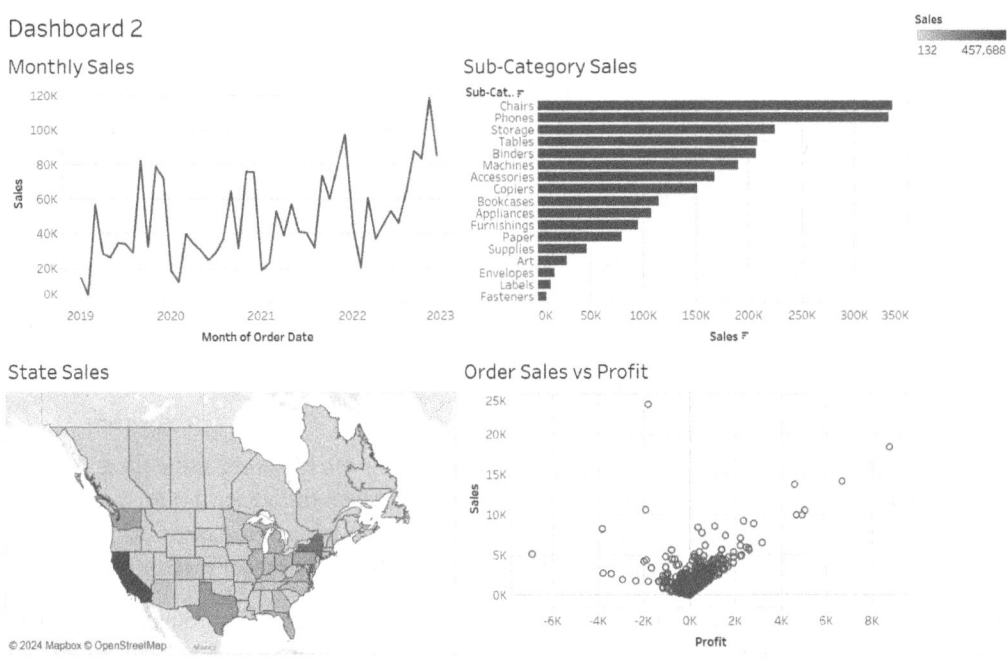

Figure 6.18: Tableau dashboard with a Tiled layout – objects are
positioned in a grid-like fashion with no overlap

If **Floating** is selected, then objects can be dragged anywhere on the canvas and will be placed wherever they are dropped. There are no shaded areas as the object will not be placed into a grid. Floating objects can be placed on top of other objects and there is a floating order that can be updated using the objects' drop-down menu options in the top right of the selected object. Selecting **Floating Order** from the menu will give the user the ability to move an object to the front or back, or move it forward or backward one place in the hierarchy.

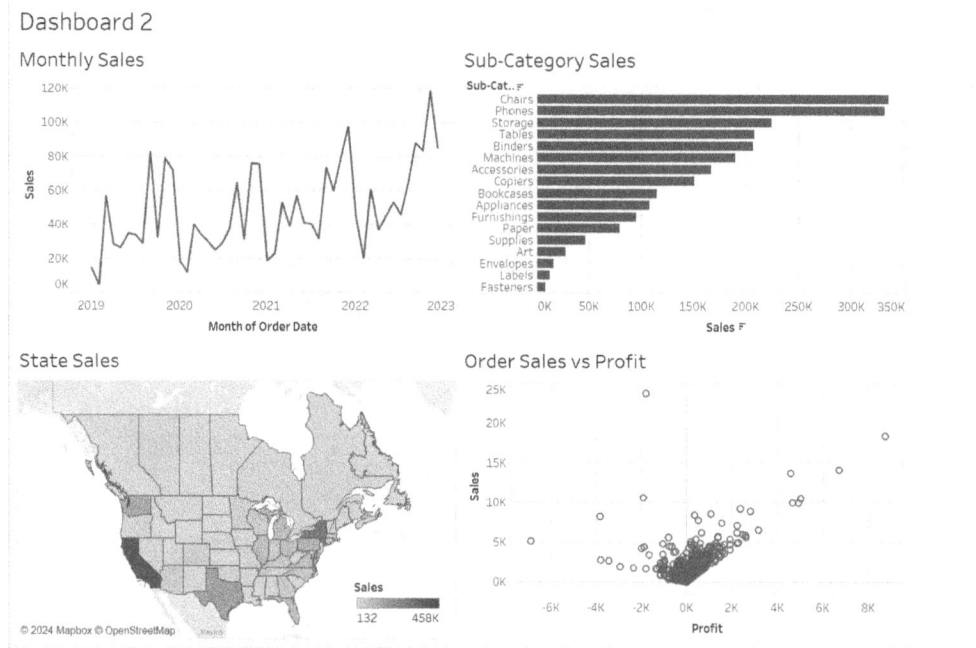

Figure 6.19: The Sales color legend is no longer a tiled object but rather floating on top of the State Sales chart

Dashboards can have both **Tiled** and **Floating** objects. All objects can be resized using the gray outline when the object is selected and if the object is tiled this will impact the adjacent objects in the grid. The size of **Floating** objects can also be updated using the **Layout** tab on the dashboard pane, which has options for increasing and decreasing the height and width of objects to a specific pixel size.

The **Layout** tab on the dashboard pane also allows users to specify an x and y coordinate on the canvas at which a floating object will be placed. This allows users to align floating objects more easily.

Tiled objects cannot have their size and position updated using the options on the **Layout** tab. If tiled objects are placed within containers, then the size can be set by fixing the height or width. Horizontal containers allow users to set a specific width for an item using the drop-down menu in the top right of the object followed by **Edit Width**. Vertical containers provide the same option but for the height as opposed to width via **Edit Height**. These options allow the user to specify a pixel width or height for the object within. Once a width/height has been fixed, the pin icon in the top right of the selected object will be colored. This can be clicked to unfix the width/height, resulting in dynamic sizing, and can also be clicked at any time to fix an object's width/height. This can also be done via the drop-down menu followed by **Fix Width/Height**.

Objects can switch between a tiled or floating state using the **Layout** tab on the dashboard pane. With the object selected, the **Floating** checkbox can be ticked or unticked to update the object status. The same option is available from the drop-down menu in the top right of a selected object; **Floating** can be either selected or deselected for any given object.

Adding Tiled Objects and Floating Objects to a Dashboard

In this exercise, you will add two new sheets to the dashboard from previous exercises using the tiled placement methodology. You will then place the legends that correspond to those charts on the dashboard using the floating methodology. This will further cement your understanding of the differences between the tiled and floating methodology in Tableau and make you a more proficient dashboard developer. Follow these steps:

1. Using the same workbook as used for *Exercises 1* to *3* in the *Dashboard Objects* section, create a new sheet.

2. Double-click **State** and place **Sales** on **color** to create a total **Sales by State** map.

3. Call the sheet `State Sales`. By default, a legend will be created for the sales color gradient.

4. Create a new sheet and place **Sales** on the rows and **Profit** on the columns and then place the order ID onto the **Detail shelf** to create an order sales versus profit scatterplot.

5. Place **Category** on color to create a color legend that breaks up order sales versus profit by various categories.

6. Call the sheet `Order Sales versus Profit`.

7. Return to the dashboard created in previous exercises and ensure that **Tiled** is selected on the dashboard pane.

8. Add the **State Sales** sheet to the view by dragging it to the bottom of the canvas until the bottom section is shaded, then dropping it.

9. Ensure that the chart is not added to the container created in the previous exercise and is instead placed below the width of the container.

10. Notice how the Sales gradient color legend has automatically been added alongside the sheet and that it is placed in its own container. Legends, filters, and parameters are placed in a single container by default.

11. Next, drag the **Order Sales versus Profit** sheet onto the view and drop it onto the right-hand side of **State Sales** and below **Sub Category Sales**. The exact area where the chart will be placed should be shaded, resulting in a grid of four charts.

12. The **Category** color legend has been added to the same container as the **Sales** gradient color legend.

13. The buttons created in the previous exercise are floating by default and can be seen overlapping the color legends. To place the color legends in more suitable positions, they can also be converted to floating objects.

14. Select the **Category** color legend and in the top-right drop-down menu, select **Floating**.

15. Do the same for the **Sales** gradient legend.

16. The container that was created to hold the legends has not automatically been removed and is taking up unnecessary space on the right side of the **Dashboard**. Right-click the container and select **Remove Container**.

17. The color legends can be dragged and resized over the relevant charts, allowing the user to more easily identify what the legend applies to.

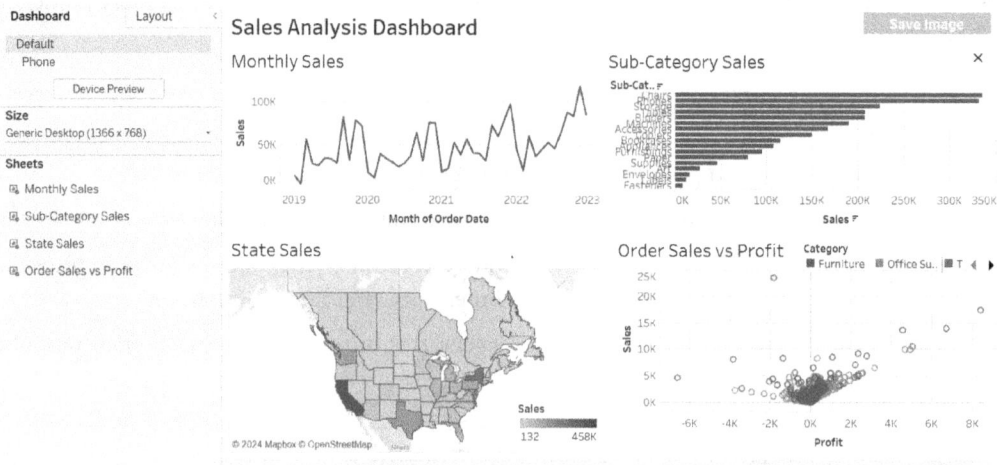

Figure 6.20: Tiled charts added to the dashboard and legends converted to floating and repositioned

In this exercise, you added two new dashboard objects using the tiled methodology. These sheets were not placed in a container and were instead tiled directly onto the dashboard canvas. You observed that sheets that contain elements such as legends (the same applies to filters and parameters) will bring those elements onto the dashboard in a designated container. You switched the placement methodology for these legends to floating and placed them overlapping the relevant charts.

Resizing the Default Canvas and Updating the Phone Layout

In this exercise, you will configure the dashboard canvas size settings to suit your personal screen size and will create a phone layout for the dashboard. It is best practice to ensure that dashboards are a suitable size for end users and phone layouts can aid in this for any user that will use the dashboard from the phone. Follow these steps:

1. The canvas is by default a fairly small size. Go to the **Size** section in the dashboard pane and resize the canvas to a size that makes sense for your machine.

2. One of the preset screen sizes can be selected or **Custom** can be selected with the specific pixel width and height defined.

3. To view the layout for phone users, select **Phone** on the dashboard pane and deselect the padlock icon to allow editing.

4. The **Show/Hide** button has been removed by default, but the other objects have been retained and are stacked on top of each other.

5. **Save Image** is not necessary for phone users, so this can be removed.

6. The title textbox may be cut off on the phone view – edit this and make the text size smaller.

7. Resize and reposition charts and legends to enhance the user experience based on what can be seen on the screen (the gray outline with bold corners identifies a screen).

8. Finally, select **Device Preview** on the dashboard pane and check through various phone and desktop layouts to ensure both phone and desktop users can utilize the dashboard.

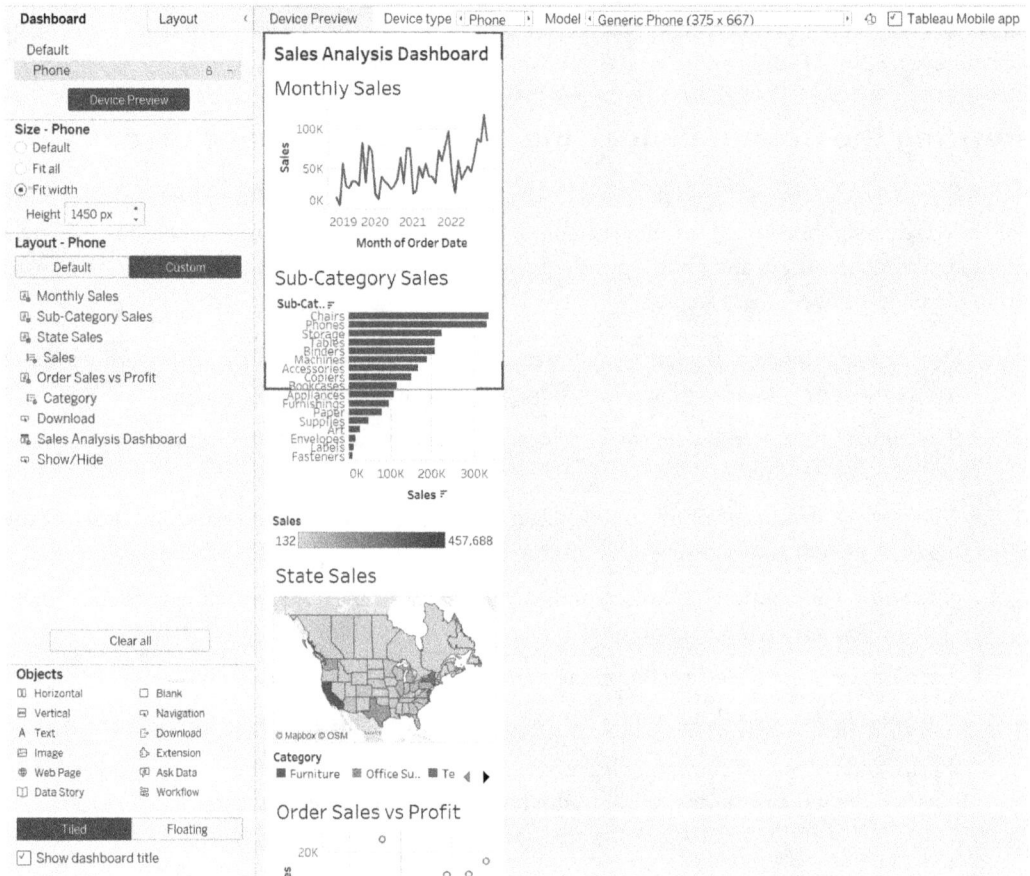

Figure 6.21: The Default and Phone layouts for the dashboard configured

In this exercise, you have customized your dashboard size by either selecting a fixed preset or setting your own canvas dimensions. You have also observed the various device layout settings and have configured a specific phone layout view.

Dashboard Interactivity

Charts in Tableau can be made to interact with each other when selected with one chart selection filtering or highlighting another. URL actions can also be set up to take users to a URL on selection or to update a web page object on a view.

Dashboard Actions

Tableau users wanting to add interactivity to charts on a dashboard can do so using dashboard actions. The dashboard actions configuration menu can be accessed from the **Dashboard** options on the top toolbar followed by **Actions**. The configuration window can then be updated to show all actions in the workbook or actions relevant to the current sheet/dashboard only.

There is an **Add Action** button that, when selected, allows the user to add a specific action type, which will open a configuration window for that action. The most commonly used actions are the top three in the list, filter, highlight, and URL actions.

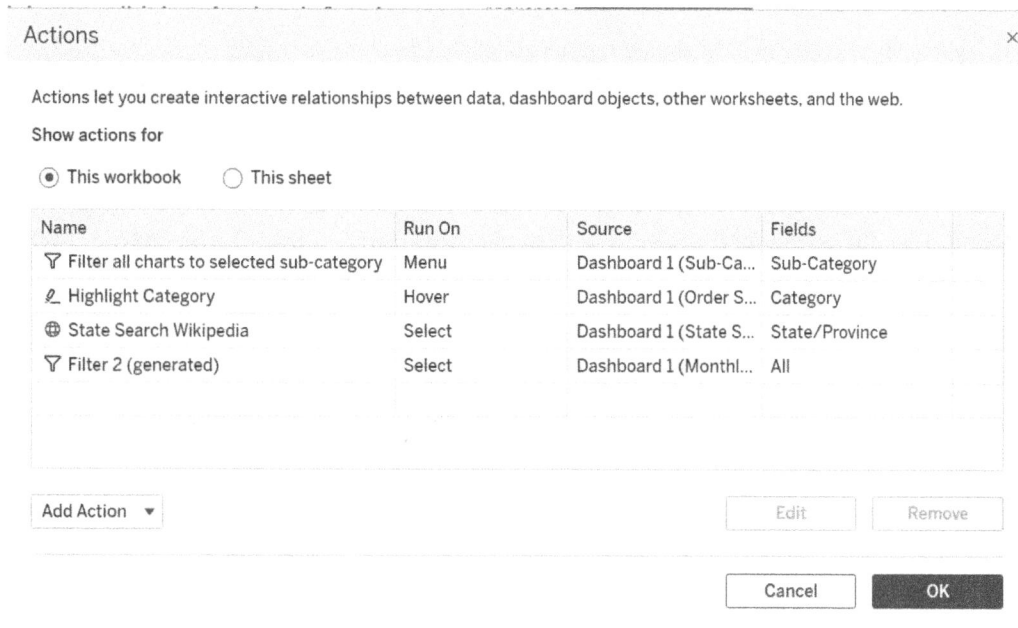

Actions				×
Actions let you create interactive relationships between data, dashboard objects, other worksheets, and the web.				
Show actions for				
⦿ This workbook ○ This sheet				

Name	Run On	Source	Fields	
▽ Filter all charts to selected sub-category	Menu	Dashboard 1 (Sub-Ca...	Sub-Category	
◢ Highlight Category	Hover	Dashboard 1 (Order S...	Category	
⊕ State Search Wikipedia	Select	Dashboard 1 (State S...	State/Province	
▽ Filter 2 (generated)	Select	Dashboard 1 (Monthl...	All	

Add Action ▾   Edit   Remove

Cancel   OK

Figure 6.22: The Actions configuration showing all of the actions in the
workbook, including two filters, a highlight, and a URL action

The **Actions** configuration window will list actions by name and will include the source (which dashboard the action is configured to take place on) and the fields that the action acts on. What the action runs on is also listed, and this can either be **Select**, **Hover**, or **Menu**. All dashboard actions provide these three options for triggering the action: **Select** refers to actions that are triggered by selecting/clicking on a chart element; **Hover** refers to actions that are triggered when the user hovers over a chart element; and **Menu** refers to actions that are triggered by a user first selecting a chart element and then selecting a hyperlink at the bottom of the corresponding tooltip. The hyperlink text for the menu action type is the name of the action itself, which is configured during the setup.

Filter Actions

Filter actions filter the elements on other charts when an element on the source chart is selected. For example, selecting a country in a map chart could filter a sales trendline chart to just sales for that country.

Filter actions can be added from the dashboard actions configuration menu by selecting **Add Action** followed by **Filter**. The **Filter** action configuration menu allows the user to name the action as well as select what to run the action on – **Hover, Select,** or **Menu**. The user can also select whether a single selection is forced or whether multiple chart elements can be selected at once to act as filters for other charts.

The **Source Sheets** section of the configuration pane allows the user to select the dashboard the action runs from as well as tick at least one sheet to filter from when interacted with. The target sheet section provides the same options allowing users to create filter relationships from multiple source sheets to multiple target sheets.

There is a section in which the user must specify what happens when the filter action is cleared, which refers to when the user stops hovering over a chart element if **Hover** is the run action on type. If **Select** or **Menu** is the run action on type then clearing the selection refers to when the user deselects the chart element being used to filter.

The first option is **Keep filtered values**, which means target sheets will only be updated when a filter selection takes place, not when it is subsequently deselected. For example, if selecting a country and filtering a sales trendline, deselecting the country in the source chart will not unfilter that country in the target trendline sheet.

The **Show all values** option means that the filter on the target sheet will be removed when the element in the source sheet is deselected. In the country and sales trendline example, the target trendline sheet would return to showing all sales.

The final option is **Exclude all values**, which results in all chart elements in the target sheet being filtered out when the source sheet element is selected. In the country and sales trendline example, the trendline would become blank until a country is selected again in the source country sheet.

By default, filter actions are set to filter target sheets using all fields from the source sheet. If, for example, a sheet has **Sales** broken down by **Category** and **Sub-Category**, then **All fields** being selected means that all target sheets would be filtered by the **Category** and **Sub-Category** values selected on the source sheet. Alternatively, **Selected fields** can be selected, which results in only the specified source fields filtering the target sheet. The selected source fields must be present on the source sheet for this to work. In the previous example, having **Category** as the only selected field for filtering would result in selections on the source sheet filtering target sheets by category only even though there is a **Sub-Category** level of detail on the sheet.

Add Filter Action   ×

Name

Filter Monthly Sales by State   Insert ▾

Source Sheets

⊞ Dashboard 1 ▾

☐ Monthly Sales   Run action on
☐ Order Sales vs Profit   ◯ Hover
☑ State Sales   ◉ Select
☐ Sub-Category Sales   ◯ Menu
                                         ☑ Single-select only

Target Sheets

⊞ Dashboard 1 ▾

☑ Monthly Sales   Clearing the selection will
☐ Order Sales vs Profit   ◯ Keep filtered values
☐ State Sales   ◉ Show all values
☐ Sub-Category Sales   ◯ Exclude all values

Filter

◉ All fields   ◯ Selected fields

☐	Source Field	Target Data Source	Target Field
☐	Click to add ▾		

Remove

Cancel   OK

Figure 6.23: Filter action set up, which filters a Monthly Sales chart by
selected States from the State Sales

Filter actions can be created quickly for a sheet on a dashboard by selecting the sheet and then clicking the filter icon in the top right so that it becomes filled. This creates a filter action that can be seen in the dashboard actions configuration menu. The filter takes the selected sheet as the only source sheet and all other sheets as the targets. It runs the action on selected, shows all values when the selection is cleared, and filters on all fields from the source sheet. The name of the action is **Filter N (generated)**, with **N** being replaced by the count of how many generated filter actions have been created by selecting the filter icon. Generated filter actions can be modified if needed from the dashboard actions menu.

Highlight Actions

Highlight actions take a selection from a source sheet and highlight the corresponding values on target sheets by fading out marks that don't match the selection. Highlight actions can be added from the dashboard **Actions** configuration menu by selecting **Add Action** followed by **Highlight**.

Highlight actions are similar to filter actions in that they take source sheets and target sheets and there can be multiple selections for both. The same **Run action on** options are available.

The only other configuration option for highlight actions is what fields should be highlighted. By default, **All Fields** is selected, which means all fields that are selected in the source sheet will be highlighted in the target sheet. If there is no field from the source sheet that the target sheet has in common, then selecting an element on the source sheet will fade out the target sheet as there is nothing to highlight.

The second option for which fields to highlight is **Dates and Times**, which will highlight only date and datetime type fields common across source and target sheets. The final highlight target option is **Selected Fields**, which allows the user to select as many fields as desired from the source sheet to highlight in the target sheet.

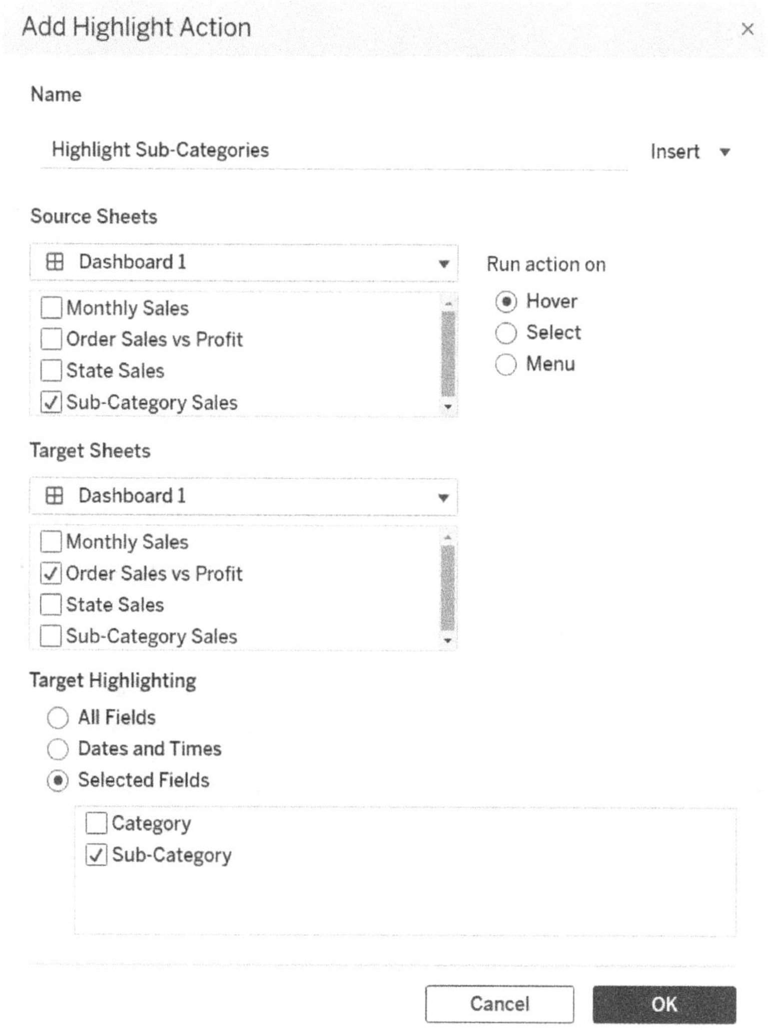

Add Highlight Action   ×

Name

Highlight Sub-Categories   Insert ▾

Source Sheets

⊞ Dashboard 1 ▾   Run action on

☐ Monthly Sales   ◉ Hover
☐ Order Sales vs Profit   ○ Select
☐ State Sales   ○ Menu
☑ Sub-Category Sales

Target Sheets

⊞ Dashboard 1 ▾

☐ Monthly Sales
☑ Order Sales vs Profit
☐ State Sales
☐ Sub-Category Sales

Target Highlighting

○ All Fields
○ Dates and Times
◉ Selected Fields

☐ Category
☑ Sub-Category

Cancel   OK

Figure 6.24: A highlight action that highlights subcategories on the Order Sales versus
Profit sheet based on the users hovering over the Sub-Category Sales sheet

URL actions

URL actions are actions that open a URL when a source sheet is interacted with. The URL can be customized to take parameters from the selected field or any parameters in the workbook. The URL can be set to open on a web page object on the dashboard or to open a new tab in the browser.

URL actions are created from the dashboard actions configuration menu by selecting **Add Action** followed by **go to URL**. The options available when configuring a URL action are the name, the source sheets to run the action off, and what action to run the URL action on: **Hover**, **Select**, or **Menu**. The user can also select whether the URL should have the **New tab if No Web Page Object Exists** option enabled, which opens a new browser tab for the URL if there is no web page object on the dashboard but otherwise opens the URL on the web page object. Alternatively, **New Browser Tab** can be selected to always open the dashboard on a new browser tab.

The final section of configuration for a URL action is the URL itself. This can take normal text input, but there is also an **Insert** button from which users can add parameters such as fields from the source sheets or parameters from the workbook. The URL that is created for the action can therefore be made dynamic based on selection, for example, putting a country field in the URL could allow the user to open a URL page relating to the country selected by the user. Below the URL configuration bar is an example of the URL as a hyperlink, which can be tested.

Advanced options for the URL configuration include **Encode data values that URLs do not support**, which converts any characters that URLs do not support to the correct equivalent. For example, a parameter might have a space in the text that would not be a recognized character in a URL. This option being selected would convert the space to a **%20**, which would allow the URL to function. **Allow multiple values via URL parameters** means that multiple selections of a parameter field in the URL will result in multiple values being placed into the URL string.

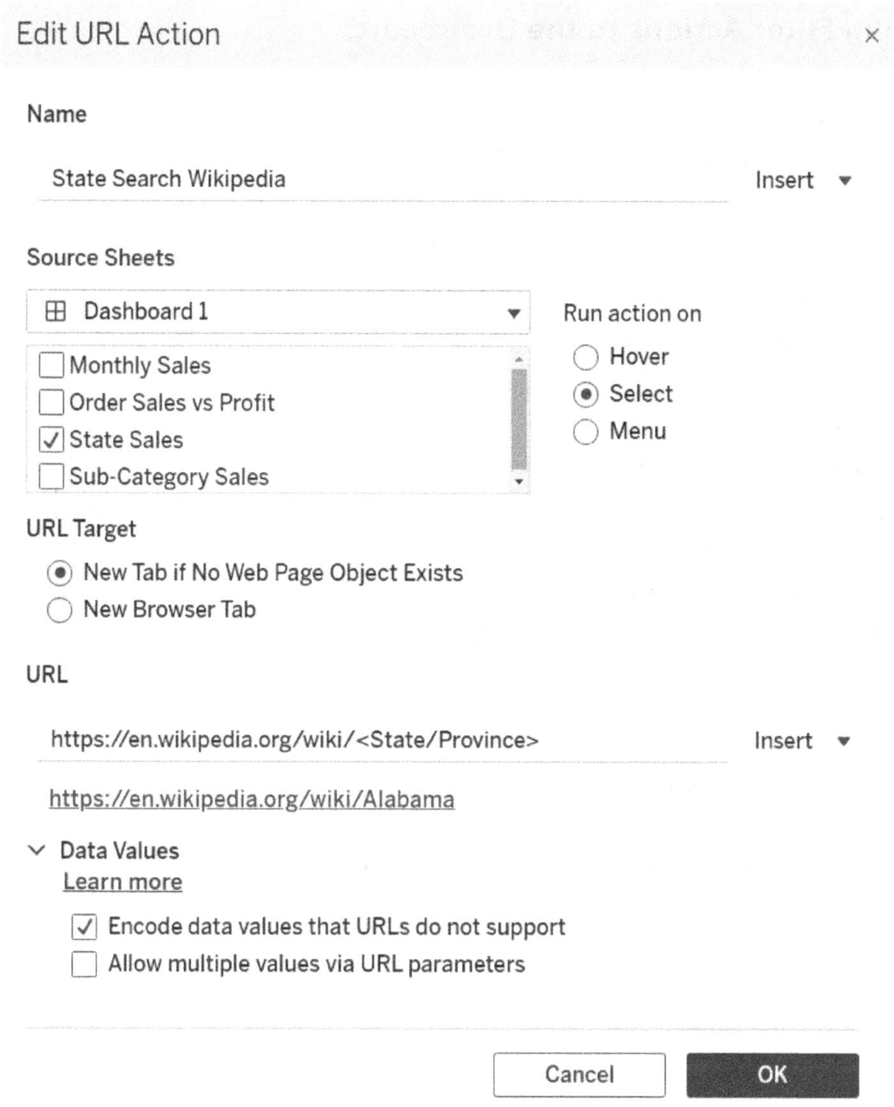

Edit URL Action   ✕

Name

State Search Wikipedia   Insert ▾

Source Sheets

⊞ Dashboard 1   ▾   Run action on

☐ Monthly Sales   ○ Hover
☐ Order Sales vs Profit   ◉ Select
☑ State Sales   ○ Menu
☐ Sub-Category Sales

URL Target
  ◉ New Tab if No Web Page Object Exists
  ○ New Browser Tab

URL

https://en.wikipedia.org/wiki/<State/Province>   Insert ▾

https://en.wikipedia.org/wiki/Alabama

∨ Data Values
   Learn more
      ☑ Encode data values that URLs do not support
      ☐ Allow multiple values via URL parameters

                                          Cancel   OK

Figure 6.25: A URL action that opens up Wikipedia articles on the selected state from the State Sales
sheet in the web page object on the dashboard or in a new browser tab if one does not exist

Adding Filter Actions to the Dashboard

In this exercise, you will add a filter action to the dashboard built in previous exercises. Adding interactivity to dashboards is a great way to allow users to unlock their own insights into the data, and selecting one chart to filter another enables this. Follow these steps:

1. On the same dashboard used in the chapter so far, open up the dashboard action configuration menu by selecting **Dashboard** followed by **Actions**.

2. Create a filter action by selecting **Add Action** and then **Filter**.

3. Configure the filter action to run on **Menu** and take **Sub-Category Sales** as the source.

4. Set all other sheets as the targets and ensure that clearing the selection shows all values on those sheets.

5. Configure the filter to filter on selected fields only and add **Sub-Category** as the field to filter on.

6. Name the action **Filter all charts to selected sub-category** so that the menu hyperlink is informative to end users.

7. Press **OK** and **OK** again and then test the filter by selecting a sub-category on the **Sub-Category Sales** chart and then selecting the hyperlink in the tooltip to filter all other charts to that sub-category.

8. Deselect the sub-category to show all values in the other charts again.

9. Create another filter quickly on the **Monthly Sales** chart by selecting the sheet followed by the filter icon in the top right.

10. The filter icon is now colored white and selecting any or multiple points on the trendline will filter all other charts on the dashboard.

11. Deselecting those points will show all values on the dashboard.

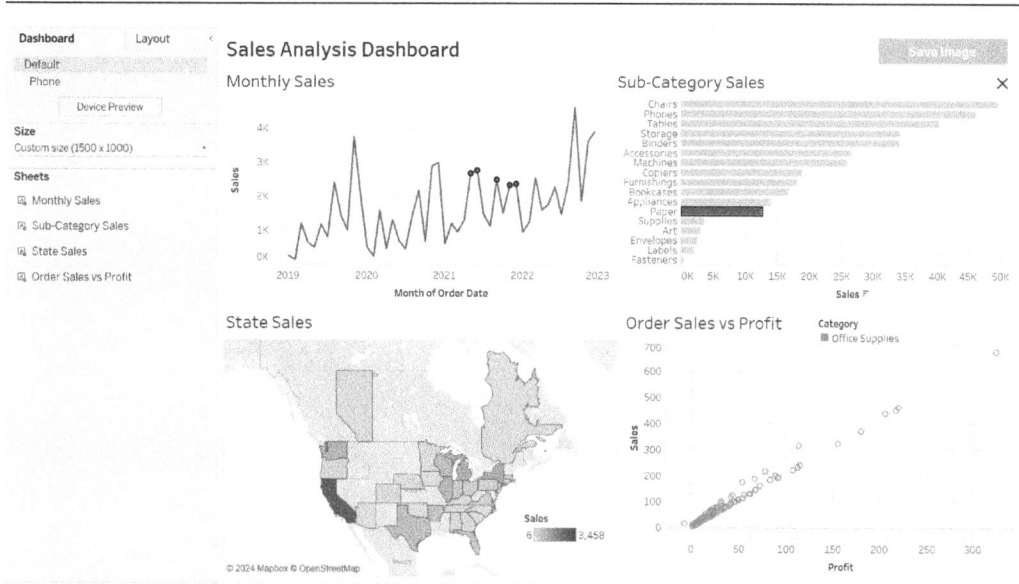

Figure 6.26: A filter action created on Sub-Category Sales as well as Monthly Sales

In this exercise, you added a filter using the quick method of selecting a filter icon and added a custom filter that runs on a specific field using the menu functionality. You observed the difference in functionality between types of filtering and now know how to customize filter actions on Tableau dashboards.

Adding a Highlight Action to the Dashboard

In this exercise, you will add an additional layer of interactivity to the dashboard via a highlight action. Highlight actions allow users to observe data points across multiple charts based on selection whilst retaining the full contextual information of all charts on the dashboard.

1. Open up the dashboard actions configuration menu again and observe that there is now a generated filter action from the **Monthly Sales** sheet.

2. Add a highlight action by selecting **Add Action** followed by **Highlight**.

3. Call the action `Highlight Category` and run the action on **Hover**.

4. Set the source sheet to **Order Sales versus Profit** and set the target sheet to **Sub-Category Sales**.

5. Set the target highlighting to **Selected Fields** and select **Category** only.

6. Press **OK** and **OK** again then test the action by hovering over points on the **Order Sales versus Profit** scatterplot.

7. The action causes the **Sub-Category Sales** chart to fade out as there is no Category field on the target sheet to highlight.

8. Navigate to the **Sub-Category Sales** sheet and add the **Category** field to detail on the marks card.

9. Test the action again and observe that the sub-categories within the relevant category are highlighted when hovering over marks on the scatterplot.

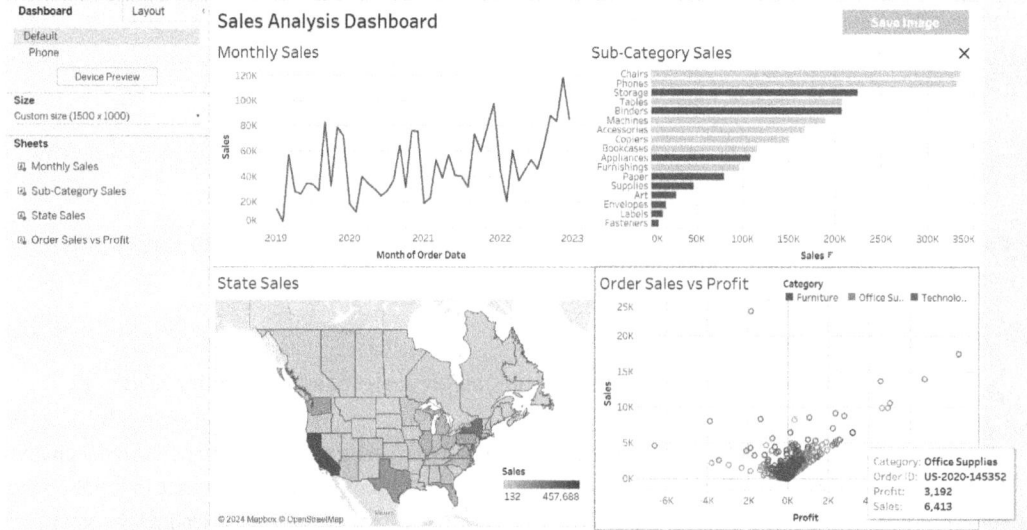

Figure 6.27: Highlight action added – hovering over a scatterplot mark
highlights the corresponding category in Sub-Category Sales

In this exercise, you learned how to add and customize highlight actions on Tableau dashboards. You also observed the requirements for highlight actions on specific fields and now know how to resolve issues with highlight actions that do not seem to function.

Adding a URL Action to the Dashboard

In this exercise, you will add a URL action to the dashboard. URL actions allow users to access web pages that can provide additional context to the analysis in the dashboard. Follow these steps:

1. Add a URL action from the dashboard configuration menu and call it `State Search Wikipedia`.

2. Set the source sheet to **State Sales** only and run the action on select.

3. Ensure the URL target is a new browser tab and type `https://en.wikipedia.org/wiki/` into the URL box.

4. Select **Insert** and **State/Province** to place the **State/Province** field into the URL as a parameter.

5. The URL can be tested here and should take the user to the Wikipedia page for Alabama (the first state in the data source).

6. Ensure **Allow multiple values via URL parameters** is deselected so that only one state is placed into the URL at a time.

7. **Encode data values that URLs do not support** can be selected to ensure no broken links are created as a result of incorrect characters.

8. Press **OK** and **OK** again and test that selecting a state on the **State Sales** chart opens a Wikipedia page for that state.

In this exercise, you added a URL action that opens a different page depending on what is selected on the chart. You can further customize URL actions with parameters and other fields. You also customized the URL action to open a new browser type and to encode any field values that are not supported by URLs.

Adding User-Guiding Sentences to Improve the End-User Experience

It can be useful to add guiding sentences when making dashboards interactive. In this exercise, you will add guiding sentences to all charts on the dashboard that will provide end users with the necessary context for how to interact with the dashboard. Follow these steps:

1. Start by editing the **Monthly Sales** chart title by double-clicking the title, and then in the line below, add the sheet name title **Click to filter by month**.

2. Make the font size smaller for this guiding sentence and use a lighter color such as gray to make it clear that it is not part of the title itself.

3. Repeat the same steps for the **State Sales** chart, adding **Click to open Wikipedia article for state**.

4. Below Order Sales versus Profit, hover over data points to highlight the order category in the **Sub-Category Sales** totals chart.

5. Below the **Sub-Category Sales** header, add **Use the tooltip hyperlink to filter all other charts by selected Sub-Category**.

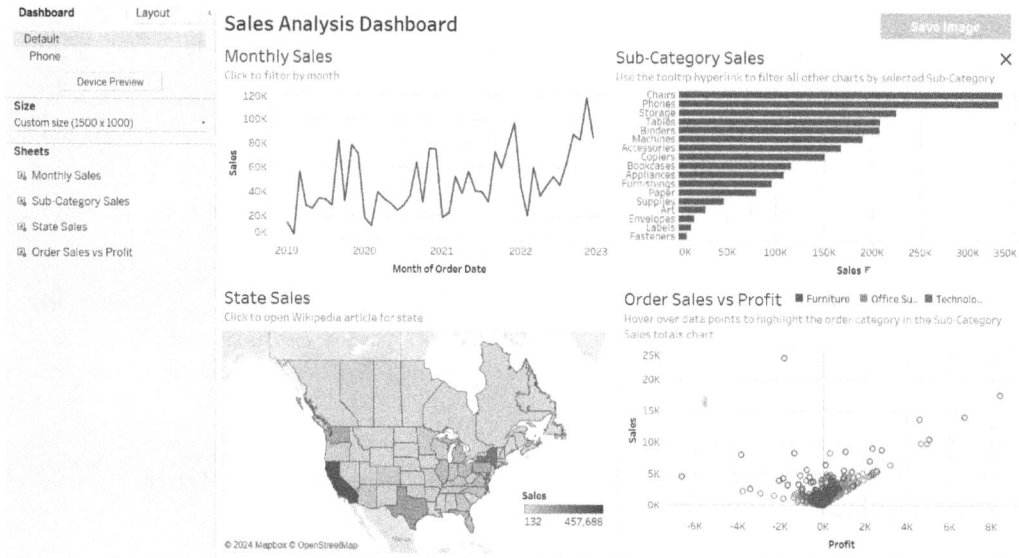

Figure 6.28: Guiding sentences added to inform end users how to interact with the dashboard

In this exercise, you added guiding sentences to the dashboard that provide end users with information that will help them better utilize the functionality on the dashboard.

Filtering Multiple Sheets

A filter created for specific sheets can be added to a dashboard but will only filter the sheet it was created on by default. Filters can be configured to filter multiple worksheets on the dashboard, and this includes sheets that did not have that field as a filter in the first place.

If a sheet has a filter on it already when being dragged onto a dashboard, then that filter will be added to the dashboard alongside the sheet in a separate container that is designated for any filters. Adding another sheet with filters on it will result in those filters being added to the same filter container. If a filter is added to a sheet after the sheet has been added to the dashboard, that filter will not automatically be added to the dashboard. To add a filter from a sheet to a dashboard, select the sheet on the dashboard, followed by the drop-down configuration menu in the top right, then **Filter**, and finally, the field to add the filter for.

Filters can be created for a sheet on a dashboard even where a filter does not exist on the sheet. Follow the same process outlined previously and then select a field listed in the filters pop-up menu. The fields available to set as a filter are fields that are used on the sheet somewhere.

Filter actions set up via the **Dashboard** configuration menu can be added as quick filter objects to the dashboard and therefore act both as filters run on the selection of chart elements and as filters interacted with directly. These filters can be found in the filter pop-up menu written in the **Action** format followed by the name of the field the action runs on in brackets.

Once filters have been added to a dashboard, they can be applied to other sheets on the dashboard by selecting the filter and then the configuration drop-down button followed by **Apply to Worksheets** and then **Selected worksheets**. This opens a configuration menu that shows the worksheets on the current dashboard. These sheets can be ticked to apply the filter to them individually or **Select all on dashboard** can be pressed if the filter should be applied to all sheets on the dashboard. **Show all worksheets in the workbook** can be checked to show all of the worksheets as opposed to just those on the dashboard. This can be useful for filters that should apply across multiple dashboards.

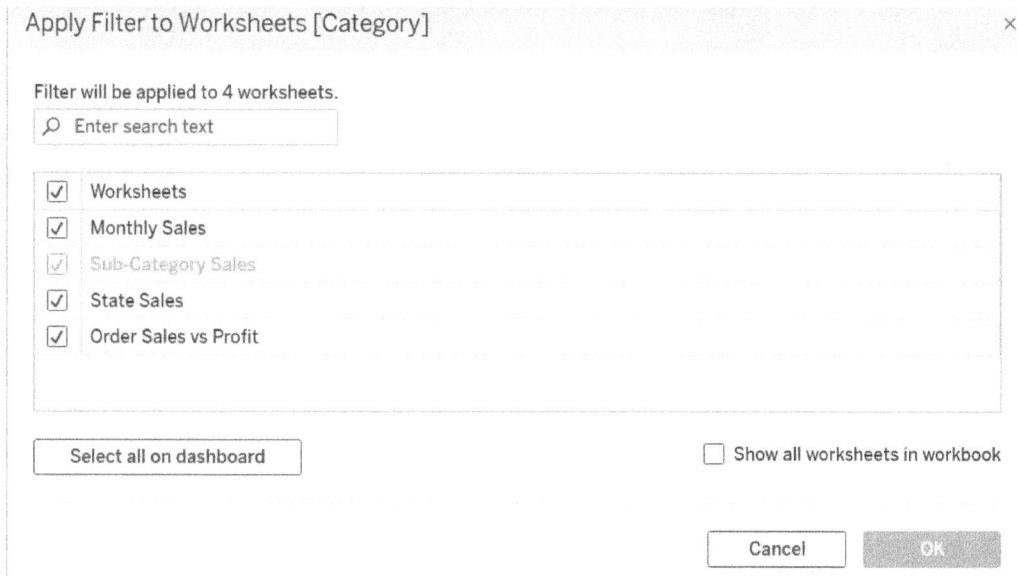

Figure 6.29: Apply filter to selected worksheets configuration

Adding a Filter to the Dashboard and Applying it to All Sheets

In this exercise, you will learn how to add a filter to dashboards. If a filter is shown on a sheet, then bringing that sheet onto the dashboard will bring the filter as well, but if the filter is added to the sheet after the dashboard has been configured, then the filter must be added manually. Similarly, if a field that is not currently a filter on a sheet is needed as a filter on the dashboard, then the filter can be added manually. It can also be useful for filters to act on all or multiple sheets on a dashboard. Here, you will apply a filter to multiple sheets. Follow these steps:

1. On the same dashboard created in previous exercises in the chapter, select the **Sub-Category Sales** sheet.

2. Use the **More Options** dropdown in the top right then select **Filters** followed by **Category** to add a category filter to the dashboard.

3. The **Category** filter has been added in its own container to the right-hand side of the dashboard.

4. Move the floating button and legend objects on the dashboard so that they fit more suitably and do not cover the filter section.

5. Test the filter by selecting **Furniture** only. Observe that only the Sub-Category Sales sheet is filtered.

6. Select the **More Options** dropdown in the top right of the **Category** filter object.

7. Select **Apply to Worksheets** followed by **Selected Worksheets** to open the apply filter to worksheets configuration popup.

8. Press **Select all on dashboard** to apply the filter to the rest of the sheets on the dashboard then press **OK**.

9. Test the filter again and observe that it now filters all charts on the dashboard.

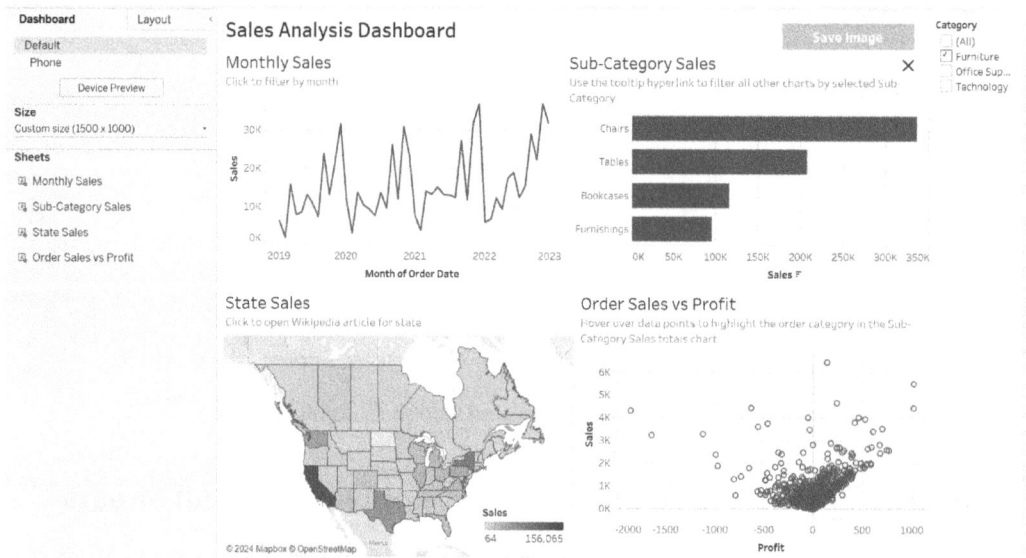

Figure 6.30: Filter added to the dashboard and applied to all sheets on the dashboard

In this exercise, you added a filter to the dashboard and configured it to filter all charts on the dashboard. Filters are a great way to enable users to explore the data on a dashboard themselves and you now know how to add and configure them.

Sheet Swapping

In Tableau, you can show a sheet on a dashboard only when that sheet is selected in a parameter. This functionality allows end users to flip between multiple views and can often be useful when there is not enough space to show all analysis on a single page or when an insight can be shown in multiple different ways.

To create a sheet swapper on a Tableau dashboard, a string parameter must first be created with the values set as the name of the sheets to be swapped in and out. An **All** or similar value can also be included to allow for the possibility of showing multiple sheets at once. Once the parameter has been created it needs to be referenced in a calculated field so that the string value is returned as a field that can be filtered on. This can be done by simply creating a calculated field and typing the parameter name in the configuration.

To apply the sheet swapper filter to the relevant sheets the calculated field that references the parameter needs to be placed on the filter shelf for each sheet that can be swapped in or out. The name of the sheet specified in the parameter should be selected as the only value accepted for that sheet. This means that when the parameter is set to that value, the sheet is not filtered, but when the parameter is set to another sheet name the sheet is filtered out. If an **All** or similar value is included in the parameter, this needs to also be included in the filter for every sheet. The easiest way to configure the filters for each sheet is to use the **Custom Value List** tab and type out the desired parameter values that should result in the sheet being displayed.

The sheets can now be placed on a dashboard and must be placed in a container. The title for each sheet must be hidden as this is not filtered out based on the parameter selection. Similarly, sheet width or height must not be fixed as this will prevent the sheet from collapsing out of view when it is filtered.

Creating a Sheet Swapper and Applying it to the Dashboard

In this exercise, you will add sheet-swapping functionality to the dashboard. Sheet swapping allows for multiple types of analysis to be shown on the same dashboard, depending on users' preferences, and can be a useful methodology for decluttering a dashboard. Follow these steps:

1. On the existing dashboard, there is a **Show/Hide** button that hides the **Sub-Category Sales** sheet, resulting in **Monthly Sales** taking up the top row of the dashboard. However, there is no functionality to hide the **Monthly Sales** and have the **Sub-Category Sales** take up the top row. To add this functionality, a sheet swapper can be created.

2. First, remove the **Show/Hide** button from the dashboard as this will no longer be needed. This can be done by pressing the **x** in the top right when the button is selected.

3. Start by navigating to the **Sub-Category Sales** sheet and creating a parameter by using the dropdown carrot in the top right of the data pane and selecting **Create Parameter**.

4. The parameter should be a string parameter called **Sheet Selector** and the allowable values should be set to **List**.

5. Type the following allowable values: **All, Line**, and **Bar**, and then press **OK**.

6. Create a calculated field called **Sheet Selector Filter** and type [**Sheet Selector**] into the configuration pane then press **OK**.

7. Drag the newly created **Sheet Selector Filter** field onto the filters shelf.

8. In the filter configuration window, select **Custom Value List**, then type All into the bar, and press + at the right end of the bar to add **All** as an allowable value that will show the sheet.

9. Repeat the previous step, but instead, type Bar and then press **OK**. The filter is now configured so that when the parameter is set to either **All** or **Bar**, the **Sub-Category Sales** sheet will display.

10. Navigate to the **Monthly Sales** sheet and drag the **Sheet Selector Filter** onto the filters shelf.

11. Repeat the previous steps using the custom value list but add the **All** and **Line** values instead.

12. The parameter selections, **All** and **Line**, will now display the **Monthly Sales** chart. This means that **All** will display both charts whereas either of the other selections will display the corresponding charts only.

13. Navigate back to the dashboard. The **Monthly Sales** and **Sub-Category Sales** sheets are already in a container so one does not need to be created. Neither chart has a fixed width so that setting does not need to be removed.

14. Both charts have titles that will prevent the sheet from collapsing when filtered. Remove the titles by right-clicking them and selecting **Hide Title**.

15. Select the **Sub-Category Sales** sheet and, using the drop-down menu in the top right, select **Parameters** followed by **Sheet Selector**.

16. The sheet selector parameter has been added to the same container as the **Category** filter. Switch the parameter value using the dropdown and observe how sheets are swapped in and out.

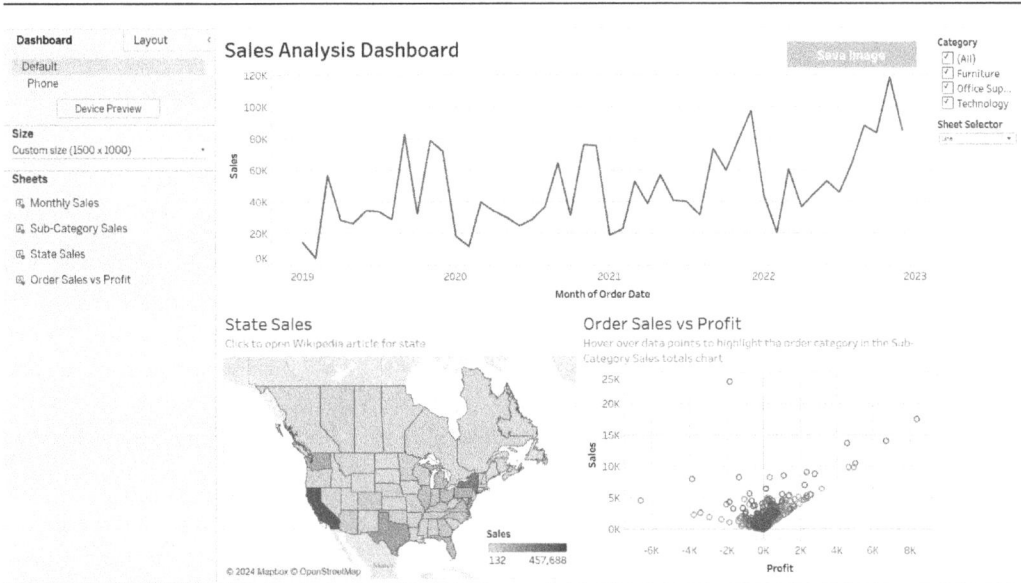

Figure 6.31: Sheet selector/swapper created and applied to the dashboard

You now know the steps required to create and customize a sheet selector in Tableau. In this exercise, the sheet selector was limited to two sheets, but as many as needed can be added.

Dashboard Best Practices

When designing a dashboard in Tableau, it is important to consider some best practices to ensure that the dashboard stays performant and so that it is best optimized for end users.

As a first step, when designing a dashboard, it is important to consider the end user. What business question is the end user trying to answer when utilizing the dashboard is a good thing to constantly consider when deciding what to include in a dashboard and where. It is also important to consider the end user's skill level and base knowledge when it comes to the data. If users are new to Tableau, then the dashboard should be made simpler; if the user is familiar with the data, then more advanced analysis can be included.

Before building a dashboard, always consider the layout. Users usually tend to start viewing a dashboard from the top left and move across then down, similar to how one would read a book. With this in mind, it often makes sense to include the title in the top left as well as the most important chart. This will allow the dashboard's context to be provided along with the key analysis. It may also make sense to start simpler in the top left and gradually become more complex in the analysis moving down to the bottom right – this will provide the end user with an experience of progression through the analysis.

It is important to consider the end user's screen size as well. Device layouts should be created where required and if end users' screen sizes are known, then the dashboard should be fixed to this size. A range of screen sizes can also be configured.

Too many views in a dashboard can make it look cluttered, be complex to understand, and degrade performance. As few views as are required to convey the necessary message should be included on a dashboard with a general rule limiting views to between two and four.

Finally, try to include interactivity in the dashboard to encourage users to explore the data themselves. This can come in the form of filters that allow users to ask questions about the data, such as "What do product sales look like in the UK?" It can also come in the form of dashboard actions, such as highlighting or filtering, that show the connection between data across multiple views, allowing for greater insight to be drawn. Guiding sentences are always a great addition to inform users of how to interact with a dashboard.

To summarize, it is always important when designing a dashboard to think about what the goal of the dashboard is with regard to the end user. There are multiple things to consider to ensure a great user experience in a Tableau dashboard, but as long as the end user is considered throughout the development, these factors should be taken care of naturally.

Stories

Tableau facilitates data-driven storytelling through sequential visualizations, both dashboards and sheets, with its Stories feature. Stories can be added the same way as sheets and dashboards using the rightmost tab option in the tab toolbar along the bottom. There is also a **Story** option on the menu toolbar from which **New Story** can be selected.

Stories allow users to create a narrative by combining sheets and dashboards into a presentation. The sheets and dashboards can be interacted with by end users in the same way as when they are accessed individually, but stories allow the developer to combine them into a sequence with each sheet/dashboard being referred to as a **story point**.

The **Story** configuration is similar to the dashboard configuration with a **Story** pane on the left-hand side that also has a **Layout** tab and the canvas in the middle. Stories differ in that they have story points, which are sequential steps in the presentation. Only a single sheet or dashboard can be placed on the canvas for any given story point. Sheets and dashboards are listed on the story pane and can be dragged onto the canvas.

New story points can be added using the **New Story Point** section on the story pane with **Blank** adding a new story point and **Duplicate** duplicating the existing story point. Textboxes can also be dragged onto the canvas, which float on top of the sheet/dashboard and can be used as annotations to enhance the narrative.

At the top of the story is the story title, which is consistent across all story points. The title can be turned on or off at the bottom of the story pane and the size of the canvas can also be configured. The canvas needs to be big enough to match the size of any dashboard used as they will not be resized to fit the story canvas.

Below the title on the story canvas are the story points. These are gray buttons by default that the user can click through to go through the story point by point. The default setting is gray rectangles called **caption boxes**, which have text that can be modified to give a brief description of the story point. The story point navigator styles can be modified in the **Layout** tab on the story pane. The caption boxes can be replaced by numbers, dots, or just arrows. There are arrows to the right- and left-hand side of the caption boxes, and numbers and dots that move the story point forward or backward one step when pressed. These can be configured on or off using the **Show arrows** checkbox.

The same sheet or dashboard can be used on different story points with differing filter configurations.

Creating a Story

In this exercise, you will use Tableau's story functionality to create a narrative analysis. Stories are great for presentations and taking users through a journey of analysis. Follow these steps:

1. Using the same workbook and dashboard created for the chapter so far, create a story by selecting **Story** on the menu toolbar followed by **New Story**.

2. Set the title of the story to `Sales Analysis Story` by double-clicking the title textbox and updating the text.

3. Drag the dashboard created in previous exercises onto the story and resize the story canvas using the custom size option at the bottom of the story pane, so that the whole dashboard is visible.

4. Type `Sales Analysis Dashboard` into the caption box. A story has now been created with the dashboard present. This is a high-level analysis but to tell a story, additional story points can be added that drill in the detail.

5. Duplicate the current story point by selecting **Duplicate** in the **New story point** section of the story pane.

6. Rename the story point `Furniture` and filter the dashboard of **Furniture** using the **Category** filter.

7. Repeat the previous steps creating story points for **Office Supplies and Technology** and update the filter accordingly.

8. There is now a story that starts at a high level showing the whole of the data and then drills the dashboard down into each **Category** level.

9. To finish the story on a more detailed piece of analysis, add a new **Blank** story point and call it `Order Analysis`.

10. Drag the **Order Sales versus Profit** sheet onto the story canvas.

11. Drag a textbox from the story pane onto the canvas and type. Hover over individual points to see the total profit and sales or individual orders in the configuration window and then press **OK**.

12. The textbox annotation now informs the user how to interact with this final piece of more detailed analysis at the end of the story. Drag the textbox to a suitable location, ensuring that none of the data points are covered.

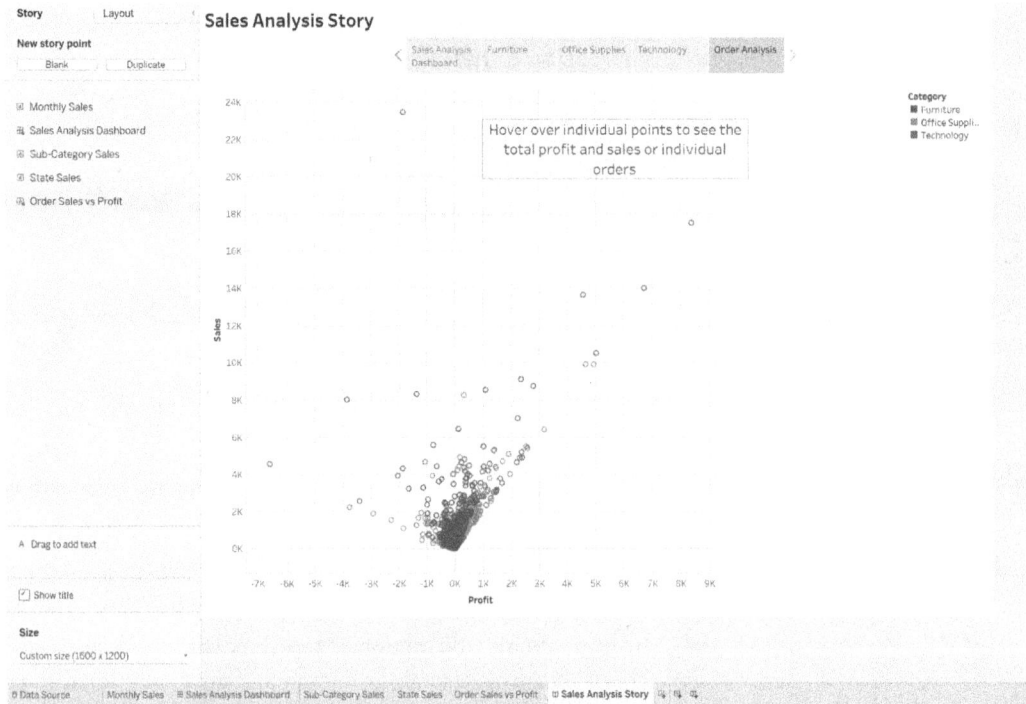

Figure 6.32: The Tableau story created that takes the user from a high-level overview through a more detailed analysis and ends with a low-level piece of analysis and some user instruction

In this exercise, you created a story that starts at a high level and walks through different steps of the analysis, providing new insights with each step. You learned how to create and customize a story with dashboards, sheets, and annotations.

Summary

In Tableau, multiple individual visualizations can be consolidated into a more advanced piece of analysis called a dashboard. Dashboards can consist of a variety of different objects, the main one being charts created on sheets. Charts can be made to interact with each other and how the user interacts with a dashboard is customizable by the developer. In addition to cross-sheet interactivity, filters can be included that apply to multiple sheets and sheets can be swapped in and out of the dashboard. Dashboards and sheets can be combined into a narrative using the Tableau stories feature.

When it comes to dashboard creation, there is a variety of objects in addition to charts that can be brought onto the canvas. These range in function from decorative to functional. How objects sit on a dashboard can also vary between tiled and floating, which each offers different benefits and drawbacks. When designing the dashboard, the canvas size can also be configured, and multiple layouts can be created for varying devices.

There are many ways to add interactivity to dashboards with actions that link charts together, such as highlighting and filtering, as well as actions that link to specific URLs. Drop-down filters can also provide interactivity across multiple worksheets and sheets can also be configured to be swapped in and out based on a parameter selection.

Stories allow users to connect dashboards and worksheets into a single cohesive narrative with annotated points and fixed pieces of analysis throughout.

The next chapter will cover formatting in Tableau dashboards.

Exam Readiness Drill – Chapter Review Questions

Apart from a solid understanding of key concepts, being able to think quickly under time pressure is a skill that will help you ace your certification exam. That is why working on these skills early on in your learning journey is key.

Chapter review questions are designed to improve your test-taking skills progressively with each chapter you learn and review your understanding of key concepts in the chapter at the same time. You'll find these at the end of each chapter.

> **How to Access these Resources**
>
> To learn how to access these resources, head over to the chapter titled *Chapter 9, Accessing the Online Practice Resources*.

To open the Chapter Review Questions for this chapter, perform the following steps:

1. Click the link – `https://packt.link/TDA_CH06`.

    Alternatively, you can scan the following **QR code** (*Figure 6.33*):

Figure 6.33: QR code that opens Chapter Review Questions for logged-in users

2. Once you log in, you'll see a page similar to the one shown in *Figure 6.34*:

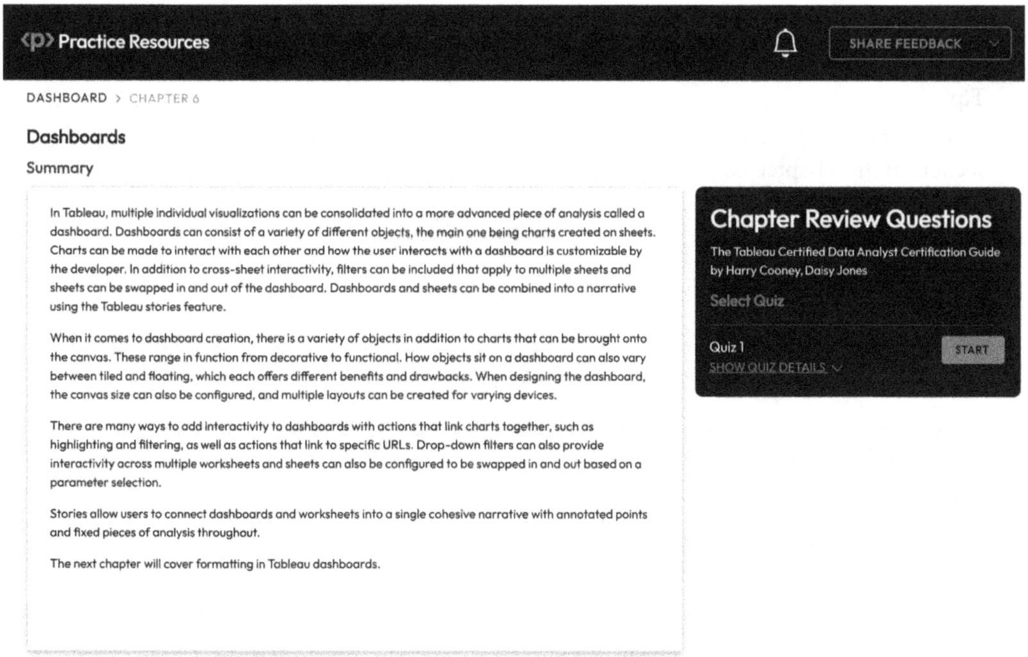
Practice Resources
SHARE FEEDBACK
DASHBOARD > CHAPTER 6
Dashboards
Summary
In Tableau, multiple individual visualizations can be consolidated into a more advanced piece of analysis called a dashboard. Dashboards can consist of a variety of different objects, the main one being charts created on sheets. Charts can be made to interact with each other and how the user interacts with a dashboard is customizable by the developer. In addition to cross-sheet interactivity, filters can be included that apply to multiple sheets and sheets can be swapped in and out of the dashboard. Dashboards and sheets can be combined into a narrative using the Tableau stories feature.
When it comes to dashboard creation, there is a variety of objects in addition to charts that can be brought onto the canvas. These range in function from decorative to functional. How objects sit on a dashboard can also vary between tiled and floating, which each offers different benefits and drawbacks. When designing the dashboard, the canvas size can also be configured, and multiple layouts can be created for varying devices.
There are many ways to add interactivity to dashboards with actions that link charts together, such as highlighting and filtering, as well as actions that link to specific URLs. Drop-down filters can also provide interactivity across multiple worksheets and sheets can also be configured to be swapped in and out based on a parameter selection.
Stories allow users to connect dashboards and worksheets into a single cohesive narrative with annotated points and fixed pieces of analysis throughout.
The next chapter will cover formatting in Tableau dashboards.
Chapter Review Questions
The Tableau Certified Data Analyst Certification Guide by Harry Cooney, Daisy Jones
Select Quiz
Quiz 1
SHOW QUIZ DETAILS
START

Figure 6.34: Chapter Review Questions for Chapter 6

3. Once ready, start the following practice drills, re-attempting the quiz multiple times.

Exam Readiness Drill

For the first three attempts, don't worry about the time limit.

ATTEMPT 1

The first time, aim for at least **40%**. Look at the answers you got wrong and read the relevant sections in the chapter again to fix your learning gaps.

ATTEMPT 2

The second time, aim for at least **60%**. Look at the answers you got wrong and read the relevant sections in the chapter again to fix any remaining learning gaps.

ATTEMPT 3

The third time, aim for at least **75%**. Once you score 75% or more, you start working on your timing.

> **Tip**
>
> You may take more than **three** attempts to reach 75%. That's okay. Just review the relevant sections in the chapter till you get there.

Working On Timing

Target: Your aim is to keep the score the same while trying to answer these questions as quickly as possible. Here's an example of how your next attempts should look like:

Attempt	Score	Time Taken
Attempt 5	77%	21 mins 30 seconds
Attempt 6	78%	18 mins 34 seconds
Attempt 7	76%	14 mins 44 seconds

Table 6.2: Sample timing practice drills on the online platform

> **Note**
>
> The time limits shown in the above table are just examples. Set your own time limits with each attempt based on the time limit of the quiz on the website.

With each new attempt, your score should stay above **75%** while your "time taken" to complete should "decrease". Repeat as many attempts as you want till you feel confident dealing with the time pressure.

7

Formatting

Introduction

This chapter will cover formatting in a Tableau dashboard. As the final element to designing a dashboard, formatting is imperative to make sure that a user can easily digest the information as it is clear and concise. From changing the size and shape of marks to highlighting key components in a chart, formatting can allow a user to quickly identify a crucial point in data for their analysis.

Correct formatting is a best practice in data analytics. This will create an easily digestible report and help drive decisions from data. This chapter will explore how to apply different formatting in a workbook, custom color palettes, shapes, and padding. Formatting is a quarter of domain 3 in the exam, which is worth 26% of the grade, so it is worth understanding how to achieve formatting in preparation for the exam.

Applying Color, Font, Shapes, and Styling

Color, font, shapes, and styling can transform a plain report into a visually appealing dashboard that tells a story about the data presented. This section will delve into each part to explain how they impact a visual and how to apply them to dashboards.

Color

Color can be used to add interest to a report and highlight key indicators to draw a user's eye to the story being told. Colors can be applied in two ways – with categorical and quantitative palettes.

Categorical palettes apply to discrete (typically multi-dimensional) fields in data. This will assign distinct colors to fields that have no particular order, and they can be used throughout the report. The colors can be changed on the **Marks** card drop-down menu or the legend drop-down menu.

Users can then select the field and choose the color they want to assign to that field.

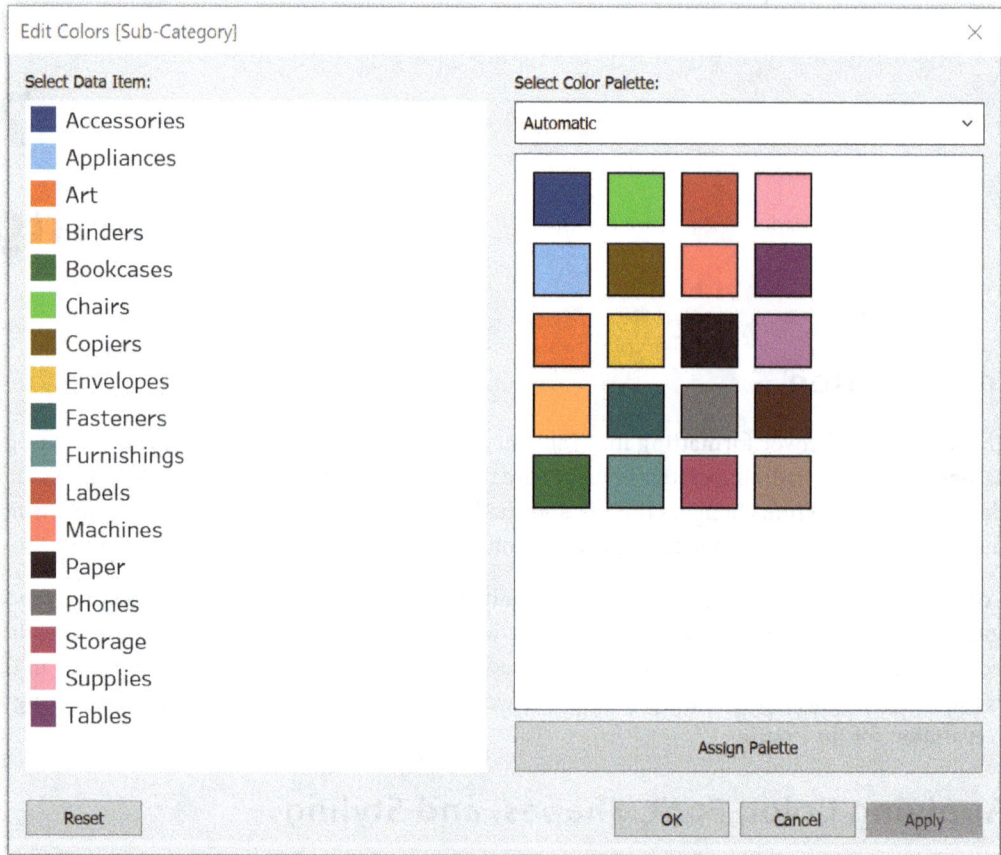

Figure 7.1: An example of a categorical palette

Quantitative palettes apply to continuous fields that will produce a color range, forming a gradient palette. Where negative values are present, two-color ranges will be applied as default, known as a diverging palette. If all values are either positive or negative, then it will return to one color, known as a sequential palette.

Although these palettes will be set as default, a user can update the palette when editing the colors.

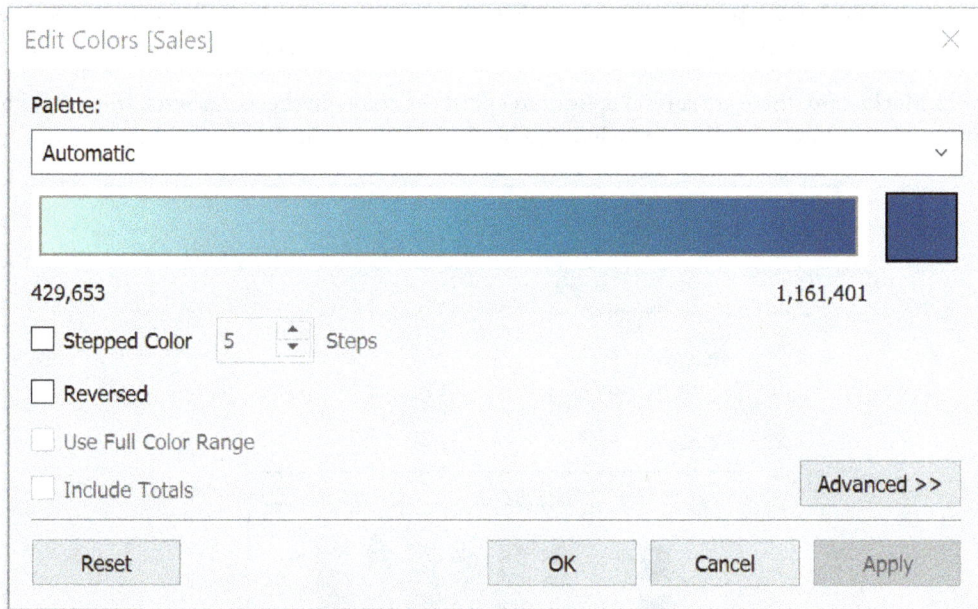

Figure 7.2: An example of a quantitative palette

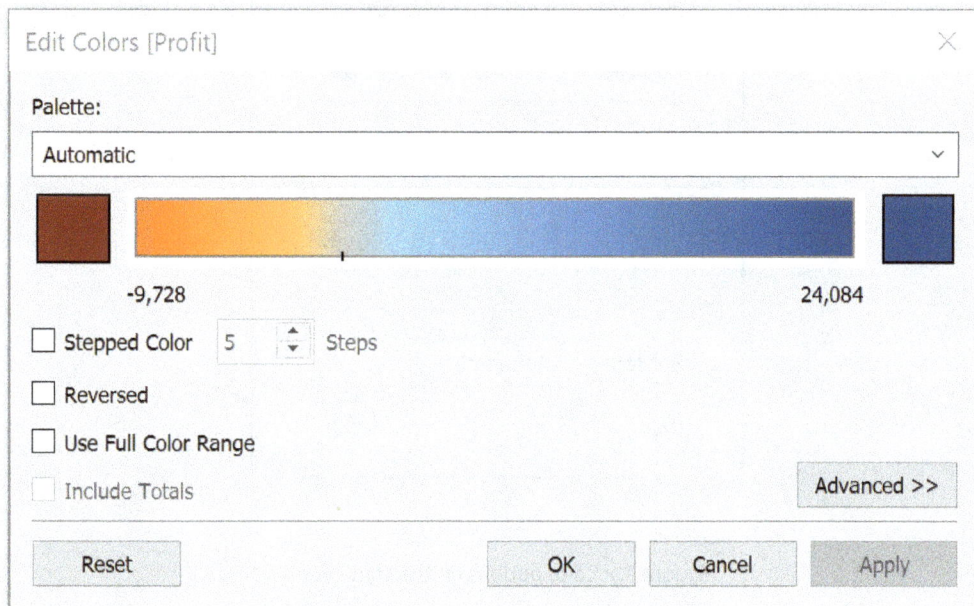

Figure 7.3: A quantitative palette with negative values

Another option that can be applied to color palettes is stepped colors, which allows the developer to control the number and range of color groups to be displayed on a quantitative palette.

On the **Marks** card, there are several options to adjust the colors in the worksheets, from changing between the palettes and opacity to including a border or halo on the data points.

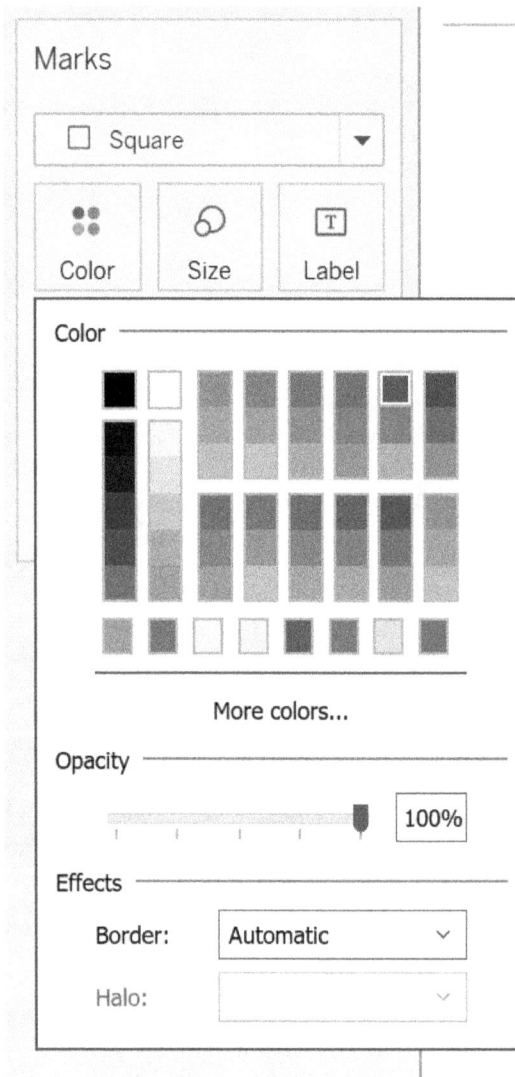

Figure 7.4: Color options on the Marks card

Opacity is a sliding option that will let the developer change the transparency of data charts. This is useful to accentuate the data points and make them stand out in a chart, such as a scatterplot, where there can be multiple points condensed together.

Borders can be applied to help separate touching or close marks. This can be used in conjunction with opacity to highlight the density of a chart.

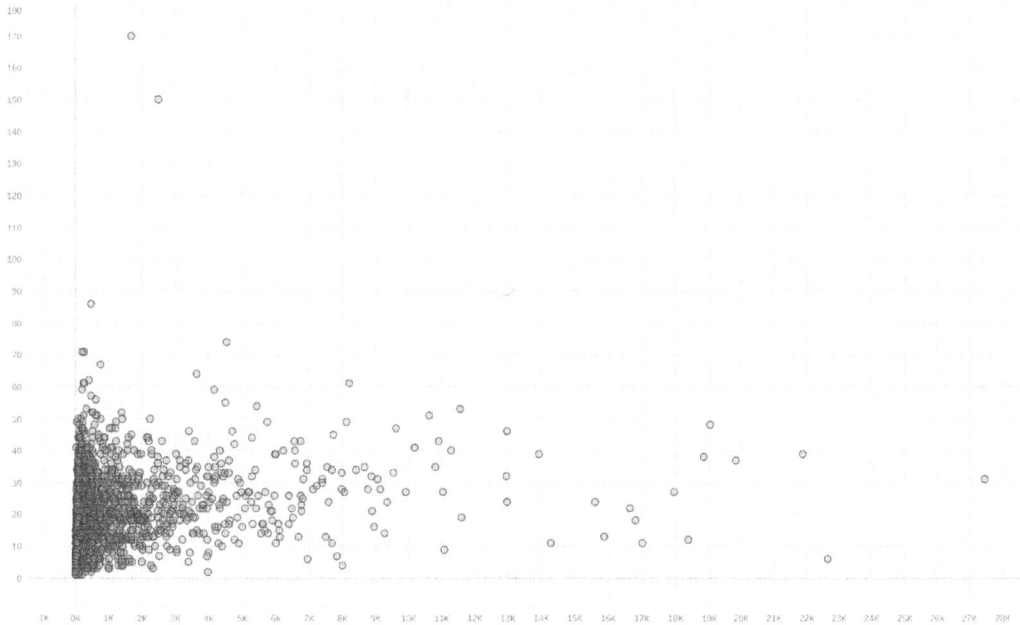

Figure 7.5: A chart with a border

Now take a look at a chart without borders on the data points:

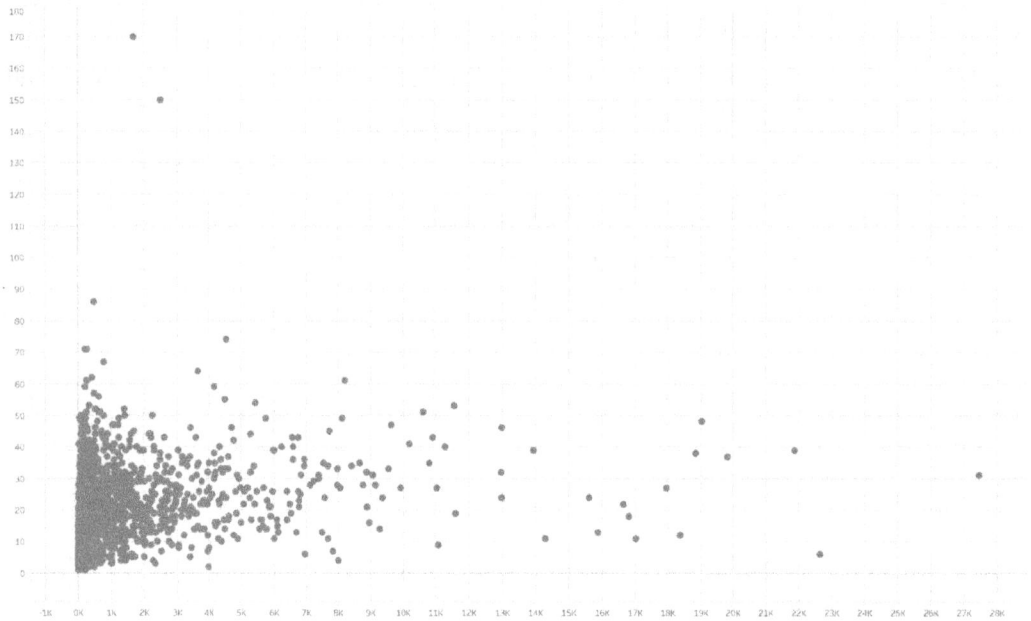

Figure 7.6: A chart without borders on the data points

Halos are available when using a background image or map, where it will halo over a solid color with another color.

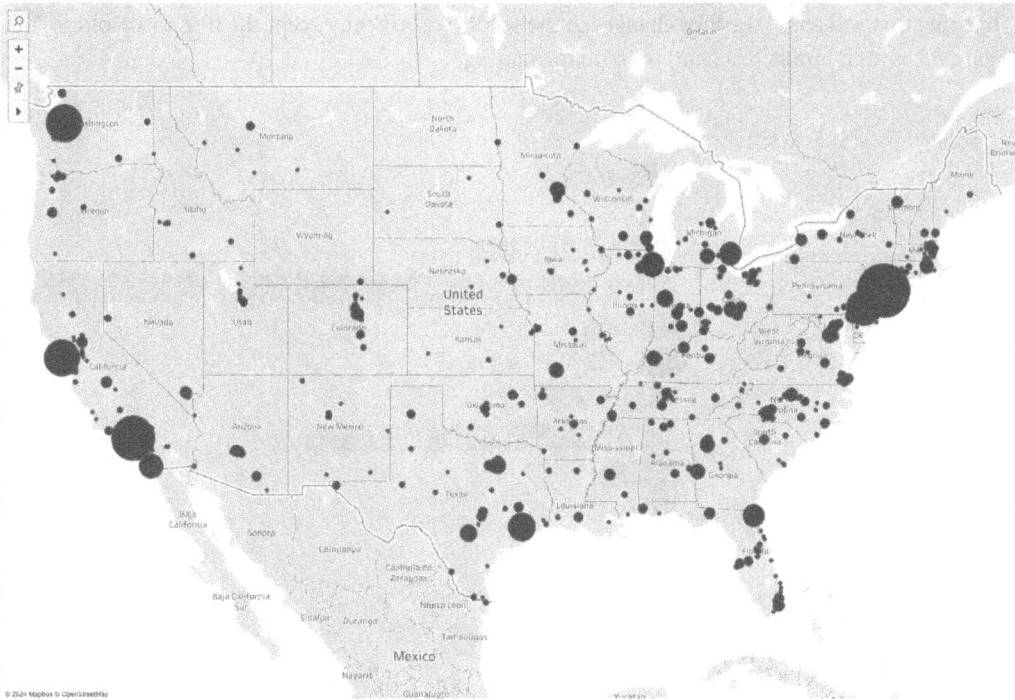

Figure 7.7: A map with halos

Next, you can find out a little about the use of fonts while working with Tableau Desktop.

Font

When working with Tableau Desktop, a developer can update the font being used by going to the **Format** tab at the top and selecting **Font**. The **Datapane** on the right-hand side will be replaced with a new pane that will show the different areas where fonts are used.

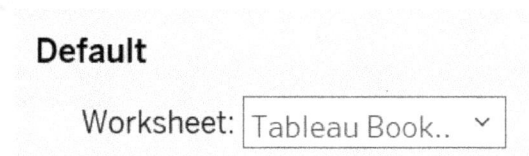

Default

Worksheet: | Tableau Book.. ∨ |

Figure 7.8: The pane showing areas where fonts are used

Changing this will apply the font changes in the whole worksheet, except the title and tooltips. Take *Figure 7.9* as an example of a chart with no formatting.

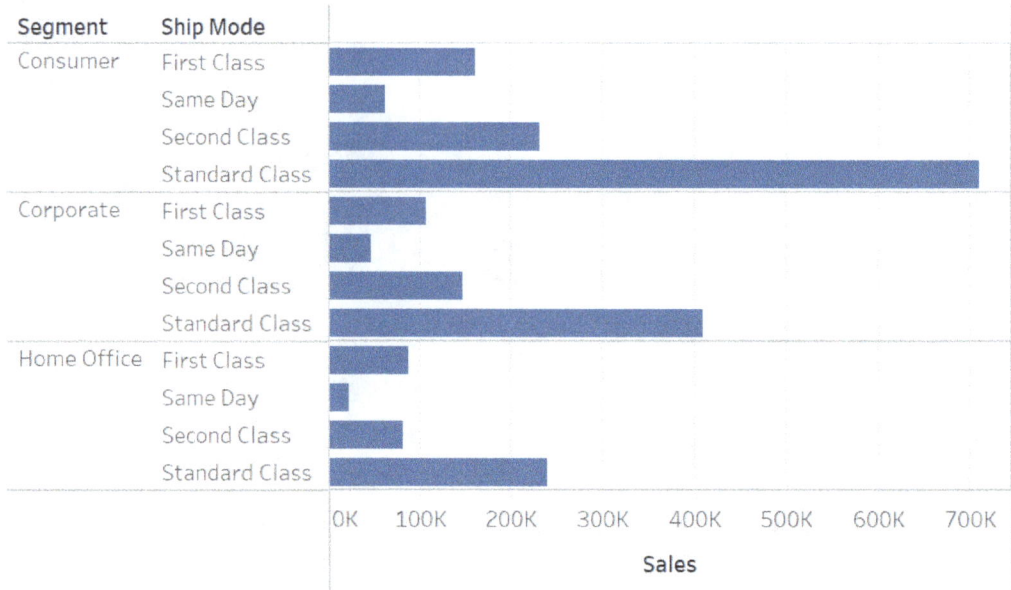

Figure 7.9: A chart with no text formatting

Now, observe how it changes when changes such as color and size are applied, as shown in *Figure 7.10*.

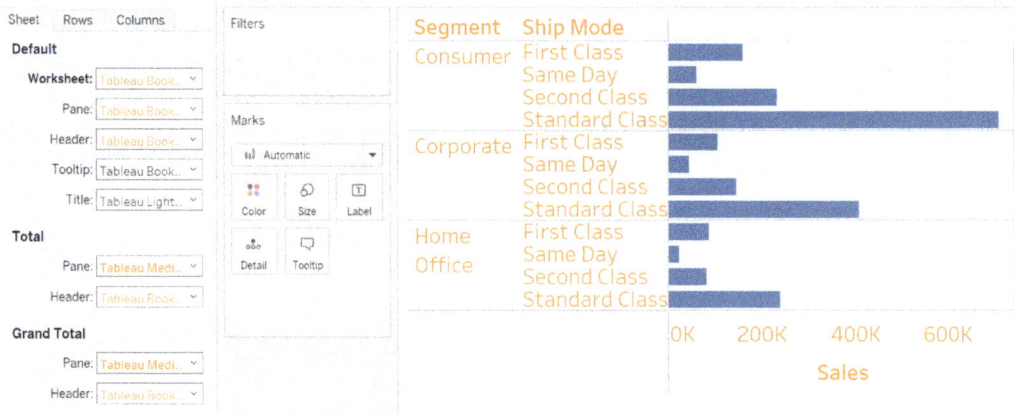

Figure 7.10: A chart with orange text and a larger font size

Note how the other options in the pane are now in the colors selected. This is because the option selected applies to all options, except tooltips and headers/titles.

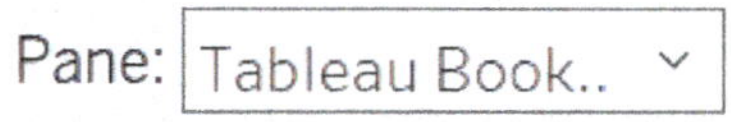

Pane:
Tableau Book..

Figure 7.11: Take a look at the Pane option

The **Pane** option allows you to change the fonts in the visual, such as the text at the end of a bar, as shown in *Figure 7.12*.

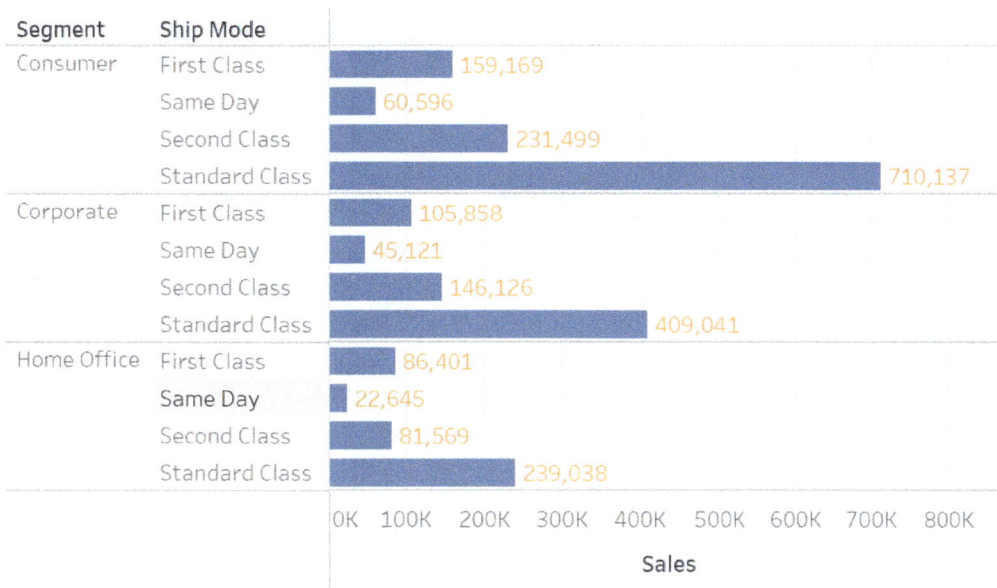

Segment
Ship Mode
Consumer
First Class
159,169
Same Day
60,596
Second Class
231,499
Standard Class
710,137
Corporate
First Class
105,858
Same Day
45,121
Second Class
146,126
Standard Class
409,041
Home Office
First Class
86,401
Same Day
22,645
Second Class
81,569
Standard Class
239,038
0K 100K 200K 300K 400K 500K 600K 700K 800K
Sales

Figure 7.12: A color and size change to the text in the chart

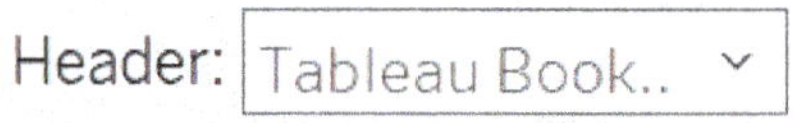

Header:
Tableau Book..

Figure 7.13: Selecting the header

The **Headers** option impacts the headers in a visual, typically the dimensions. This will essentially change the axes headers on the edges of the chart.

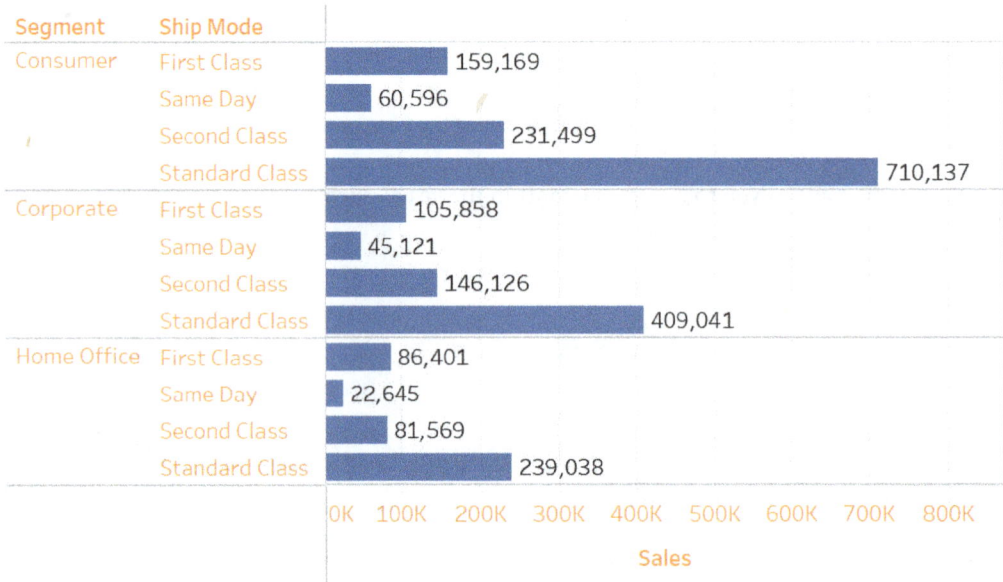

Segment	Ship Mode	
Consumer	First Class	159,169
	Same Day	60,596
	Second Class	231,499
	Standard Class	710,137
Corporate	First Class	105,858
	Same Day	45,121
	Second Class	146,126
	Standard Class	409,041
Home Office	First Class	86,401
	Same Day	22,645
	Second Class	81,569
	Standard Class	239,038

0K 100K 200K 300K 400K 500K 600K 700K 800K

Sales

Figure 7.14: Color changes in the headers

Tooltip: | Tableau Book.. ∨ |

Figure 7.15: Changing the tooltips

Tooltip will impact the tooltips on the visuals. This will change the default color of the value that is set for tooltips when building a chart. If a developer needs more flexibility to change the format of the tooltip, this can be done in the **Marks** card.

Segment	Ship Mode	
Consumer	First Class	159,169
	Same Day	60,596
	Second Class	231,499
	Standard Class	710,137
Corporate	First Class	105,858
	Same Day	45,121
	Second Class	146,126
	Standard Class	409,041
Home Office	First Class	
	Same Day	2.
	Second Class	
	Standard Class	

Segment: Corporate
Ship Mode: Second Class
Sales: 146,126

0K   100K   200K   300K   400K   500K   600K   700K   800K

Sales

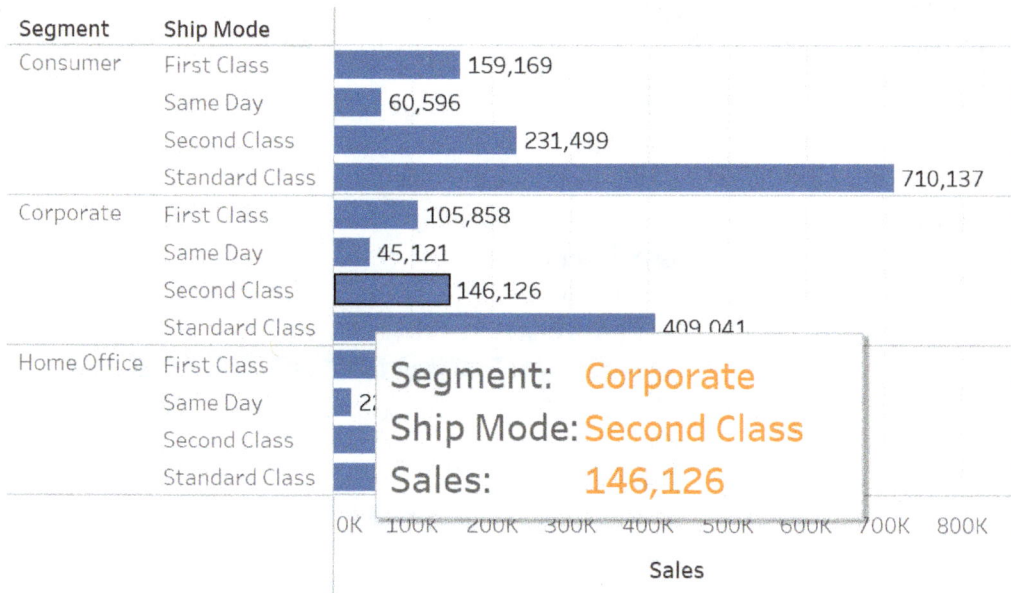

Figure 7.16: A tooltip color change

Title: | Tableau Light.. ⌄ |

Figure 7.17: Updating the title of a worksheet

Title will allow the developer to update the title of a worksheet, and the whole title will change. If a user wanted more options for changing the title color and font, they would have to apply this in the **Title** text editor.

Ship Mode by Segment

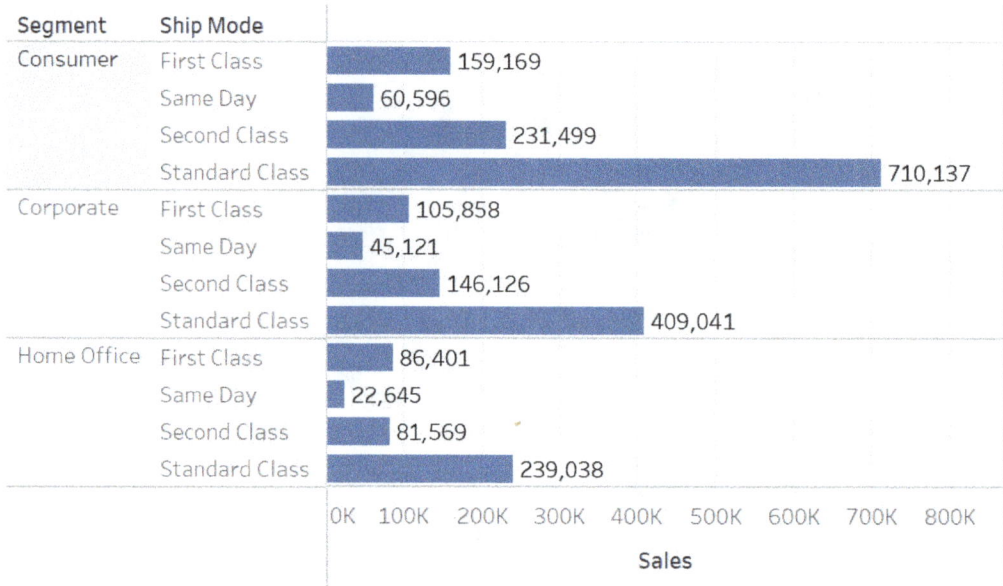

Segment	Ship Mode	
Consumer	First Class	159,169
	Same Day	60,596
	Second Class	231,499
	Standard Class	710,137
Corporate	First Class	105,858
	Same Day	45,121
	Second Class	146,126
	Standard Class	409,041
Home Office	First Class	86,401
	Same Day	22,645
	Second Class	81,569
	Standard Class	239,038

0K   100K   200K   300K   400K   500K   600K   700K   800K

Sales

Figure 7.18: A title change

Total

Pane: | Tableau Medi.. ⌄ |

Figure 7.19: The Total pane

This applies to a chart that includes a **total**, created by using the **Analytics** tab in the Tableau worksheet. This option will change the font of the text displaying the total.

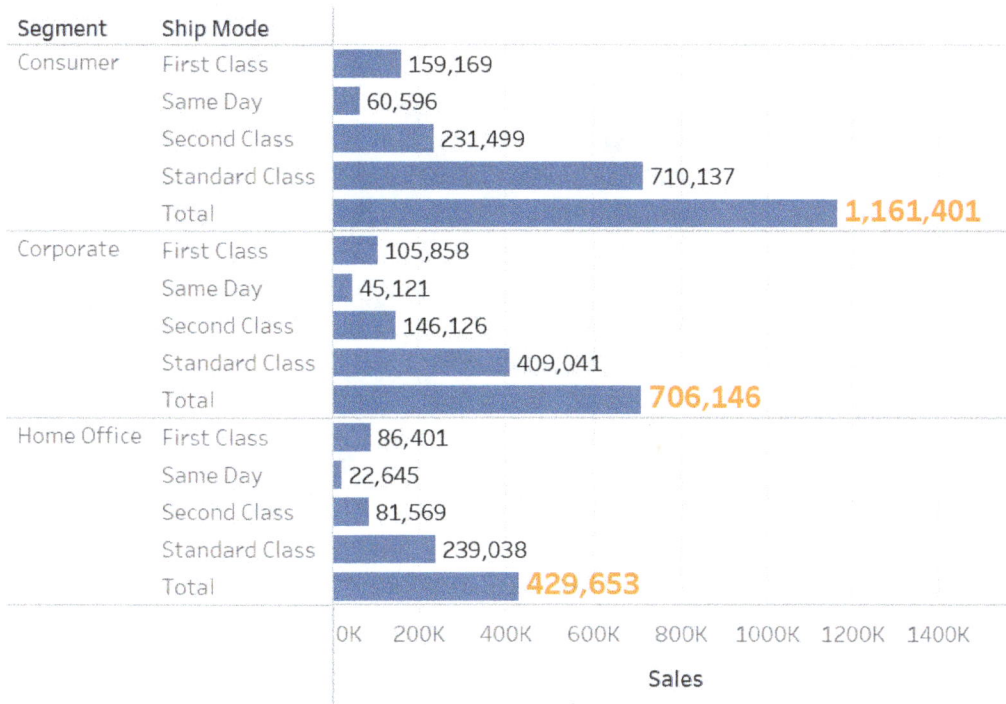

Segment	Ship Mode	Sales
Consumer	First Class	159,169
	Same Day	60,596
	Second Class	231,499
	Standard Class	710,137
	Total	1,161,401
Corporate	First Class	105,858
	Same Day	45,121
	Second Class	146,126
	Standard Class	409,041
	Total	706,146
Home Office	First Class	86,401
	Same Day	22,645
	Second Class	81,569
	Standard Class	239,038
	Total	429,653

Figure 7.20: Total changes

Total

Pane: Tableau Medi.. ⌄

Header: Tableau Book.. ⌄

Figure 7.21: Making changes to Pane and Header

As shown in the preceding figure, this will change the **Total** section, but for **Header**, it will only update the header and not the value.

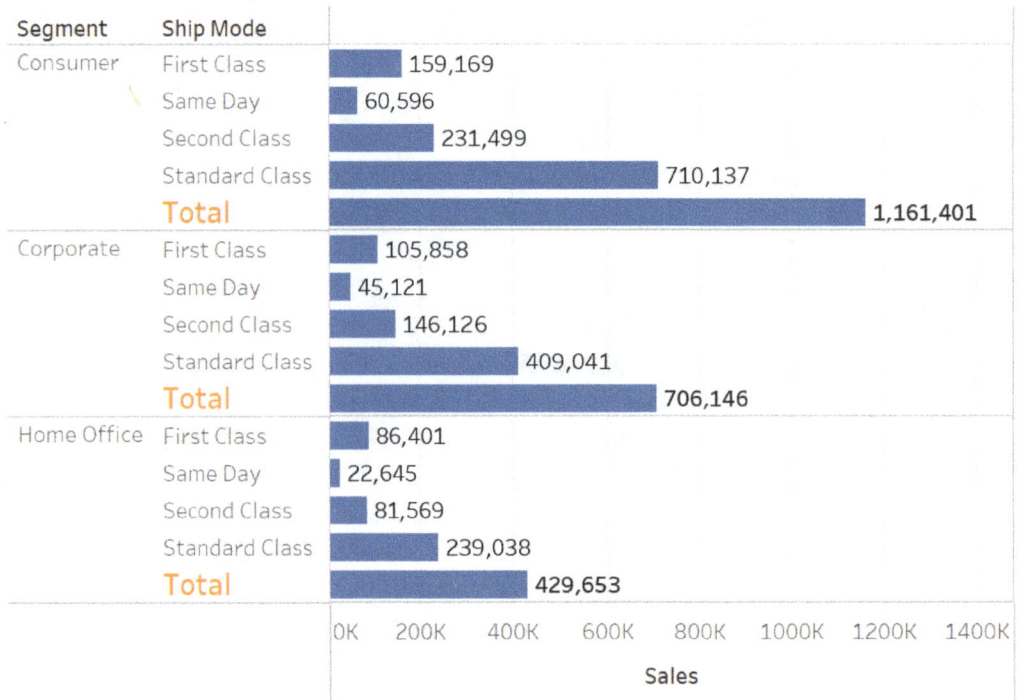

Segment	Ship Mode	Sales
Consumer	First Class	159,169
	Same Day	60,596
	Second Class	231,499
	Standard Class	710,137
	Total	1,161,401
Corporate	First Class	105,858
	Same Day	45,121
	Second Class	146,126
	Standard Class	409,041
	Total	706,146
Home Office	First Class	86,401
	Same Day	22,645
	Second Class	81,569
	Standard Class	239,038
	Total	429,653

Figure 7.22: Total header updates

Grand Total

Pane: Tableau Medi.. ⌄

Header: Tableau Book.. ⌄

Figure 7.23: Using Grand Total

Using **Grand Total** is exactly the same as with **Total**, but it only applies to the grand total.

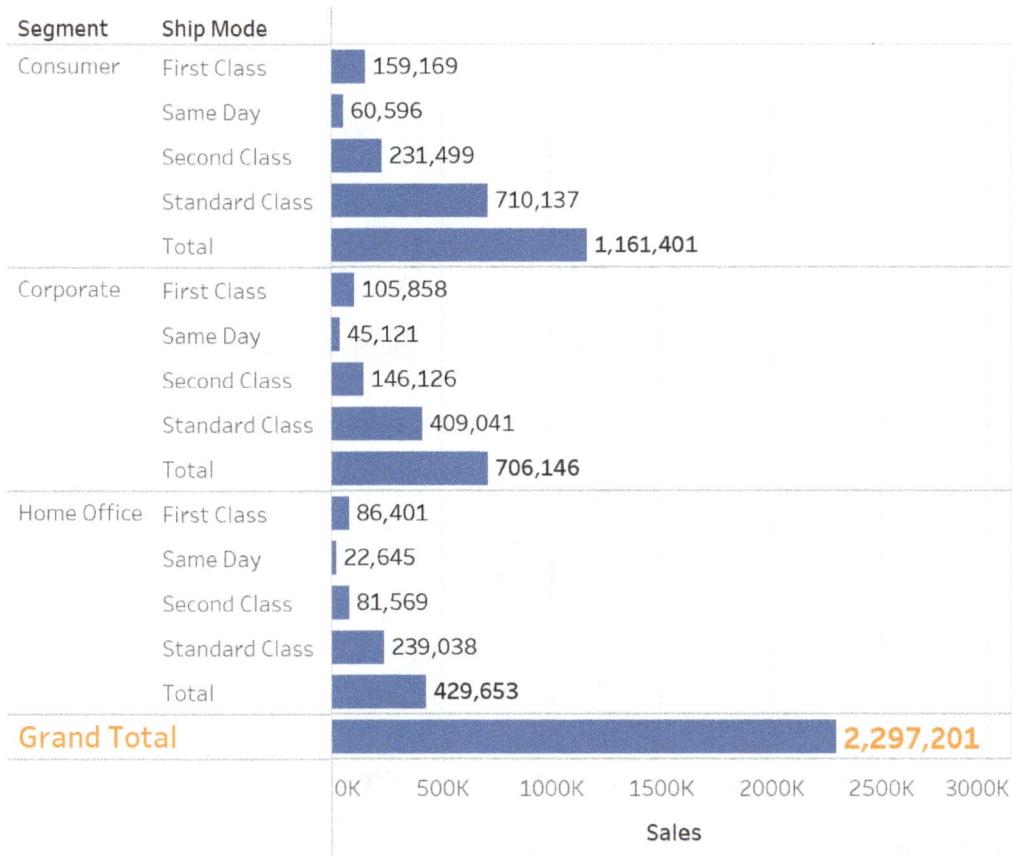

Segment	Ship Mode	Sales
Consumer	First Class	159,169
	Same Day	60,596
	Second Class	231,499
	Standard Class	710,137
	Total	1,161,401
Corporate	First Class	105,858
	Same Day	45,121
	Second Class	146,126
	Standard Class	409,041
	Total	706,146
Home Office	First Class	86,401
	Same Day	22,645
	Second Class	81,569
	Standard Class	239,038
	Total	429,653
Grand Total		2,297,201

Figure 7.24: A Grand Total update

Alignment

In most cases, Tableau is able to define what would be best in terms of alignment, but this can be adapted by the developer. Go to **Format | Alignment**, and the **Data pane** will change to allow changes to the alignment of the worksheet.

As shown previously, this is broken into three sections – **Default**, **Totals**, and **Grand Totals**. There are four ways to adjust alignment:

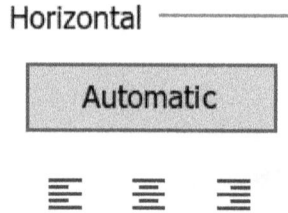

Figure 7.25: Changing alignment using Horizontal

Horizontal will change where the text sits to either the left, right, or center.

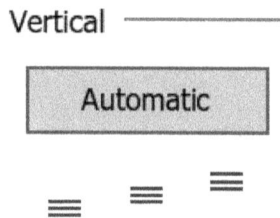

Figure 7.26: Changing alignment using Vertical

Vertical will change where the text sits on a vertical axis to the top, middle, or bottom.

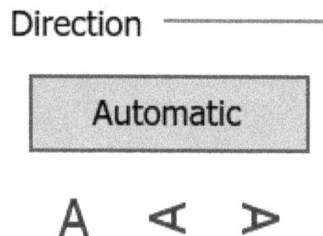

Figure 7.27: Changing alignment using Direction

Direction will rotate the text in the direction visualized in the options.

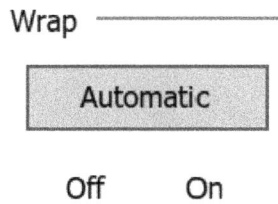

Wrap ————————————

Automatic

Off   On

Figure 7.28: Changing alignment using Wrap

Wrap will allow the text to wrap to the next line or display abbreviated.

Tooltips

As mentioned previously, tooltips can have their font adjusted, but this can be limited to only the value part of a tooltip. A tooltip can be formatted either by the **Marks** card or in the **Worksheet** tab at the top.

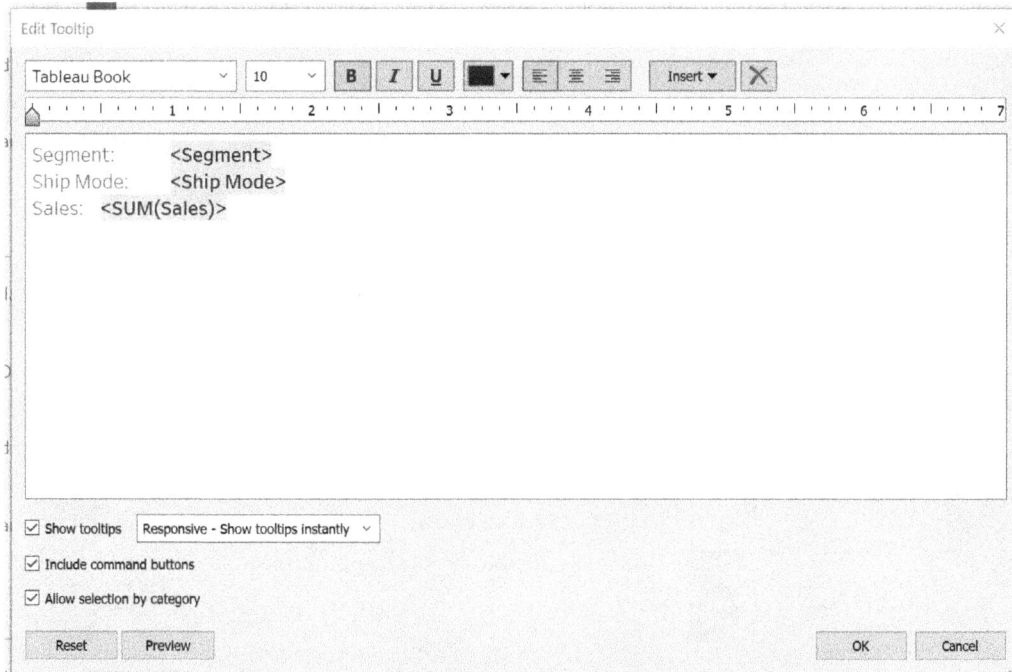

Figure 7.29: The tooltip editor

Here, the developer has the flexibility to edit the tooltip completely, such as changing the size of certain points or even inserting additional fields in the tooltip to provide more information about a data point. Check out the example in *Figure 7.30* to see how a tooltip can be reinvented.

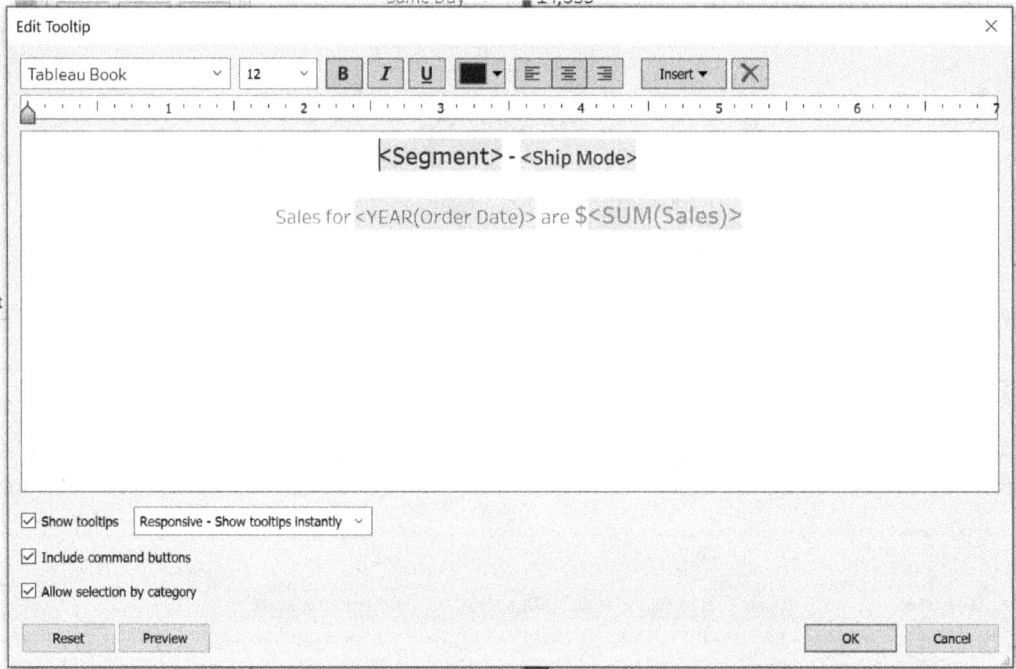

Figure 7.30: A tooltip example

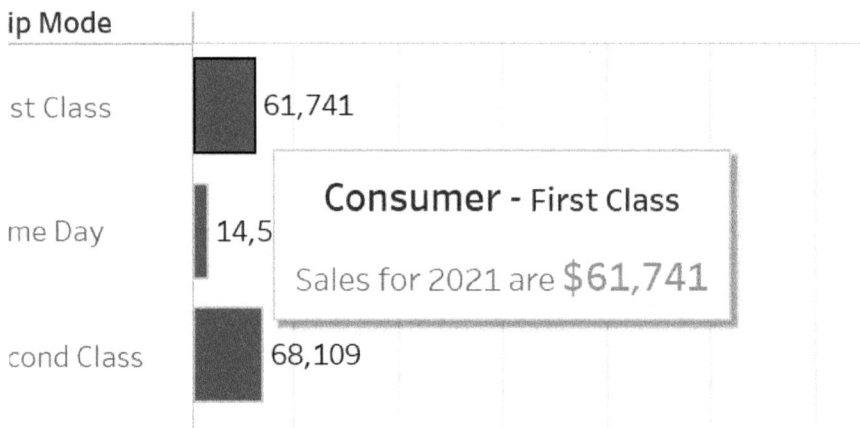

Figure 7.31: The tooltip example result

Shapes

One of the options in the **Marks** card is **Shape**, which means a developer can choose a shape to represent a category within a data point or multiple points in a chart. This can help identify items as belonging to certain categories, which an end user can use to spot any trends in data.

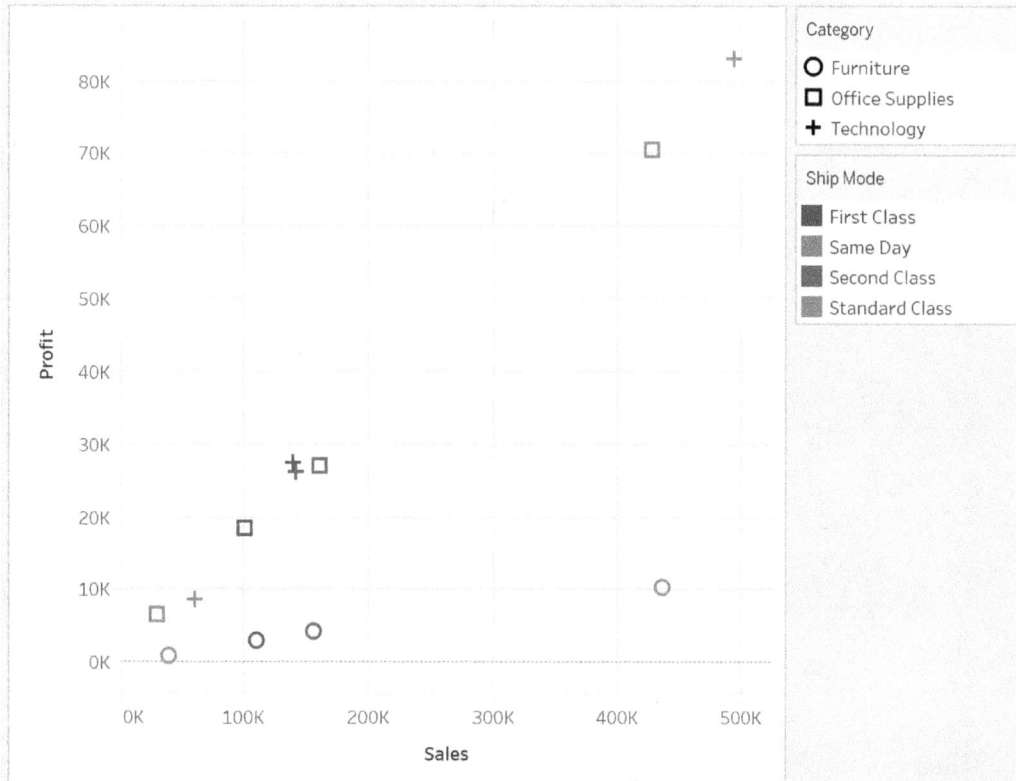

Figure 7.32: Shapes applied to a scatter graph

Dragging in a dimensional field to the **Shape** shelf will assign a unique shape per category. If a measure is used, this will convert the measure to a discrete measure.

Tableau, by default, has a variety of unique shapes available to use. These will be automatically assigned once the user places a field onto **Shape**. Editing these shapes is as simple as selecting the **Shape** tab in the **Marks** card and choosing between the provided palettes.

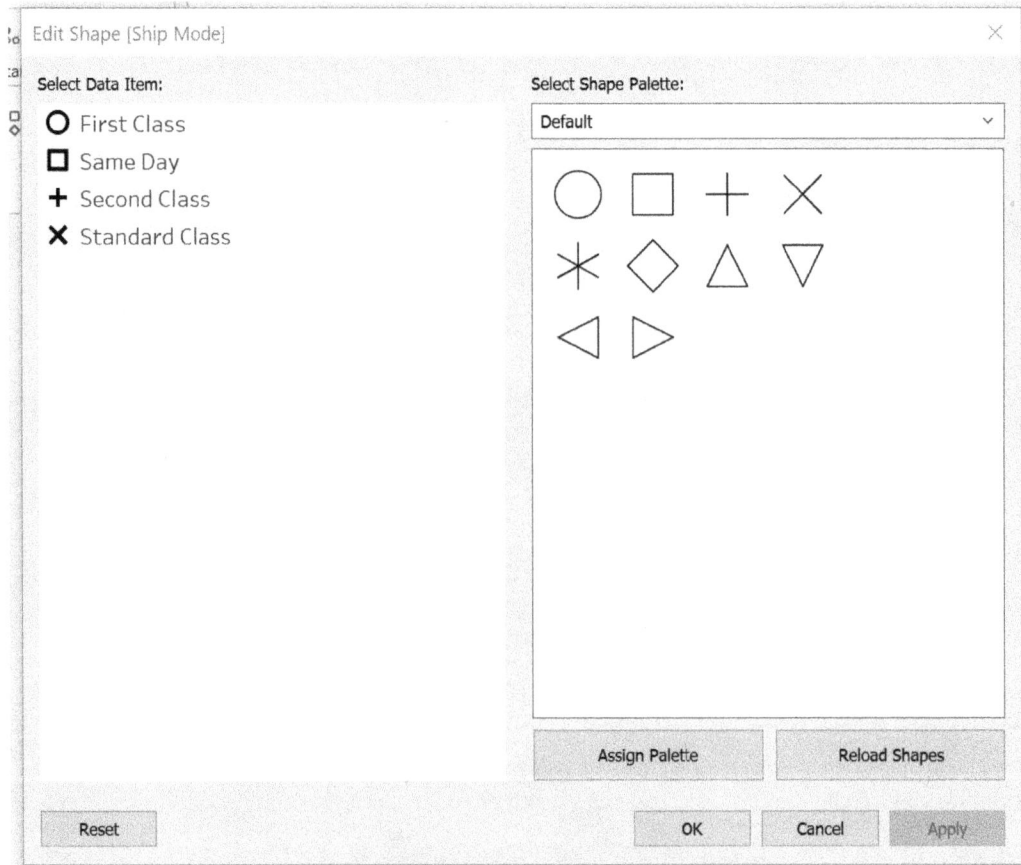

Figure 7.33: The shape editor in Tableau

The developer can manually assign each item a shape or simply allow Tableau to assign a palette that will apply to the items. If there are more items than what the palette offers, there will be duplicates if the developer allows it.

Adding Custom Shapes and Color Palettes

Although Tableau offers a great number of unique shapes and colors, a developer can include their own styles in a report to match a company color palette or simply for a personalized touch.

To include custom shapes, a developer will need to navigate to the Tableau repository to the Documents folder. When the developer uses these custom shapes, they will be saved in a workbook so that if the workbook is shared, the recipients are still able to view the report with the custom shapes.

Before diving into creating custom shapes, a location for them needs to be created. This can be simply achieved by creating a new folder in the Shapes folder in the Tableau Repository.

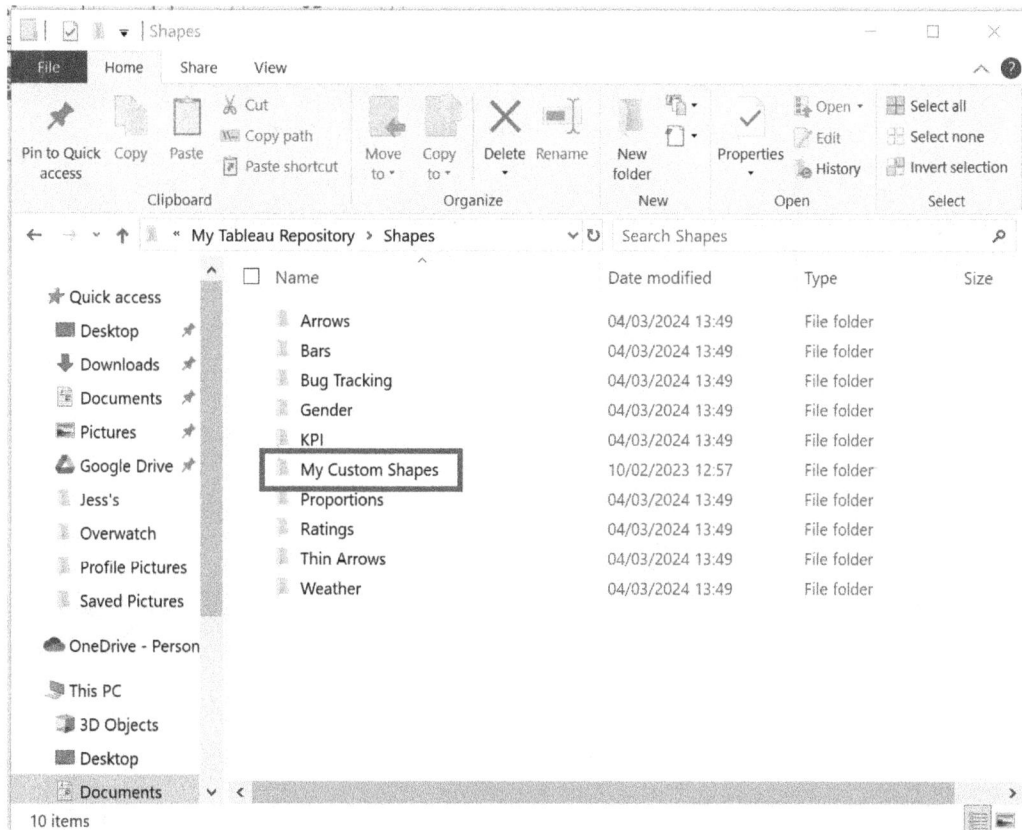

Figure 7.34: The file directory of the custom shapes folder

Once the shapes have been saved in the folder, go to Tableau and open **Edit Shape** in the **Marks** card. The new palette of custom shapes should be available for use. If they are not present straight away, click **Reload Shapes**.

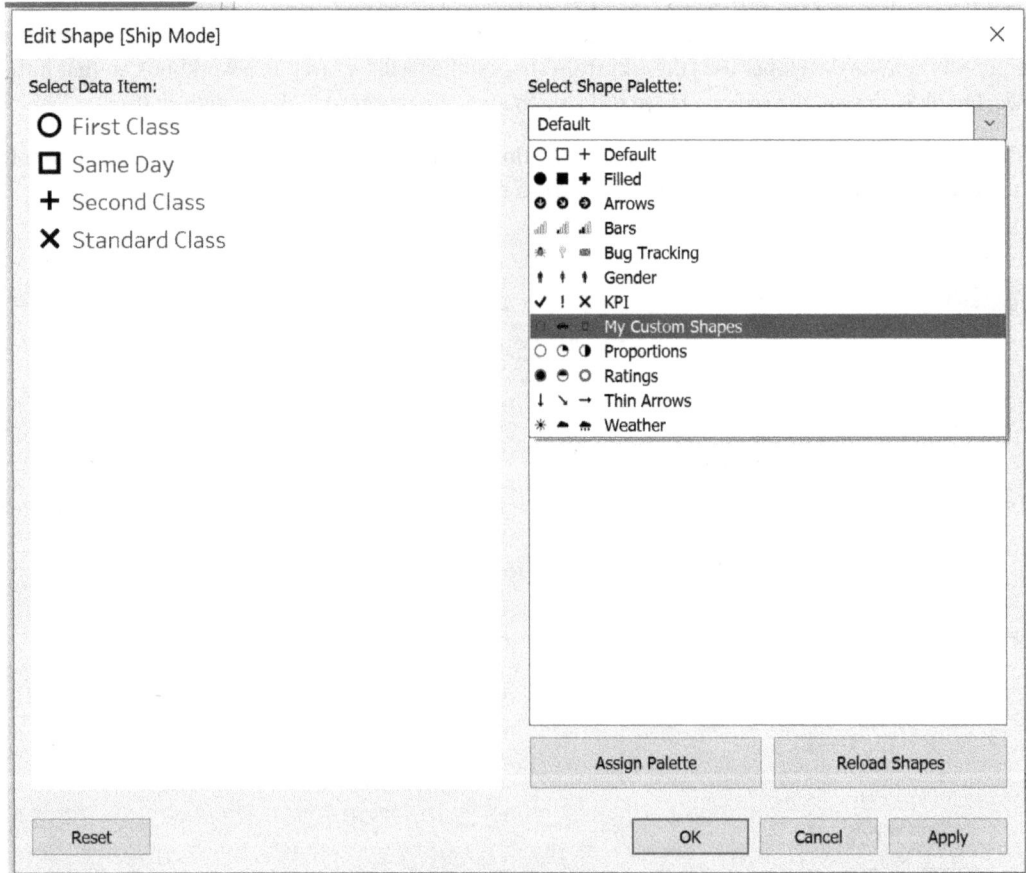

Figure 7.35: The result of the new custom shapes

Try to use a transparent background for the shapes; otherwise, there will be a square border surrounding the shape.

Updating a color palette in Tableau requires a little more editing of the `Preferences.tps` file that is included in Tableau Desktop. This file is where custom color palettes, which a developer can use whenever they open Tableau, are included.

In the `Preferences.tps` file, there is no limit to the number of palettes that can be included. However, it is important to note that the **Tableau editor** will only be able to display up to 20 colors. To resolve this, create multiple palettes containing 20 or fewer colors.

All new colors will need to use the HTML format to input the exact shade, and Tableau will need to be restarted to be able to use the palette.

To add the new palette, open the `Preferences.tps` file in a text editor, such as Notepad, to update the file. To input the new palette, type in an opening and closing `<preferences>` statement. It should look something like this:

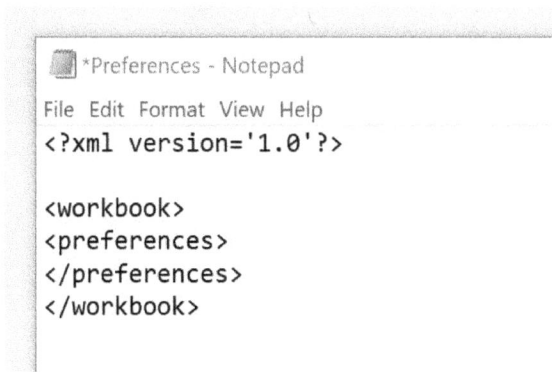

```
*Preferences - Notepad
File  Edit  Format  View  Help
<?xml version='1.0'?>

<workbook>
<preferences>
</preferences>
</workbook>
```

Figure 7.36: The preferences notepad

The line between each `<preferences>` instance will define the name and type of the palette desired. It will be something like the following:

```
<color-palette name=["Palette Name"] type=["Type of Palette]" >
```

Note

Make sure to use straight brackets when typing in the text editor; otherwise, there will be an error.

For categorical color palettes, use the `regular` type. This palette will contain several unique colors that will be applied to a discrete dimension. The file should look like what is shown in *Figure 7.37*.

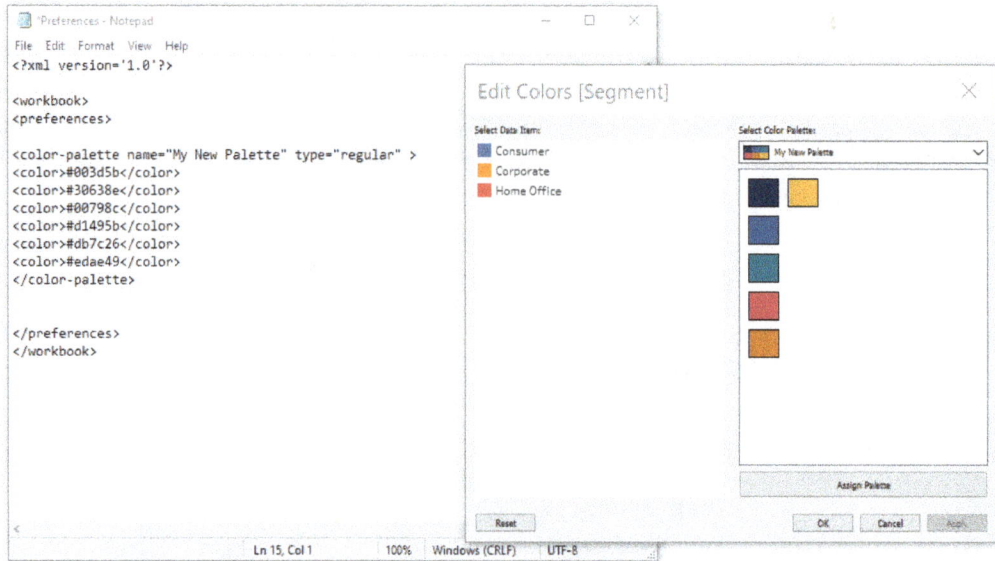

Figure 7.37: An example of a quantitative color palette and the Preferences text

When creating a sequential palette, use `ordered-sequential`. A sequential palette will typically show a singular color that will show a gradient of intensity.

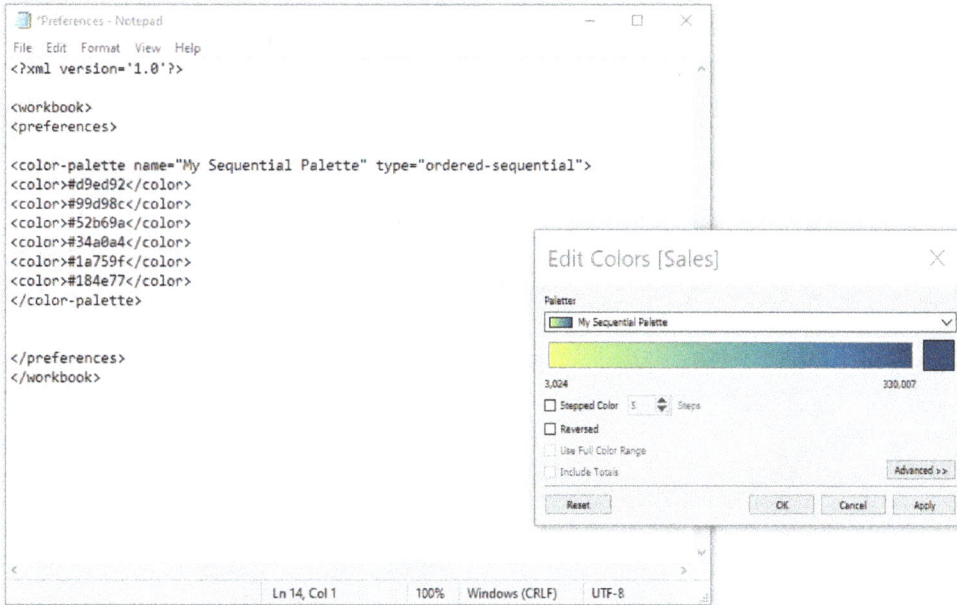

Figure 7.38: An example of a sequential color palette and the Preferences text

A diverging palette will use two opposing colors to show the contrast between two measures. These are commonly used to show positive versus negative values. To define this palette, use `ordered-diverging`.

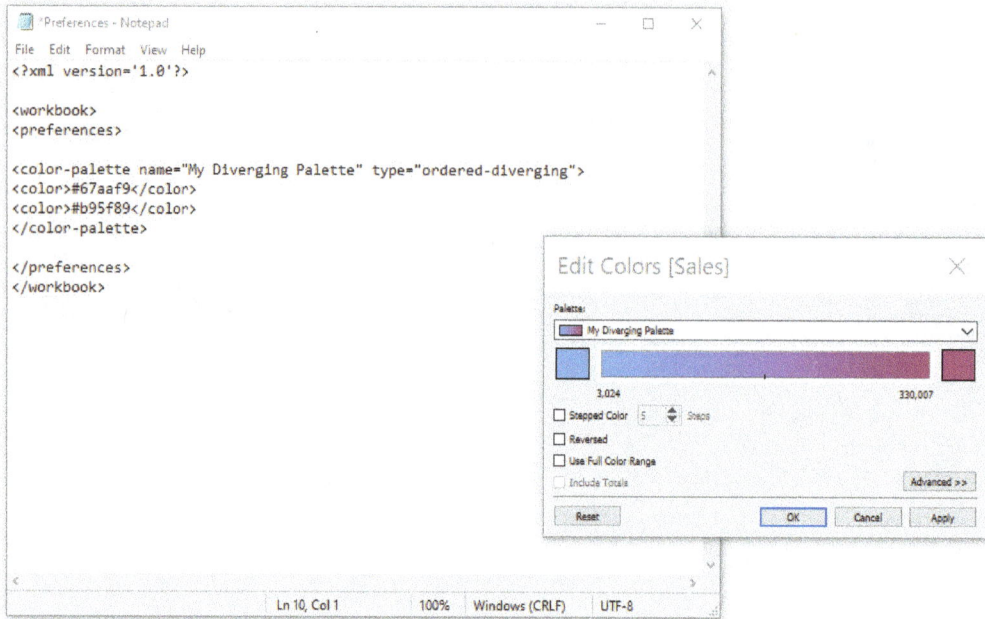

Figure 7.39: An example of a diverging color palette and the Preferences text

Adding annotations

Annotations allow a developer to draw the user's eye to a key point in a dashboard and include additional information about that point.

Take *Figure 7.40*, for example:

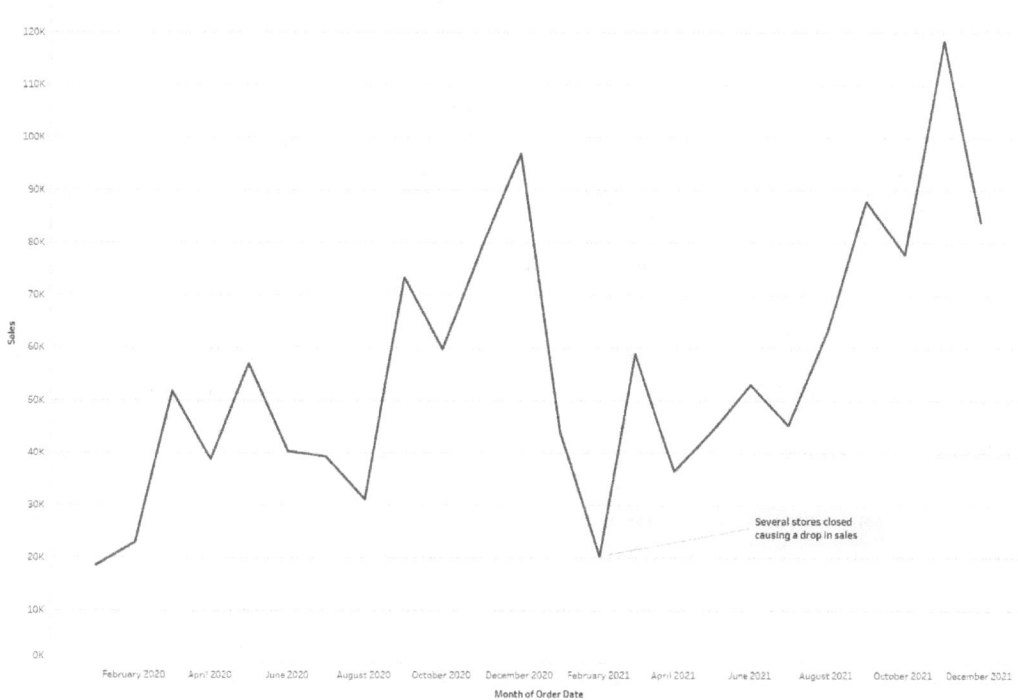

Figure 7.40: An example of an annotation in a chart

The trend in this chart is clearly shown, but where there are dips in data, it can be hard to distinguish why. Including an annotation can help clarify the reason for the dips that contributed to the data result.

To create these annotations, go to the worksheet of the chart and right-click on the point where you need to annotate. There are three options for annotations – **Mark**, **Point**, and **Area**:

- **Mark**: This will create an annotation on a specific mark selected on the worksheet. This option will only apply to a selected mark; selecting multiple points will create individual annotations for each mark.

- **Point**: Similar to marks, this will create an annotation point around a selected area. Selecting multiple marks will create only one annotation point.

- **Area**: This will create an annotation in a general area where a developer can highlight a point of interest in a report.

When selecting the annotation type, a window will open in which a developer can adjust the font and style of the annotation, along with the information they wish to display. By using the **Insert** menu, dynamic options can be included in the annotation.

If a developer needs to update the annotation, this can be done by right-clicking the annotation and selecting the option to annotate.

When using annotations, a developer has the option to move, resize, and adjust the line and move the text to fit in the chart appropriately. This can be achieved by dragging the item around and using the size anchors to change the width. Lines can be adjusted with the point anchors to make sure that the annotation points at the correct outlier.

For an area, the size anchors can be used to create an area that encompasses the required points.

Formatting can be applied to annotations that will include some customization of the points. These formatting options can change the form of the box, the text, the shading, and the corners, as well as the style of the line, such as turning it into an arrow or changing how the line ends.

Adding tooltips

Tooltips allow users to view additional information about a chart element. Hovering over a mark or more will show the tooltip that is filtered to the affected areas.

Automatically, the tooltip will generate a result based on the fields that have appeared on the chart. This text can be edited in the **Marks** card to create hardcoded text, or by selecting **Insert**, developers can create a dynamic tooltip with the text.

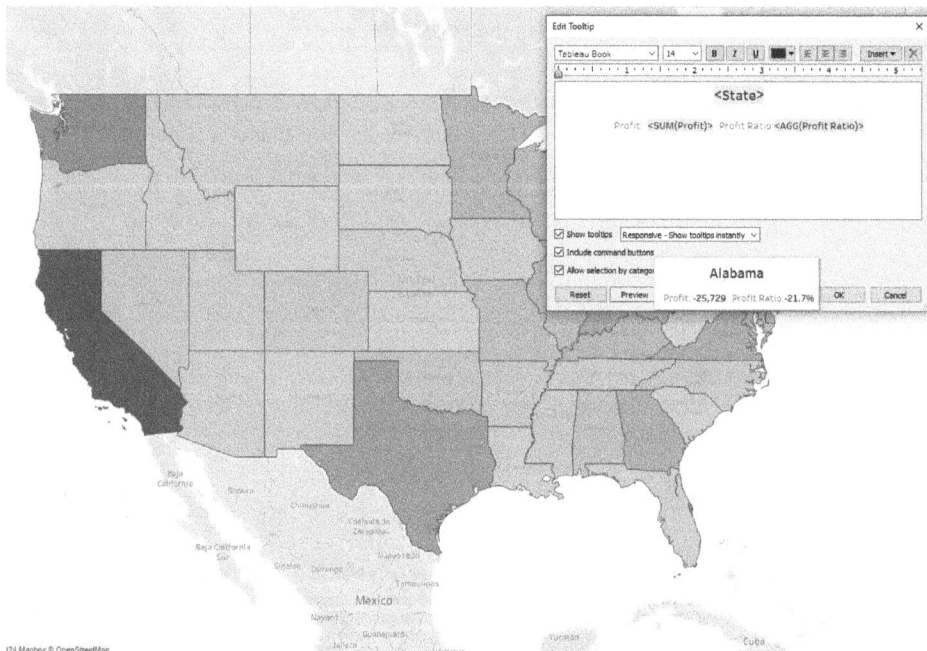

Figure 7.41: The tooltip editor and preview

Developers can include command buttons that give users options to keep data only, exclude data, and view data on the selected data point. If the command button is deselected, users will not get to see a summary of the results at the top of the bar.

Applying padding

Padding helps create uniform spacing between charts, objects, and containers in a dashboard. By using padding, all elements of the dashboard will not be clustered.

To add padding, go to the **Layout** tab on the left-hand side. There will be two padding options – inner padding and outer padding. Take a look at *Figure 7.42* to understand the difference:

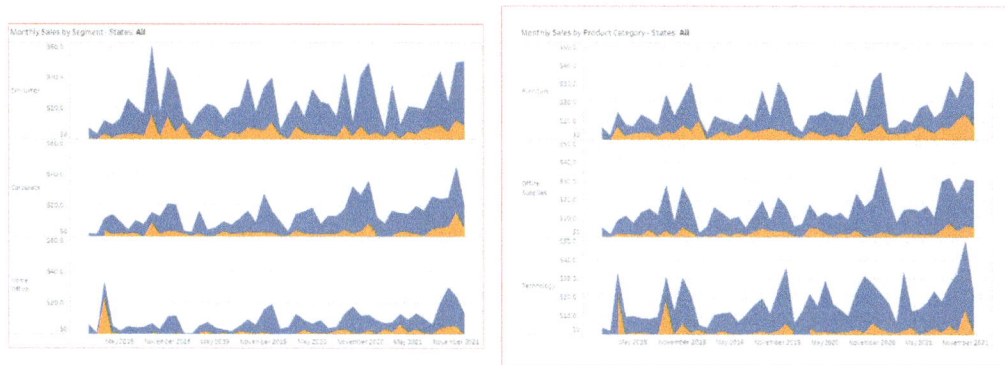

Figure 7.42: Examples of outer and inner padding

The chart on the left represents outer padding while the chart on the right represents inner padding.

All sheets will have a border around the visual. Although this may not be immediately apparent this is where padding can impact the sheet depending on which area is chosen. If a developer uses outer padding, this will create space on the outside of the border and from other sheets. Conversely, using inner padding will create space on the inside of the border and the chart.

How is this useful? The reason for using padding is to evenly and precisely space out charts so that the dashboard can be perceived clearly. This gives a developer full control of the report and how it is styled for any user to understand clearly.

Removing Gridlines, Row-Level and Column-Level Bands, and Shading

Using gridlines can help guide the eye in a chart so that a user can understand where the points lie and the value they present. A developer can control the appearance of these lines by going to **Format | Lines**.

In *Figure 7.43*, the lines have been changed to orange from light gray.

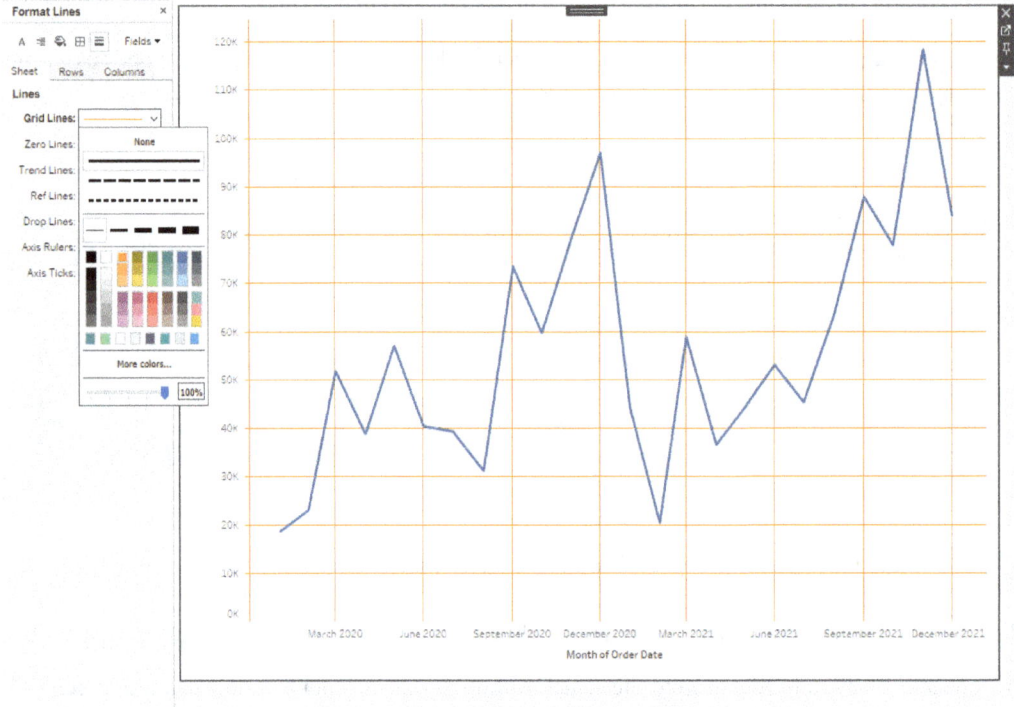

Figure 7.43: The gridlines changed to orange

There are more options that let you change the gridlines from dashed to dotted, and even the opacity of the lines.

Shading can be applied to headers to control the background of a worksheet. This can be accessed with **Format | Shading**.

One example of shading is using it for column or row banding. In the following example, take a look at one table without any banding and one with banding. This helps the viewer distinguish each row.

State	Profit	Sales	State	Profit	Sales
Connecticut	3,511	13,384	Connecticut	3,511	13,384
Delaware	9,977	27,451	Delaware	9,977	27,451
District of Colu..	1,060	2,865	District of Columbia	1,060	2,865
Illinois	-12,608	80,166	Illinois	-12,608	80,166
Indiana	18,383	53,555	Indiana	18,383	53,555
Iowa	1,184	4,580	Iowa	1,184	4,580
Kansas	836	2,914	Kansas	836	2,914
Maine	454	1,271	Maine	454	1,271
Maryland	7,031	23,706	Maryland	7,031	23,706
Massachusetts	6,786	28,634	Massachusetts	6,786	28,634
Michigan	24,463	76,270	Michigan	24,463	76,270
Minnesota	10,823	29,863	Minnesota	10,823	29,863
Missouri	6,436	22,205	Missouri	6,436	22,205
Nebraska	2,037	7,465	Nebraska	2,037	7,465
New Hampshire	1,707	7,293	New Hampshire	1,707	7,293
New Jersey	9,773	35,764	New Jersey	9,773	35,764
New York	74,039	310,876	New York	74,039	310,876
North Dakota	230	920	North Dakota	230	920
Ohio	-16,971	78,258	Ohio	-16,971	78,258
Oklahoma	4,854	19,683	Oklahoma	4,854	19,683
Pennsylvania	-15,560	116,512	Pennsylvania	-15,560	116,512
Rhode Island	7,286	22,628	Rhode Island	7,286	22,628
South Dakota	395	1,316	South Dakota	395	1,316
Texas	-25,729	170,188	Texas	-25,729	170,188
Vermont	2,245	8,929	Vermont	2,245	8,929
West Virginia	186	1,210	West Virginia	186	1,210
Wisconsin	8,402	32,115	Wisconsin	8,402	32,115

Figure 7.44: Before and after row banding

There are three options that can help change the banding:

- **Pane and Header**: This is the color of the bands
- **Band Size**: This helps change how thick the bands are
- **Level:** Depending on whether there are multiple-dimensional fields, the banding style will change to include either the different categories or individual rows

Summary

Now, you should be feeling confident with regard to formatting a workbook, from incorporating custom color palettes and shapes to formatting gridlines and text in a dashboard.

You are encouraged to experiment with their own reports to familiarize themselves with the steps and visualize the impact that formatting makes on their dashboards.

Exam Readiness Drill – Chapter Review Questions

Apart from a solid understanding of key concepts, being able to think quickly under time pressure is a skill that will help you ace your certification exam. That is why working on these skills early on in your learning journey is key.

Chapter review questions are designed to improve your test-taking skills progressively with each chapter you learn and review your understanding of key concepts in the chapter at the same time. You'll find these at the end of each chapter.

> **How to Access these Resources**
>
> To learn how to access these resources, head over to the chapter titled *Chapter 9, Accessing the Online Practice Resources*.

To open the Chapter Review Questions for this chapter, perform the following steps:

1. Click the link – `https://packt.link/TDA_CH07`.

   Alternatively, you can scan the following **QR code** (*Figure 7.45*):

Figure 7.45: QR code that opens Chapter Review Questions for logged-in users

2. Once you log in, you'll see a page similar to the one shown in *Figure 7.46*:

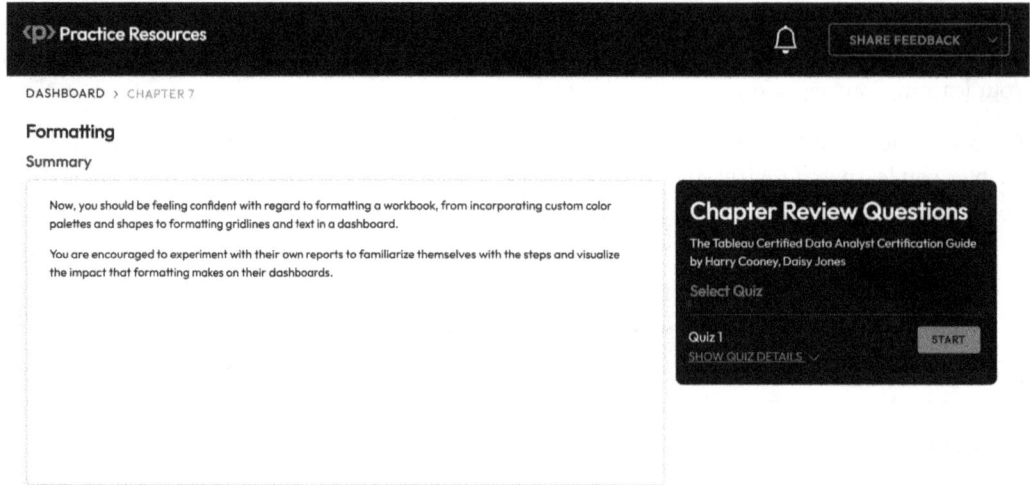

Figure 7.46: Chapter Review Questions for Chapter 7

3. Once ready, start the following practice drills, re-attempting the quiz multiple times.

Exam Readiness Drill

For the first three attempts, don't worry about the time limit.

ATTEMPT 1

The first time, aim for at least **40%**. Look at the answers you got wrong and read the relevant sections in the chapter again to fix your learning gaps.

ATTEMPT 2

The second time, aim for at least **60%**. Look at the answers you got wrong and read the relevant sections in the chapter again to fix any remaining learning gaps.

ATTEMPT 3

The third time, aim for at least **75%**. Once you score 75% or more, you start working on your timing.

> **Tip**
>
> You may take more than **three** attempts to reach 75%. That's okay. Just review the relevant sections in the chapter till you get there.

Working On Timing

Target: Your aim is to keep the score the same while trying to answer these questions as quickly as possible. Here's an example of how your next attempts should look like:

Attempt	Score	Time Taken
Attempt 5	77%	21 mins 30 seconds
Attempt 6	78%	18 mins 34 seconds
Attempt 7	76%	14 mins 44 seconds

Table 7.1: Sample timing practice drills on the online platform

> **Note**
>
> The time limits shown in the above table are just examples. Set your own time limits with each attempt based on the time limit of the quiz on the website.

With each new attempt, your score should stay above **75%** while your "time taken" to complete should "decrease". Repeat as many attempts as you want till you feel confident dealing with the time pressure.

8

Publishing and Managing Content

Introduction

This chapter will cover the steps that need to be taken to publish content to Tableau Server and Tableau Cloud. Tableau Server is a self-hosted service for sharing Tableau content with end users and working collaboratively with other developers, and Tableau Cloud is the same but is hosted on the cloud. Along with publishing content, exporting content from the server/cloud will also be covered in the chapter. Managing content to ensure that data is kept up to date and end users are notified about content if needed will also be discussed.

Tableau Server and Tableau Cloud are, for the most part, the same and function as a way for Tableau developers, who will either have the role of a Creator or an Explorer, to share content with end users who may be Explorers or Viewers.

Tableau Server and Cloud are organized by **projects**, which function essentially as folders to group collections of content together. Content can be Tableau workbooks, data sources, and Tableau Prep flows. Permissions to access projects can be set both for specific users and for user groups. User groups are a functionality that allows admins to group many users together, for example, a sales department, and manage their permissions collectively. Permissions can also be specified at the content level.

This chapter will cover the process that is followed when a developer wants to publish content to Tableau Server or Cloud. You will also learn how to publish Tableau workbooks, including embedding data sources within and limiting the published workbook to specific views. You will look at publishing data sources to Tableau Server and Cloud as well. The process for end users of dashboards who want to export the dashboard to their local machine, for example, as a PDF, will also be covered.

To ensure that dashboards and insights are accurate and relevant, data must be kept up to date on Tableau Server and Cloud. To do this, users can schedule data refreshes if they are direct extracts. Tableau Prep flows can also be scheduled and can take data from multiple sources, manipulate and transform it, and then output the up-to-date data to a published data source location.

Workbooks published on Tableau Server and Cloud can be configured with thresholds on relevant data points. If the thresholds are passed, alerts can be sent out to a user or group of users. Subscriptions can also be sent out on a regular basis to keep end users up to date on published content.

Sharing Tableau Content

From Tableau Desktop, there are options available to publish both the workbook and any of the data sources being used. Throughout the rest of this chapter, Tableau Cloud will be referred to, as this is what will be used for the demo, but the functionalities of Tableau Server and Tableau Cloud are the same. While interactive dashboards are the recommended method of sharing Tableau content, static exports such as PDFs and images can also be created from Tableau workbooks.

Publishing Workbooks

Once a dashboard or a series of dashboards has been developed, they can be shared with end users by uploading the Tableau workbook to Tableau Cloud.

The first step when publishing a Tableau workbook to Tableau Cloud is to ensure that Tableau Desktop is connected to Tableau Cloud. This can be checked by opening up **Server** on the top menu. The top result in the menu will either say **Sign In…** or will provide the name of the server the user is currently signed in to. If the menu says **Sign In…**, then Tableau Desktop is not currently connected to any Tableau Server/Cloud. If this is the case, then the user should select **Sign In…**.

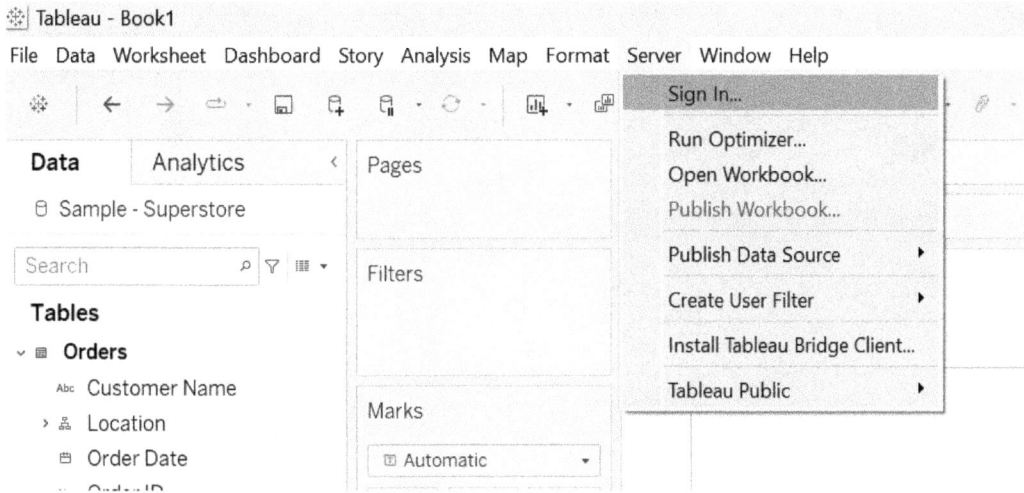

Figure 8.1: The Tableau Desktop Server Sign In… location

If the server name listed is incorrect, then the user should select the server name followed by **Sign in** to another Server. Both options will result in the **Tableau Server Sign In** pop-up box appearing, allowing the user to input the correct Tableau Cloud URL. After the URL has been inputted and **Connect** pressed, another popup will appear that will allow the user to sign in to the Tableau Cloud site.

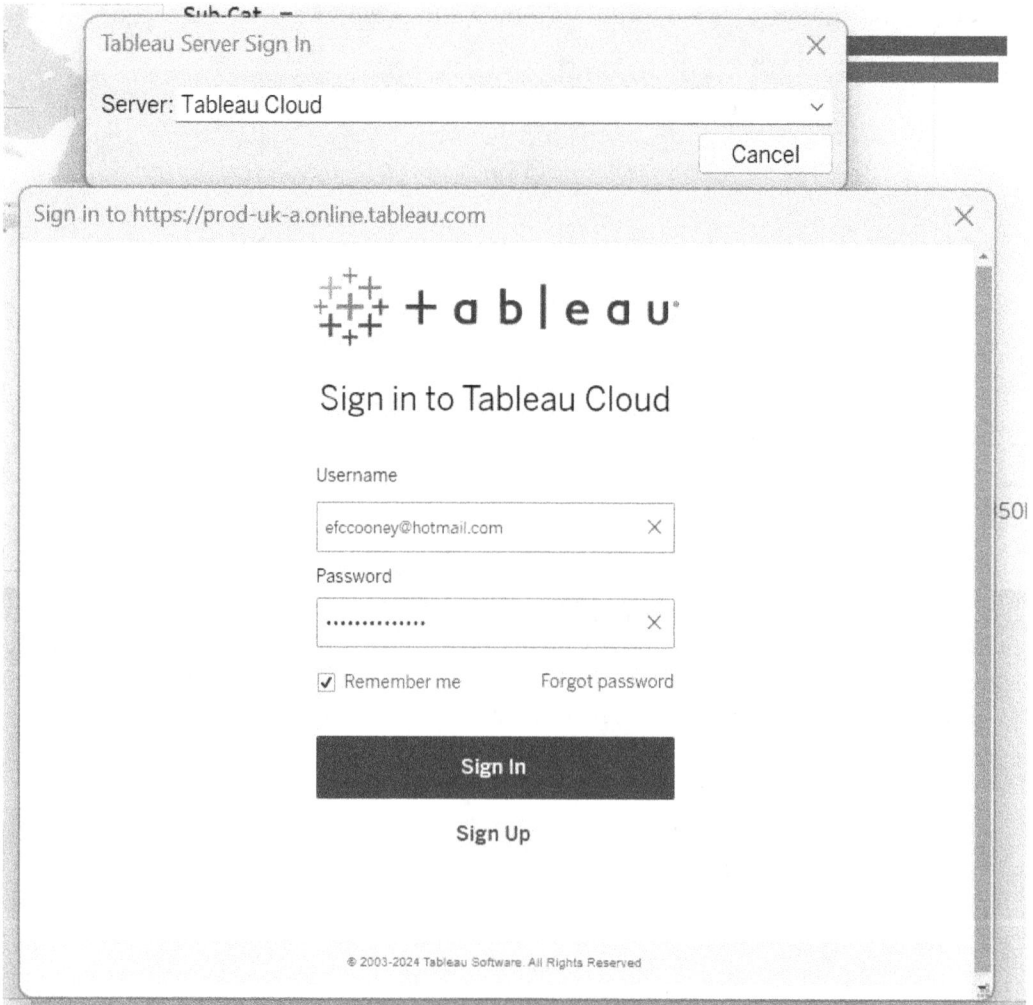

Figure 8.2: The Tableau Server Sign In pop-up box and the subsequent Sign in to Tableau Cloud popup

Once Tableau Cloud is connected and the user is signed in, the workbook can be published. To publish a Tableau workbook, select **Server** on the top menu, followed by **Publish Workbook**. This will open the **Publish Workbook** popup that first asks the user to select a project to upload the workbook to. There is a **Default** project on every Tableau site that will always be the default publish location. A name must also be given for the published workbook, and if there is already a workbook with that name in the selected project, red text will show below, indicating that publishing the workbook will overwrite the existing content.

Location

Samples

Name

Superstore

Workbook name is already in use. Publishing will overwrite the existing workbook.

Figure 8.3: A workbook called Superstore already exists in the Samples project – publishing the workbook will overwrite the existing Superstore workbook on the Tableau site

A description can optionally be added to provide end users with some context about the dashboards within the workbook. Tags can also be added and can be useful for end users searching for content on a Tableau site. By default, all sheets in a workbook are published, but if only certain sheets/dashboards need to be displayed to the end user, there is an option to select and deselect as required. An **Only Dashboards** shortcut button can be selected to ensure that only dashboards from the workbook will be displayed to the end user.

Permissions default to those set for the project that the workbook is being published to. If **Edit** is selected, then custom permissions can be set for the workbook, both for individual users and for permission groups.

The **Data Sources** section of the configuration allows users to select whether to embed any local files within the published workbook or publish the data source as a separate data source on Tableau Server, which the workbook will then be connected to. If a data source is an extract that needs to be refreshed, then **Embed password** followed by **Allow refresh access** must be selected. Passwords can also be embedded for any connection to external databases. If the password is not embedded, then the end user will be asked for the password when opening the dashboard.

The final configuration options under **More Options** allow the user to select some of the following options: **Show sheets as tabs**, **Show selections**, and **Include external files**. If selected, **Show sheets as tabs** will list every sheet/dashboard in the workbook as a tab along the top of the screen, allowing the end user to navigate easily through the workbook. **Show selections** saves any selection on Tableau Desktop, for example, an insightful data point on a key chart, and shows them as selected in the published workbook. **Include external files** will package up external files such as Excel, Access, Text, Data Extract, and image files with a Tableau workbook upon publishing.

Publishing a Tableau Workbook

In this exercise, you will learn how to publish a Tableau workbook to Tableau Cloud. Analyses are shared with end users on Tableau Cloud or Desktop, so the ability to publish workbooks is vital for anyone who wants to share their content to Tableau Server/Cloud and has signed in from Tableau Desktop. Follow these steps to publish a Tableau workbook to Tableau Cloud:

1. In a Tableau workbook, create a dashboard that is ready for consumption by end users.

2. For example, create one sheet with **Total Sales** on rows by **Discrete Year** and **Discrete Month** using the **Order Date** field on columns.

3. On another sheet, show **Total Sales** by sub-category as a bar chart colored by category and sorted by **Sales** descending.

4. On a final sheet, create a map of **Total Sales by City** with the circles sized as per the number of **Sales**.

5. Combine all three sheets on a dashboard using containers and add a title.

6. Now that the dashboard is ready for end users, it can be published by selecting **Server** on the top menu followed by **Publish Workbook...**.

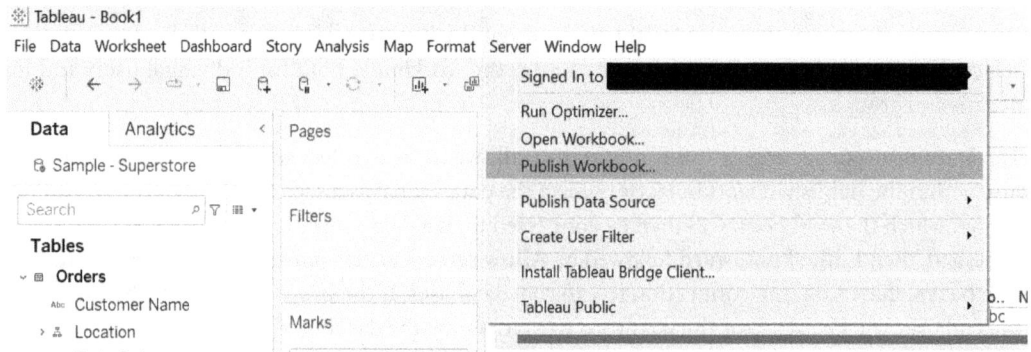

Figure 8.4: The Publish Workbook... menu option

7. In the popup, select the project you want to publish the workbook to.

8. Either create a project on your Tableau site or select **Default**.

9. Name the workbook **Sales Analysis Dashboard**.

10. Provide a brief description such as Dashboard breaking down sales by month, sub-category, and city.

11. In the sheets section, select **Only Dashboards** to ensure the end user only sees the **Sales Analysis Dashboard** and not the individual sheets that are used to make it.

12. Permissions can be left the same as the project and the data source can be left to be embedded within the workbook.

13. None of the **More Options** options need to be ticked as there is only one dashboard, there are no relevant selections, and there are no external files to include.

14. Click **Publish** to publish the workbook.

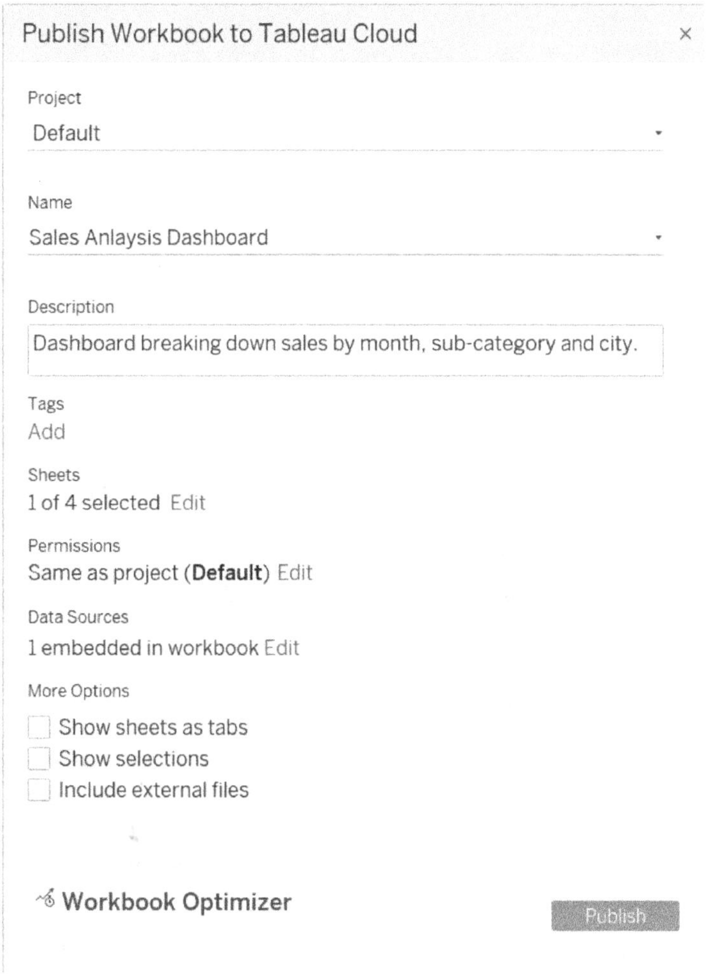

Publish Workbook to Tableau Cloud   ✕

Project

Default   ▾

Name

Sales Anlaysis Dashboard   ▾

Description

Dashboard breaking down sales by month, sub-category and city.

Tags
Add

Sheets
1 of 4 selected   Edit

Permissions
Same as project (**Default**) Edit

Data Sources
1 embedded in workbook Edit

More Options
☐ Show sheets as tabs
☐ Show selections
☐ Include external files

⚙ **Workbook Optimizer**   [Publish]

Figure 8.5: The Publish Workbook to Tableau Cloud configuration

15. Once the workbook has finished publishing, the Tableau site will open with the workbook selected. Click on the dashboard thumbnail to open it and view how the end user will interact with the published dashboard.

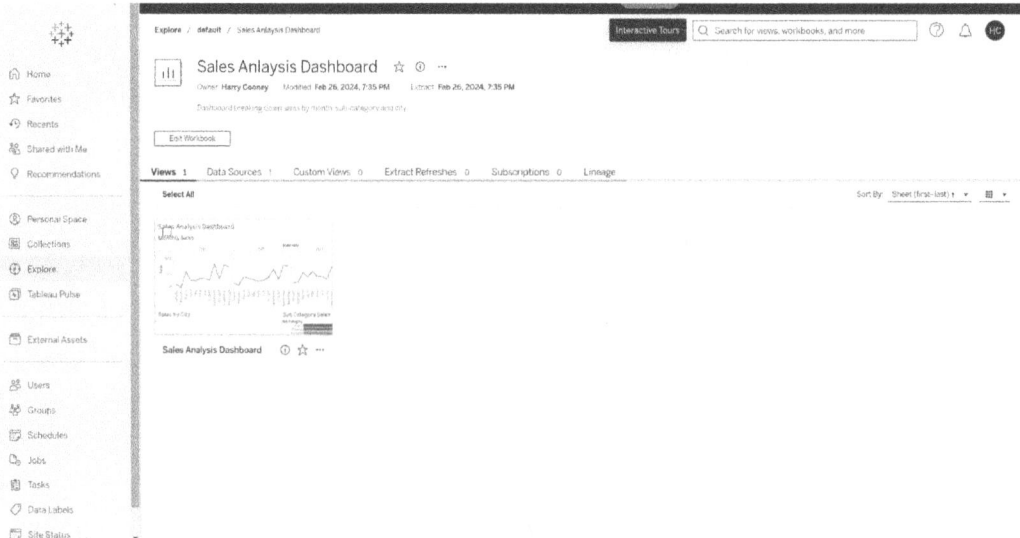

Figure 8.6: A Tableau workbook published to Tableau Cloud

In this exercise, you have created a basic dashboard and have published it to Tableau Cloud. You now know how to publish and share dashboards in Tableau and also know what customization options are available when publishing.

Publishing Data Sources

In addition to publishing workbooks for consumption by end users, data sources can be published to Tableau Cloud. Published data sources can be accessed by end users who want to ask questions about data sources or create specific metrics on them. Published data sources can also be connected to by other Tableau developers and are a great way to create a single source of truth that multiple different pieces of analysis can be connected to.

To publish a Tableau data source, open the **Server** menu and select **Publish Data Source**. Then, select the data source in the workbook you want to publish. The **Publish Data Source** pop-up menu is similar to the **Publish Tableau Workbook** menu. The first option is the project on the Tableau site where the data source should be published to. This is followed by the name; if a data source of the same name exists in the project, red text will show, informing the user that the existing data source will be overwritten. A description and tags can also be provided.

Permissions are similar to publishing a Tableau workbook, in that, by default, they are inherited from the project being published to but can be manually configured for specific users and permission groups.

If the data source being published is an extract, then there will be an option to set up a refresh schedule. This allows the user to set up specific times when the data will be refreshed. For the refresh to be possible, there must be a connection between the Tableau site and the data source. If a data source is on-premises, then to set up a connection between the Tableau Cloud site and the data source, an additional piece of software called **Tableau Bridge** has to be configured.

If the data source is a database connection, then there will be an authentication section that allows for the data source to either prompt end users for a password whenever they try and access it or have the credentials currently used embedded within. If the credentials are embedded, then end users will be able to access the data source without having to input credentials.

The final section of the configuration window has **More Options**. This includes an **Include external files option**; this option is similar to the **Publish Workbook…** option that allows users to embed local file data and images. The **Update workbook to use the published data source** option replaces the existing data source in the workbook with the newly published version.

Publishing a Data Source

In this exercise, you will publish a data source to Tableau Cloud. Tableau-published data sources are a great way to ensure that an organization has one source of truth when it comes to data. If all users use the same published data source, then the risk of misaligned data sources is nullified. Follow these steps to publish a data source:

1. In the same workbook from the previous exercise, select **Server**, followed by **Publish Data Source** and **Sample - Superstore**.

2. In the pop-up menu, select the project you want to publish the data source to; in this case, the **Default** project can be selected.

3. Name the data source **Sample - Superstore** and give a brief description such as **Tableau's default sample data source**.

4. No tags need to be added and the permissions can be kept the same as the project that is being published to.

5. The next section says **Tableau Bridge required for on-premises data**. This is because the file is being published from your local drive and the Tableau Cloud site does not have access to this location. Tableau Bridge is an additional piece of software that can facilitate a connection but is not required for this practice example. This means the data will not be able to be updated. If data needs to be kept fresh, then a data source that Tableau Cloud can access must be set up or Tableau Bridge must be configured.

6. In **More options**, **Include external files** can be left unticked, but **Update workbook to use published data source** can be selected as we want to update the data source in the workbook. Select **Publish** to publish the data source.

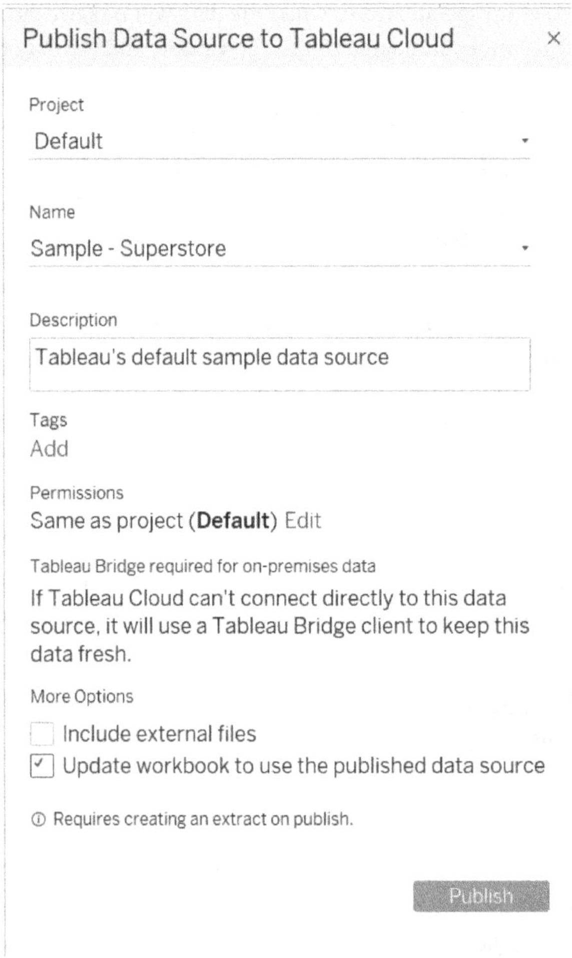

Publish Data Source to Tableau Cloud   ×

Project
Default   ▾

Name
Sample - Superstore   ▾

Description
Tableau's default sample data source

Tags
Add

Permissions
Same as project (**Default**) Edit

Tableau Bridge required for on-premises data
If Tableau Cloud can't connect directly to this data source, it will use a Tableau Bridge client to keep this data fresh.

More Options
☐ Include external files
☑ Update workbook to use the published data source

ⓘ Requires creating an extract on publish.

Publish

Figure 8.7: Publish data source configuration

7. Once the data source has been published, the Tableau site will open in the browser with a publishing complete message. The data source will be open on the Tableau site.

8. Navigate back to the Tableau Desktop workbook and open one of the sheets.

9. The data source in the top left, **Sample - Superstore**, can no longer be observed with a data source icon but with a Tableau logo icon, indicating the data source is a published Tableau data source.

10. Right-click the data source and select **Edit Data Source**. This opens the data source configuration menu and now there are no options to configure relationships; instead, a list of data sources on the Tableau Cloud site can be selected from.

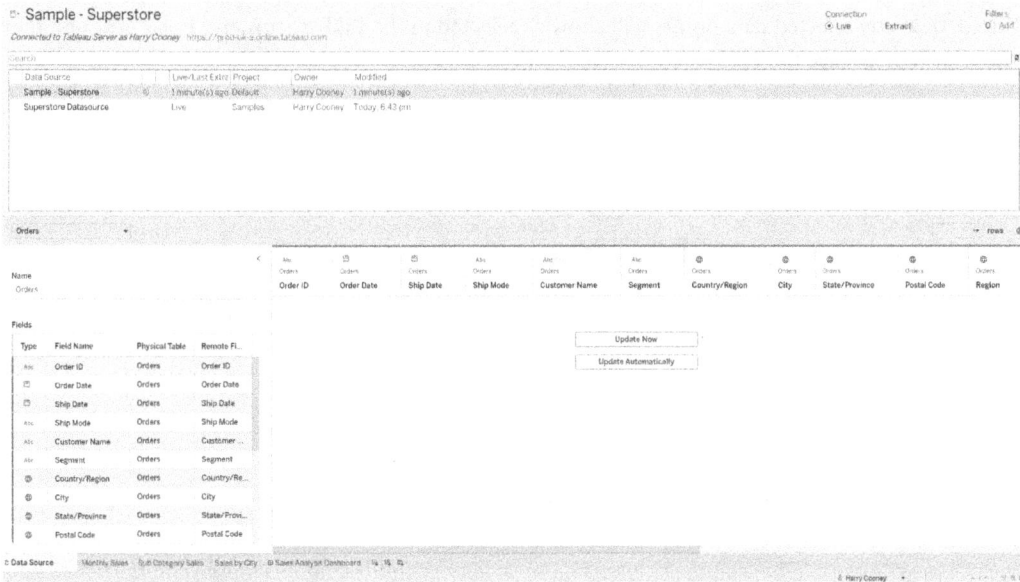

Figure 8.8: The Tableau data source has been published to the cloud site and the workbook data source has been updated to a connection to the published data source

In this exercise, you have successfully published a data source to Tableau Cloud. You have observed how data sources are stored on the cloud site, and you have also seen the impact of publishing on the workbook, with the original data source now being a connection to the published data source on the site.

Exporting Content

To share sheets and dashboards from Tableau workbooks as static images or to include them in presentations, they can be exported directly from Tableau Desktop as a PDF or an image file. Sheets and dashboards can also be sent directly to a printer if a hard copy is required. Data can also be exported from Tableau Desktop.

To export a sheet or dashboard to a PDF, select **File** from the top menu followed by **Print to PDF**. The pop-up configuration menu allows the user to select to export the whole workbook to PDF, the active sheet that the user is currently displaying, or selected sheets. The PDF paper size and orientation can be configured, and whether to view the PDF after printing can also be selected. If **Show selections** is ticked, then any selected data points will show as selected in the PDF document. Press **OK**, and then choose where to save the PDF document and with what name.

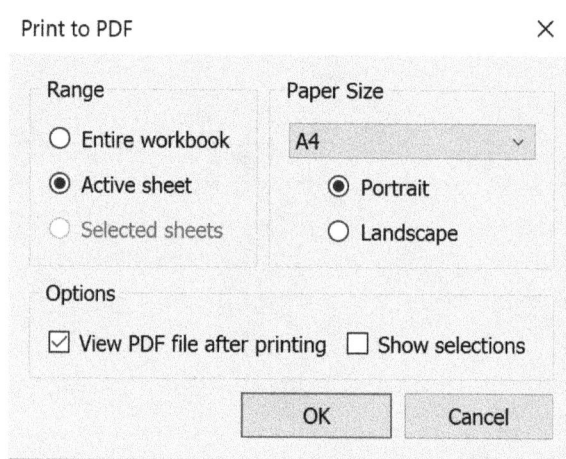

Figure 8.9: The Print to PDF configuration

To customize how individual sheets/dashboards are displayed when exporting to PDF, go to **File** and **Page Setup** for each sheet that needs to be customized. The **Page setup** configuration window contains multiple options for customization of the page such as whether to show titles, views, captions, or legends, where to place legends on the page or margin sizes, and how to fit the page. These page configuration options apply to PDF printing and printing to hard copy.

To print a hard copy of the workbook, select **File** followed by **Print**. In the print configuration popup, a printer and the number of copies you want can be selected, and here you can decide whether you will print the entire workbook, the active sheet, or the selected sheets.

To export the workbook to PowerPoint, select **File** followed by **Export As PowerPoint**. The pop-up window defaults to printing just the current displayed view, but the **Include** dropdown can be updated to **Specific sheets from this workbook**, which allows the user to select the relevant sheets. Click **Export**, then name and save the PowerPoint file.

There are two options when it comes to exporting a Tableau Desktop sheet or dashboard to an image. The current view can be copied to the clipboard as an image that can be pasted into another application, or it can be saved as an image file. To copy the view to the clipboard, select **Worksheet**, followed by **Copy**, and then **Image**. The configuration popup allows the user to select which elements to copy as well as where to position any legends.

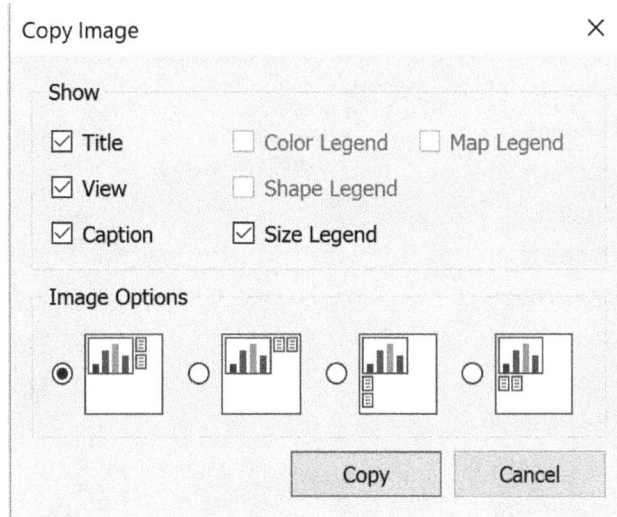

Figure 8.10: Copy image to clipboard configuration options

To save a Tableau view as an image, select **Worksheet**, followed by **Export**, and then **Image**. The same configuration options when copying the image are provided but, instead of selecting **Copy**, **Export** is selected, followed by naming and choosing the destination to save the file.

Similar to copying and exporting images, the data in the current view can be either copied or exported. Both menus are accessed the same way via **Worksheet**, and then either **Copy** or **Export**. The options are then either **Data** or **Crosstab to Excel**. Selecting **Copy** and **Data** will copy the selected data points in the view to the clipboard, ready to be pasted into another application. If no data points are selected, then all data points in the view will be copied. Selecting **Copy** followed by **Crosstab to Excel** copies the data in a format that is suited to Excel when pasting.

When exporting data, there are the same **Data** and **Crosstab to Excel** options. The **Data** option allows the user to save the data points to a Microsoft Access database. The **Crosstab to Excel** option opens up an Excel file with the data points cross-tabulated.

	A	B	C	D
1			Category	
2	Sub-Category	Furniture	Office Supplies	Technology
3	Chairs	335,768		
4	Phones			331,843
5	Storage		224,645	
6	Tables	208,020		
7	Binders		207,355	
8	Machines			189,925
9	Accessories			167,380
10	Copiers			150,745
11	Bookcases	115,361		
12	Appliances		108,213	
13	Furnishings	95,598		
14	Paper		79,541	
15	Supplies		46,725	
16	Art		27,659	
17	Envelopes		16,528	
18	Labels		12,695	
19	Fasteners		8,532	

Figure 8.11: Category and Sub-Category Sales cross-tabbed to Excel

Exporting to PDF

In this exercise, you will export a dashboard to PDF. While dashboards are the preferred method for sharing analyses in Tableau, as these facilitate interactivity, users may ask for a PDF version of the analysis. Follow these steps to export a PDF:

1. Open the **Sales Analysis Dashboard** that you created during the first exercise of the chapter.
2. Select **File** followed by **Print to PDF**.
3. Set the range to **Active sheet** so that only the dashboard is printed.

4. Set the paper size to **A4** and the orientation to **Landscape**.

5. Ensure that selections are not shown when printing and that the PDF is viewed after printing, and then press **OK**.

6. Name the PDF export and save it to your machine.

7. The PDF export will now open on your machine using whichever PDF reader is available.

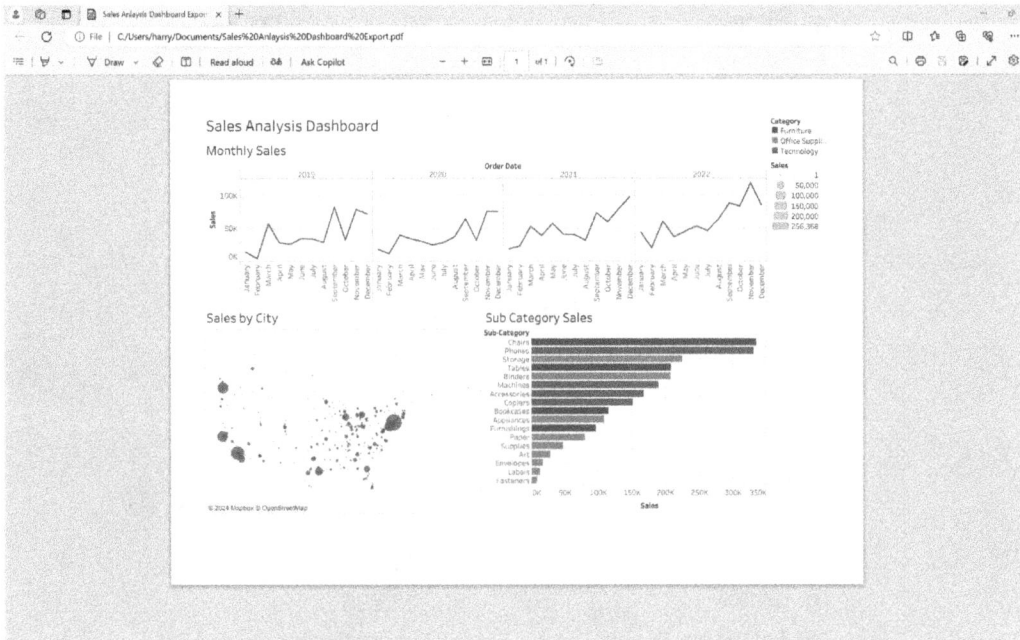

Figure 8.12: Tableau dashboard exported to PDF

In this exercise, you have exported a single dashboard from your workbook to a PDF and have saved it locally. You now know how to export PDFs from Tableau and are aware of the configuration options available when doing so.

Cross-Tab Selected Data to Excel

In this exercise, you will learn how to export data directly from Tableau to Excel in a cross-tabulated format. It can sometimes be useful to get the data related to a specific view in Tableau to share with others or for personal reference. Cross-tabbing allows users to extract view-specific data points from a data source, preventing the need to manually dive into the data in Excel. Follow these steps to crosstab selected data to Excel:

1. On the **Sub-Category Sales** worksheet created in the first exercise of the chapter, select **Worksheet** on the top menu followed by **Export**, and finally **Crosstab to Excel**.

2. An Excel document has opened up with the data from the worksheet displayed crosstabbed.

	A	B	C	D
1			Category	
2	Sub-Category	Furniture	Office Supplies	Technology
3	Chairs	335,768		
4	Phones			331,843
5	Storage		224,645	
6	Tables	208,020		
7	Binders		207,355	
8	Machines			189,925
9	Accessories			167,380
10	Copiers			150,745
11	Bookcases	115,361		
12	Appliances		108,213	
13	Furnishings	95,598		
14	Paper		79,541	
15	Supplies		46,725	
16	Art		27,659	
17	Envelopes		16,528	
18	Labels		12,695	
19	Fasteners		8,532	

Figure 8.13: Sub-Category, Category, and Sales data crosstabbed to Excel

In this exercise, you have crosstabbed data from Tableau into Excel. You now know how to export view-specific data from Tableau into Excel.

Scheduling Data Updates

Once content has been published to a Tableau site, the data needs to be kept fresh so that the insights on the dashboards are relevant. Live connections are kept constantly up to date, but extracts contain a point-in-time snapshot of the data. Extracts can be configured to refresh on a schedule. Tableau Prep flows can also be created to combine and transform data before outputting to a Tableau data source on a Tableau site. These prep flows can be scheduled so that the resulting data sources are kept fresh.

Extract Refreshes

Sometimes, live connections to data are not feasible and Tableau extracts are also more performant than live connections. To refresh the data on a regular basis, an extract refresh can be configured. On Tableau Server, the schedules available to run the refresh are configured by the server admin. On Tableau Cloud, the refresh frequency is fully configurable by the user setting up the extract refresh.

When publishing a data source that is an extract or when publishing a workbook that contains an extract, the configuration menu contains the necessary options to set up a refresh schedule. When publishing to Tableau Server, a specific refresh schedule can be selected, and when publishing to Tableau Cloud, an hourly, daily, weekly, or monthly refresh can be set up to run at selected times on selected days. The refresh type can also be set to a full refresh or an incremental refresh. Incremental refreshes will only add rows that are new since the previous refresh.

Once a data source has been published, it can be opened on the Tableau site, and you can find a specific tab called **Extract Refreshes**. From this tab, **New Extract Refresh** can be selected and the same configurations based on time and date can be configured. Existing extract refresh schedules can also be configured from this menu.

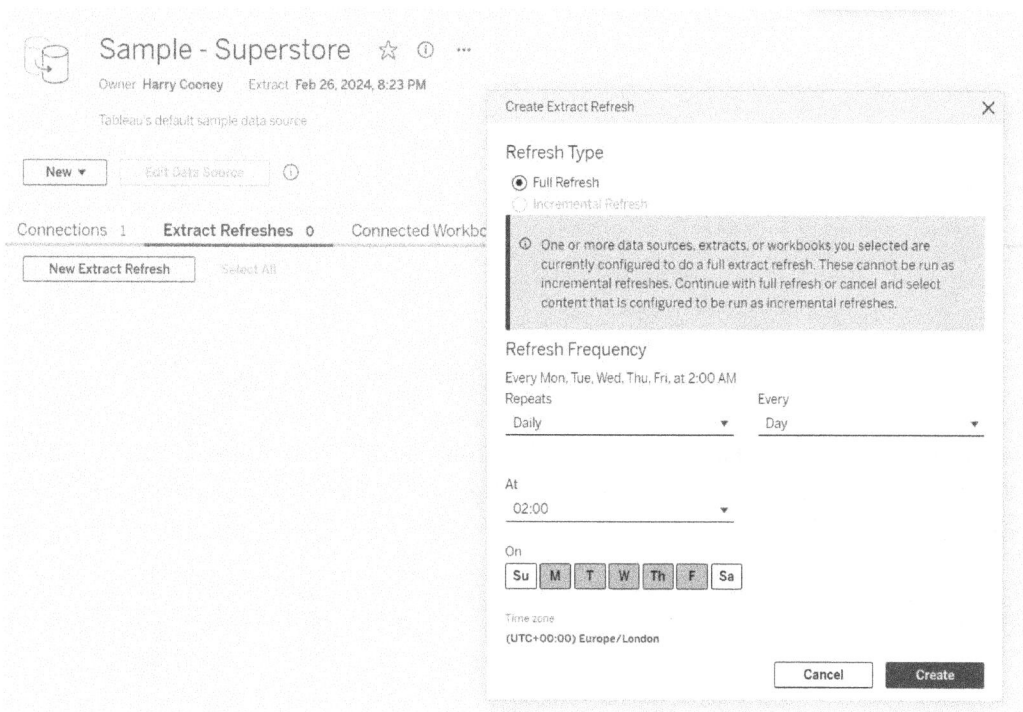

Figure 8.14: Extract refresh configured to update the data on weekdays at 2 a.m.

Scheduling an Extract Refresh

In this exercise, you will add a scheduled refresh time to the data source that you published earlier in the chapter. For insights to be meaningful to end users, the data in the analysis must be kept up to date. If a published data source is an extract, then it can be manually refreshed or refreshed on a schedule. Usually, it makes the most sense to schedule updates to extracts as this allows for a consistent refresh time. Follow these steps to schedule an extract refresh:

1. On Tableau Cloud, open the **Sample - Superstore** data source that you previously published in this chapter.

2. Open the **Extract Refreshes** tab and select **New Extract Refresh**.

3. Configure the refresh to run weekly on a Sunday at 11.30 p.m. and then select **Create**.

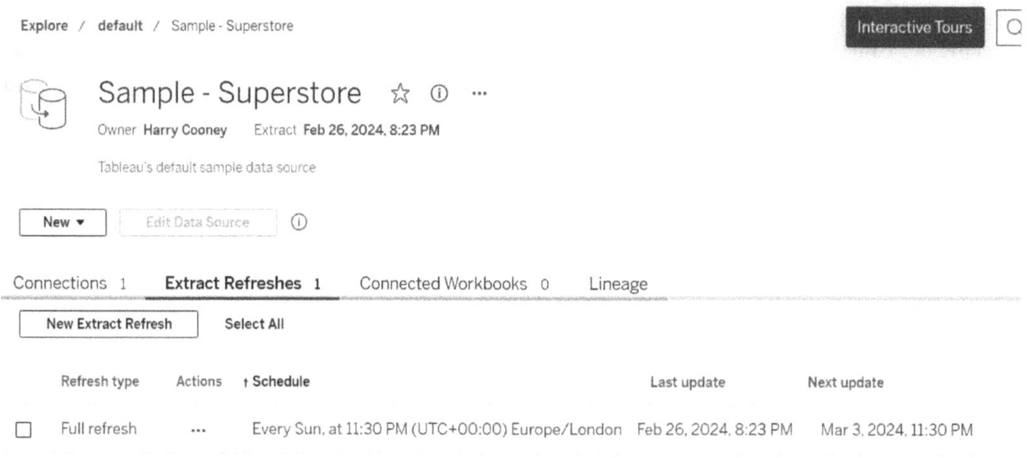

Figure 8.15: Extract refresh schedule created and configured in Tableau Cloud

In this exercise, you have added a weekly refresh schedule to a published data source. You have also observed the other configuration options available when adding the schedule, including options for daily and monthly refreshes.

Tableau Prep Flows

Tableau Prep flows can also be published to a Tableau site, where they can take input data sources and output to published data sources on the site. Therefore, these Tableau Prep flows can also be used to schedule data source refreshes.

To add a Tableau Prep flow refresh on a schedule, first open the Tableau Prep flow on the Tableau site. Open the **Scheduled Tasks** tab and select **New Task**. In the **Single Task** tab, a schedule can be selected using the **Select a schedule** dropdown. The available schedules are preconfigured and range from running on the first day of the month at 1 a.m. to running every hour of the day. Each schedule name is descriptive and describes exactly when the flow will run. Once the task has been created, it can also be edited from the same location.

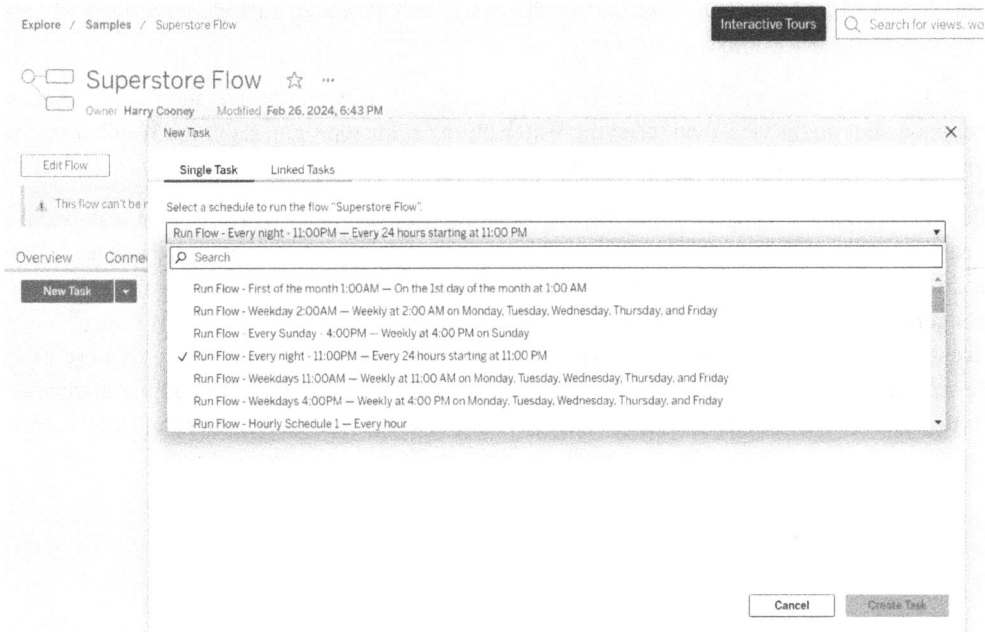

Figure 8.16: Adding a Tableau Prep flow scheduled task to run every night at 11 p.m.

Managing Published Workbooks

Once content has been published to a Tableau site, it can be managed with additional functionality to engage users. Data-driven alerts can be set up such that when a preconfigured threshold on a view is surpassed, an email, Slack notification, or Tableau site notification is sent to the desired end users. Screenshots or PDF exports of a specific view or a whole workbook can also be configured to be sent out on a regular basis.

Alerts

As data is refreshed in a published Tableau workbook, the charts within will update accordingly. End users may not be aware of the updated data and may, as a result, miss out on key insights. Data-driven alerts allow for thresholds to be specified on charts, which, if exceeded during data refresh, will result in an alert being sent out to specific users.

To create an alert, open a published workbook and navigate to the view the alert should run off. Select a continuous axis on the view then select the **Watch** button at the top right. From the **Watch** dropdown, select **Alerts**, then in the configuration pane on the right, select **Create**.

The alert configuration popup allows the user to set the alert to trigger when the data is above, above or equal to, below, below or equal to, or equal to a specified threshold. The threshold will be displayed on the selected axis as a red line with light red shading above. A subject line for the email can be configured, as well as when to send the alert. The options for when to send the alert range from only once when the threshold is first met to as frequently as possible. Recipients for the alert are Tableau site accounts and can be typed into the **Recipients** box. The final configuration option is whether or not to make the alert visible to others. If the alert is made visible, then any user can add themself to the alert so that they also receive the email/notification when the threshold is met.

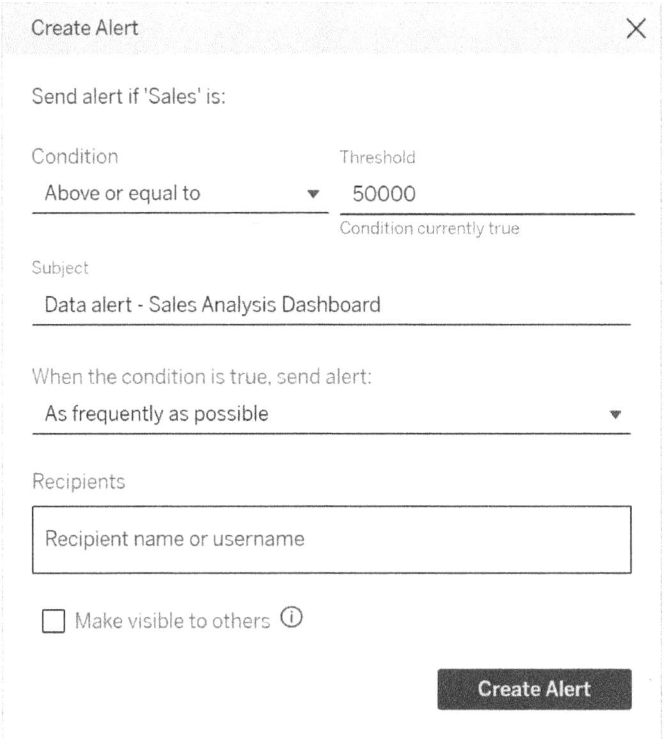

Figure 8.17: Alert configured to trigger as frequently as possible whenever monthly sales are over 50k

Creating a Data-Driven Alert

In this exercise, you will add a data-driven alert to the dashboard that you published earlier in the chapter. Data-driven alerts are a great way to let users know whether something noteworthy has occurred in the data and nudge them to look into it on the dashboard. Follow these steps to create a data-driven alert:

1.   Open the workbook published to Tableau Cloud earlier in the chapter and select the thumbnail to open the **Sales Analysis Dashboard**.

2.   Select the **Sales** axis on the **Monthly Sales** chart followed by the **Watch** button at the top right and then **Alerts**. Select **Create** in the **Alerts** pane to open up the alert configuration menu.

3.   In the alert configuration popup, set the threshold to 90,000 and the condition to **Above or equal to**. Notice how the red threshold indicator in the chart behind moves up accordingly and now only a few data points surpass the threshold.

4.   Set the notification subject line to **Data alert - Sales Analysis Dashboard** and have the alert run **Daily at most**.

5.   Set yourself as the recipient of the alert then select **Save Alert**.

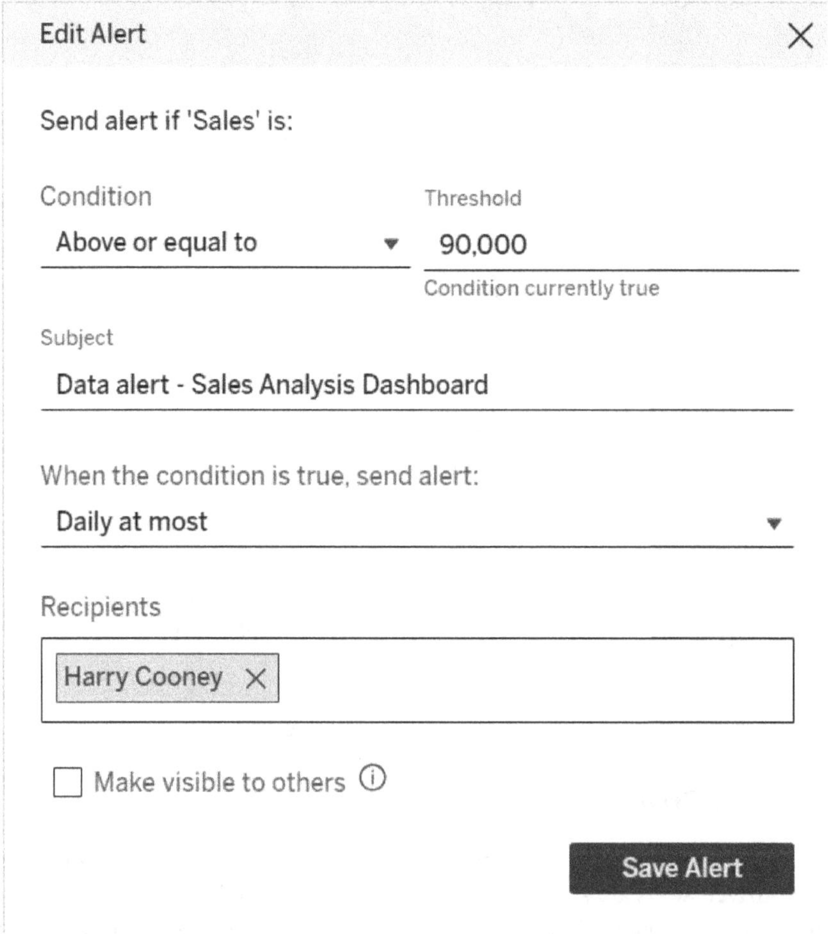

Edit Alert   ✕

Send alert if 'Sales' is:

Condition   Threshold
Above or equal to   ▼   90,000
                                   Condition currently true

Subject
Data alert - Sales Analysis Dashboard

When the condition is true, send alert:
Daily at most   ▼

Recipients
┌───┐
│ Harry Cooney ✕   │
└───┘

☐ Make visible to others ⓘ

                                          Save Alert

Figure 8.18: Alert configuration

6.   Check your email – you should have received a data-driven alert from Tableau, informing you
     that the sales are above or equal to 90,000, providing a link to the dashboard and a screenshot
     of the dashboard with the threshold visible.

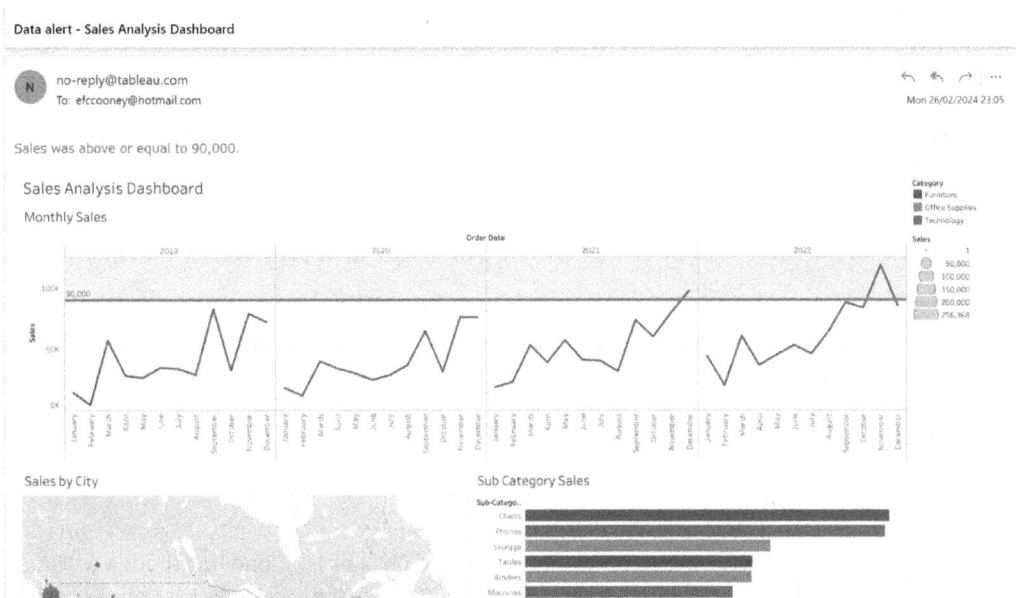

Figure 8.19: Data-driven alert created and email received

In this exercise, you added a data-driven alert to a dashboard and saw the configuration options available. You added an email alert to your own address and received the email that provided a screenshot of the dashboard and a link to it.

Subscriptions

Subscriptions can be set up to email selected users on a regular basis with either PDF or image exports of a published dashboard.

To create a subscription, open the workbook you want to set the subscription for on the Tableau site. Select **Watch** at the top right followed by **Subscriptions**. The **Subscriptions** configuration pane allows you to first type in the users or permission groups you want to send the regular exports to, and there is also a **Subscribe me** checkbox to subscribe yourself.

The currently opened view can be selected as the sole export to send to users or the whole workbook can be selected in the **Include** section. There is also an option to not send the email if the view is empty (for example, if there was an issue with a data source refresh or connection).

The **Format** section allows the user to select whether to export the view as a PDF, image, or both. The PDF is shared as an attached file and the image is embedded into the email body. Paper size and orientation allow the user to configure the formatting of the exports.

A subject line for the email must be configured and a message can be provided for the email body.

Finally, a frequency can be set for when to send the email out. If the frequency is set to **On Selected Schedule**, then a predefined schedule can be selected on Tableau Server or a configurable schedule can be selected on Tableau Cloud. These schedules function in the same way as schedules for extract refreshes with the same options available. If **When Data Refreshes** is selected, then the email will be sent out when the data source extract refreshes.

Creating a Subscription

In this exercise, you will create a subscription to a dashboard on Tableau Cloud that sends a scheduled email to your own email account with a PDF attachment of the dashboard. Subscriptions are a great way to keep users up to date with the analysis on the published dashboard. Follow these steps to create a subscription:

1. Open the **Sales Analysis Dashboard** on the Tableau site that you created earlier in the chapter and click the **Watch** button at the top right, followed by **Subscriptions**.

2. In the **Subscriptions** configuration popup, select **Subscribe me** to add yourself to the email subscription for the dashboard.

3. Include **This view** only and send the export in PDF and image format with a paper size of **A4** and a **Landscape** orientation.

4. Set the subject line to **Sales Analysis Dashboard** and optionally provide a message for the email body, such as `Hi, check out my Sales Analysis Dashboard!`.

5. Set the frequency to **On Selected Schedule** and then configure the schedule to run hourly between the closest time coming up and for five minutes after for the current day only.

6. Once complete, select **Subscribe**.

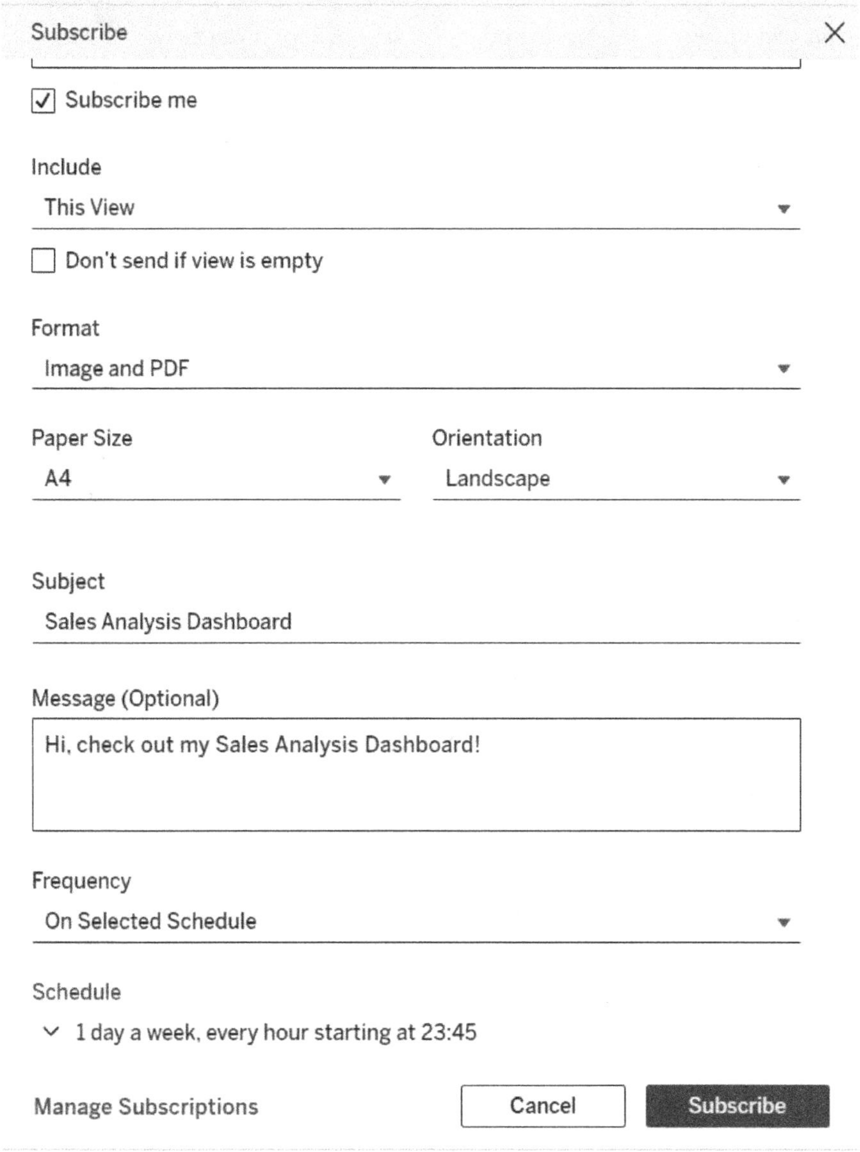

Figure 8.20: Subscription configuration

7.   When it hits the scheduled time, you will receive an email containing the PDF and image exports, along with the specified subject line and email body.

8.   To prevent further emails, the subscription can be deleted if you navigate back to the **Default** project, open the **Sales Analysis Dashboard**, and then select the **Subscriptions** tab.

9.   Press **...** and select **Unsubscribe** from the dropdown.

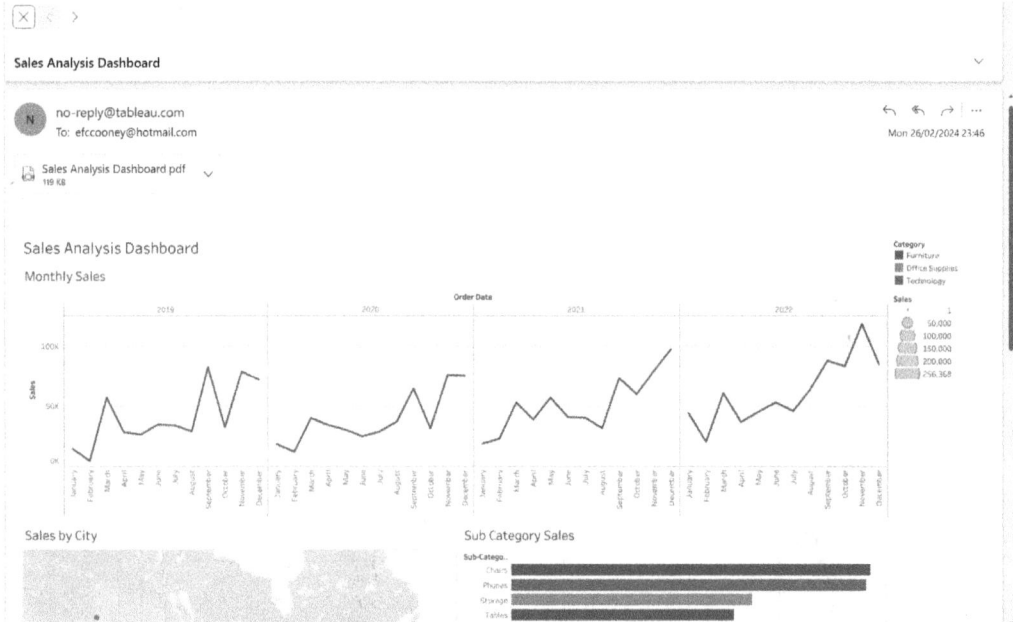

Figure 8.21: Subscription created and email received, containing both an image and PDF export

In this exercise, you have added a subscription to a published dashboard and have customized both the email content and the attachment included. You have also received the subscription email and now know what end users will receive when they are subscribed to a dashboard.

Summary

When it comes to publishing and managing content on a Tableau site, there is a wide variety of functionality. End users can be provided with both workbooks and data sources. Workbooks can be published whole or as specific sheets that are relevant to end users. Data sources can be published and set to replace the existing data sources in a Tableau workbook.

Tableau Server and Cloud are not the only means of sharing Tableau content. PDF files, images, and data extracts in the form of Access databases and Excel crosstabs can all be exported directly from Tableau. Hard copies of Tableau workbooks can also be printed directly from Tableau Desktop.

When it comes to keeping data fresh on a Tableau site, extracts can be configured to run a full or incremental refresh on a regular basis. Tableau Prep flows can also be configured to run transformations on data, combine multiple sources, and output a consolidated data source from which reports are built. These Tableau Prep flows can also be configured to be run on a regular basis, resulting in up-to-date data in the end reports.

To keep users aware of data updates, subscriptions can be set up to run whenever a data source is refreshed. Subscriptions send an email with an image or PDF file attachment exported from the workbook. Subscriptions are highly customizable and can also be configured to run at regular intervals. Alerts can also be configured on published workbooks so that if data is updated and a pre-set threshold condition is met, specific users will be notified.

Exam Readiness Drill – Chapter Review Questions

Apart from a solid understanding of key concepts, being able to think quickly under time pressure is a skill that will help you ace your certification exam. That is why working on these skills early on in your learning journey is key.

Chapter review questions are designed to improve your test-taking skills progressively with each chapter you learn and review your understanding of key concepts in the chapter at the same time. You'll find these at the end of each chapter.

> **How to Access these Resources**
>
> To learn how to access these resources, head over to the chapter titled *Chapter 9, Accessing the Online Practice Resources*.

To open the Chapter Review Questions for this chapter, perform the following steps:

1. Click the link –`https://packt.link/TDA_CH08`.

    Alternatively, you can scan the following **QR code** (*Figure 8.22*):

Figure 8.22: QR code that opens Chapter Review Questions for logged-in users

2. Once you log in, you'll see a page similar to the one shown in *Figure 8.23*:

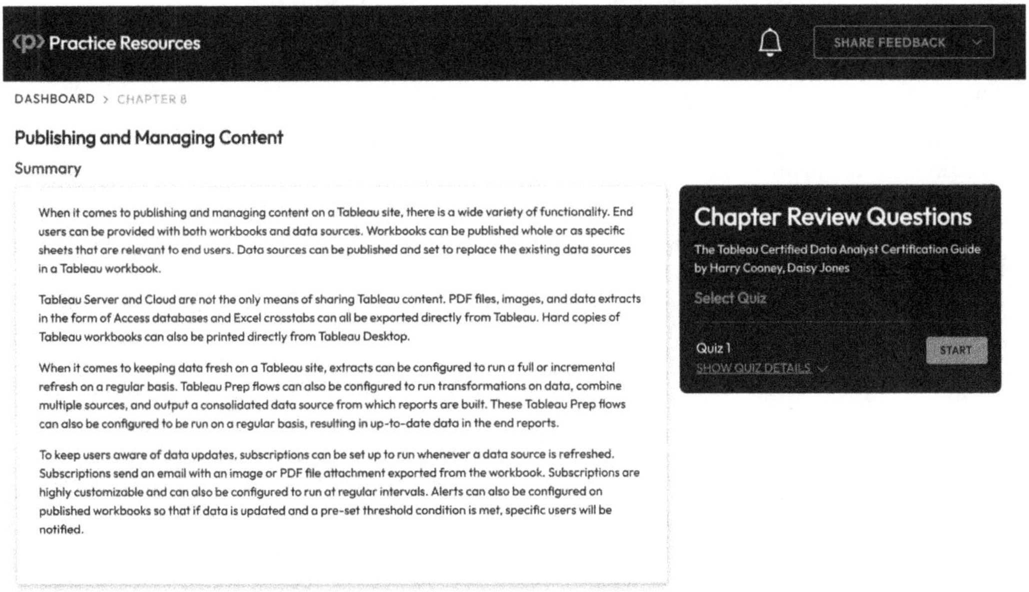

Practice Resources   SHARE FEEDBACK

DASHBOARD > CHAPTER 8

Publishing and Managing Content

Summary

When it comes to publishing and managing content on a Tableau site, there is a wide variety of functionality. End users can be provided with both workbooks and data sources. Workbooks can be published whole or as specific sheets that are relevant to end users. Data sources can be published and set to replace the existing data sources in a Tableau workbook.

Tableau Server and Cloud are not the only means of sharing Tableau content. PDF files, images, and data extracts in the form of Access databases and Excel crosstabs can all be exported directly from Tableau. Hard copies of Tableau workbooks can also be printed directly from Tableau Desktop.

When it comes to keeping data fresh on a Tableau site, extracts can be configured to run a full or incremental refresh on a regular basis. Tableau Prep flows can also be configured to run transformations on data, combine multiple sources, and output a consolidated data source from which reports are built. These Tableau Prep flows can also be configured to be run on a regular basis, resulting in up-to-date data in the end reports.

To keep users aware of data updates, subscriptions can be set up to run whenever a data source is refreshed. Subscriptions send an email with an image or PDF file attachment exported from the workbook. Subscriptions are highly customizable and can also be configured to run at regular intervals. Alerts can also be configured on published workbooks so that if data is updated and a pre-set threshold condition is met, specific users will be notified.

Chapter Review Questions

The Tableau Certified Data Analyst Certification Guide by Harry Cooney, Daisy Jones

Select Quiz

Quiz 1   START
SHOW QUIZ DETAILS

Figure 8.23: Chapter Review Questions for Chapter 8

3. Once ready, start the following practice drills, re-attempting the quiz multiple times.

Exam Readiness Drill

For the first three attempts, don't worry about the time limit.

ATTEMPT 1

The first time, aim for at least **40%**. Look at the answers you got wrong and read the relevant sections in the chapter again to fix your learning gaps.

ATTEMPT 2

The second time, aim for at least **60%**. Look at the answers you got wrong and read the relevant sections in the chapter again to fix any remaining learning gaps.

ATTEMPT 3

The third time, aim for at least **75%**. Once you score 75% or more, you start working on your timing.

> **Tip**
> You may take more than **three** attempts to reach 75%. That's okay. Just review the relevant sections in the chapter till you get there.

Working On Timing

Target: Your aim is to keep the score the same while trying to answer these questions as quickly as possible. Here's an example of how your next attempts should look like:

Attempt	Score	Time Taken
Attempt 5	77%	21 mins 30 seconds
Attempt 6	78%	18 mins 34 seconds
Attempt 7	76%	14 mins 44 seconds

Table 8.1: Sample timing practice drills on the online platform

> **Note**
> The time limits shown in the above table are just examples. Set your own time limits with each attempt based on the time limit of the quiz on the website.

With each new attempt, your score should stay above **75%** while your "time taken" to complete should "decrease". Repeat as many attempts as you want till you feel confident dealing with the time pressure.

9

Accessing the Online Practice Resources

Your copy of *Tableau Certified Data Analyst Certification Guide* comes with free online practice resources. Use these to hone your exam readiness even further by attempting practice questions on the companion website. The website is user-friendly and can be accessed from mobile, desktop, and tablet devices. It also includes interactive timers for an exam-like experience.

How to Access These Resources

Here's how you can start accessing these resources depending on your source of purchase.

Purchased from Packt Store (packtpub.com)

If you've bought the book from the Packt store (`packtpub.com`) eBook or Print, head to `https://packt.link/tdaunlock`. There, log in using the same Packt account you created or used to purchase the book.

Packt+ Subscription

If you're a *Packt+ subscriber*, you can head over to the same link (`https://packt.link/tdapractice`), log in with your `Packt ID`, and start using the resources. You will have access to them as long as your subscription is active.

If you face any issues accessing your free resources, contact us at `customercare@packt.com`.

Purchased from Amazon and Other Sources

If you've purchased from sources other than the ones mentioned above (like *Amazon*), you'll need to unlock the resources first by entering your unique sign-up code provided in this section. **Unlocking takes less than 10 minutes, can be done from any device, and needs to be done only once**. Follow these five easy steps to complete the process:

STEP 1

Open the link `https://packt.link/tdaunlock` OR scan the following **QR code** (*Figure 9.1*):

Figure 9.1: QR code for the page that lets you unlock this book's free online content.

Either of those links will lead to the following page as shown in *Figure 9.2*:

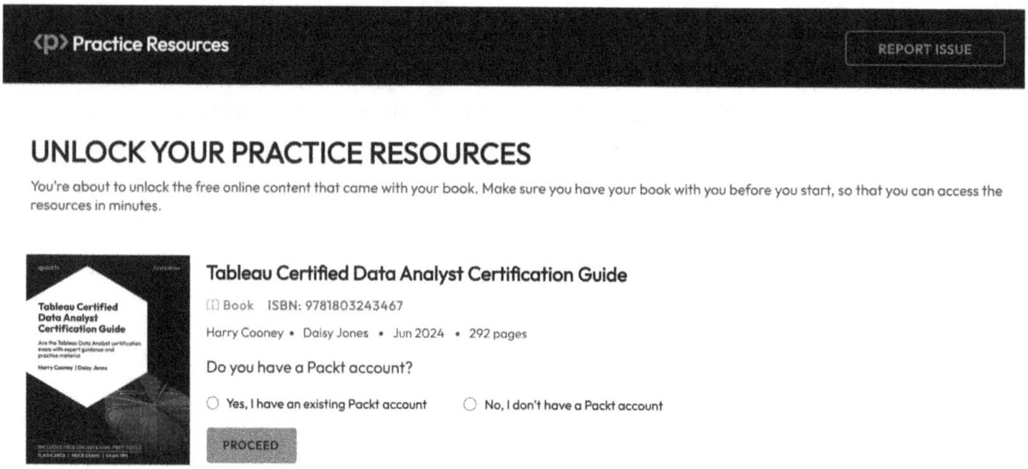

Figure 9.2: Unlock page for the online practice resources

STEP 2

If you already have a Packt account, select the option `Yes, I have an existing Packt account`. If not, select the option `No, I don't have a Packt account`.

If you don't have a Packt account, you'll be prompted to create a new account on the next page. It's free and only takes a minute to create.

Click `Proceed` after selecting one of those options.

STEP 3

After you've created your account or logged in to an existing one, you'll be directed to the following page as shown in *Figure 9.3*.

Make a note of your unique unlock code:

`PSV9397`

Type in or copy this code into the text box labeled 'Enter Unique Code':

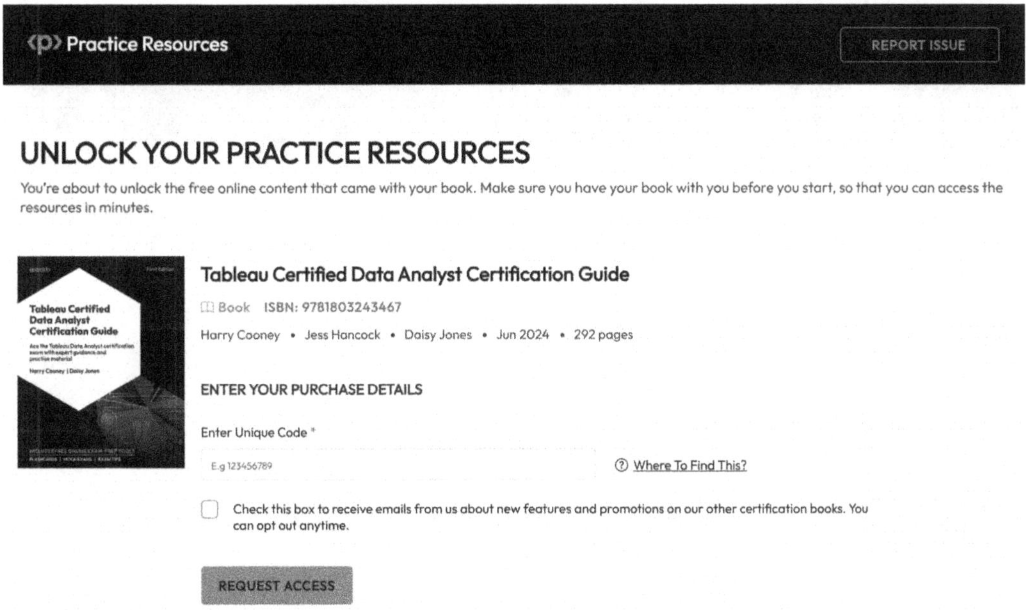

Figure 9.3: Enter your unique sign-up code to unlock the resources

> **Troubleshooting Tip**
>
> After creating an account, if your connection drops off or you accidentally close the page, you can reopen the page shown in *Figure 9.2* and select `Yes, I have an existing account`. Then, sign in with the account you had created before you closed the page. You'll be redirected to the screen shown in *Figure 9.3*.

STEP 4

> **Note**
>
> You may choose to opt into emails regarding feature updates and offers on our other certification books. We don't spam, and it's easy to opt out at any time.

Click `Request Access`.

STEP 5

If the code you entered is correct, you'll see a button that says, OPEN PRACTICE RESOURCES, as shown in *Figure 9.4*:

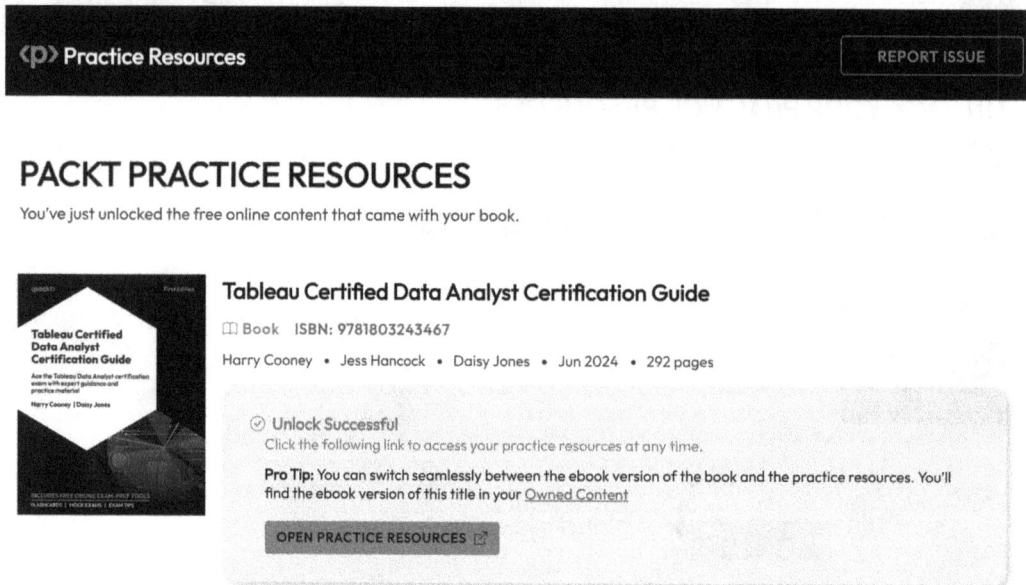

Figure 9.4: Page that shows up after a successful unlock

Click the OPEN PRACTICE RESOURCES link to start using your free online content. You'll be redirected to the Dashboard shown in *Figure 9.5*:

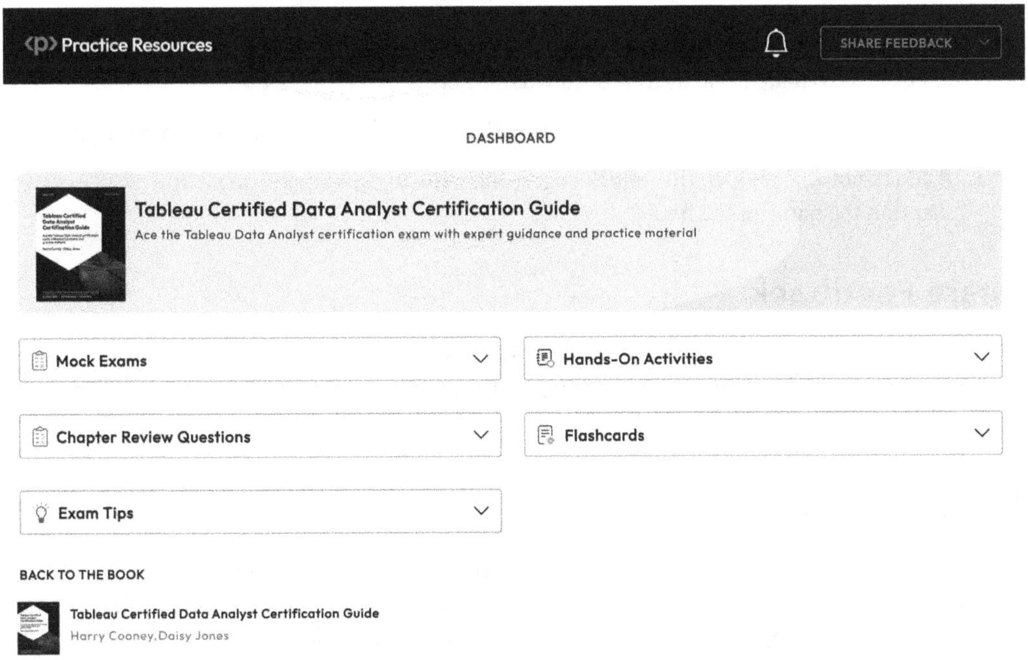

Figure 9.5: Dashboard page for Tableau Certified Data Analyst practice resources

Bookmark this link

Now that you've unlocked the resources, you can come back to them anytime by visiting `https://packt.link/tdapractice` or scanning the following QR code provided in *Figure 9.6*:

Figure 9.6: QR code to bookmark practice resources website

Troubleshooting Tips

If you're facing issues unlocking, here are three things you can do:

- Double-check your unique code. All unique codes in our books are case-sensitive and your code needs to match exactly as it is shown in *STEP 3*.

- If that doesn't work, use the `Report Issue` button located at the top-right corner of the page.

- If you're not able to open the unlock page at all, write to `customercare@packt.com` and mention the name of the book.

Share Feedback

If you find any issues with the platform, the book, or any of the practice materials, you can click the `Share Feedback` button from any page and reach out to us. If you have any suggestions for improvement, you can share those as well.

Back to the Book

To make switching between the book and practice resources easy, we've added a link that takes you back to the book (*Figure 9.7*). Click it to open your book in Packt's online reader. Your reading position is synced so you can jump right back to where you left off when you last opened the book.

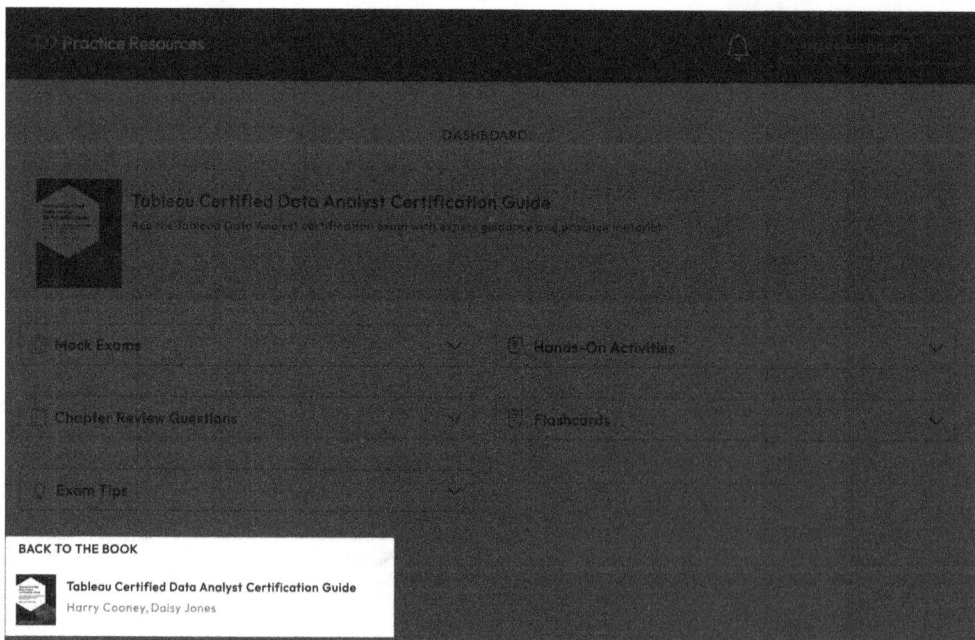

Figure 9.7: Dashboard page for Tableau Certified Data Analyst practice resources

Index

D

W

‹packt›

www.packtpub.com

Subscribe to our online digital library for full access to over 7,000 books and videos, as well as industry leading tools to help you plan your personal development and advance your career. For more information, please visit our website.

Why subscribe?

- Spend less time learning and more time coding with practical eBooks and Videos from over 4,000 industry professionals

- Improve your learning with Skill Plans built especially for you

- Get a free eBook or video every month

- Fully searchable for easy access to vital information

- Copy and paste, print, and bookmark content

At www.packtpub.com, you can also read a collection of free technical articles, sign up for a range of free newsletters, and receive exclusive discounts and offers on Packt books and eBooks.

Other Books You May Enjoy

If you enjoyed this book, you may be interested in these other books by Packt:

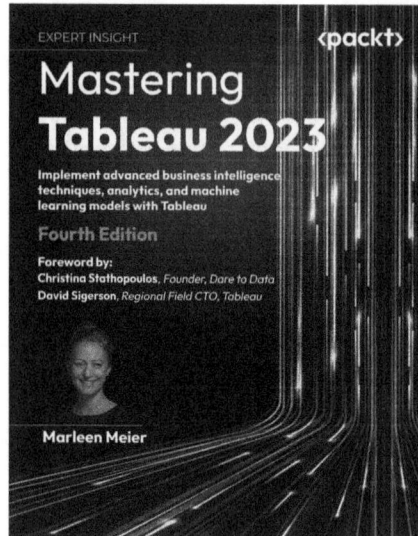

Mastering Tableau 2023

Marleen Meier

ISBN: 978-1-80323-376-5

- Learn about various Tableau components, such as calculated fields, table calculations, and LOD expressions
- Master ETL (Extract, Transform, Load) techniques using Tableau Prep Builder
- Explore and implement data storytelling with Python and R
- Understand Tableau Exchange by using accelerators, extensions, and connectors
- Interact with Tableau Server to understand its functionalities
- Study advanced visualizations and dashboard creation techniques
- Brush up on powerful self-service analytics, time series analytics, and geo-spatial analytics
- Find out why data governance matters and how to implement it

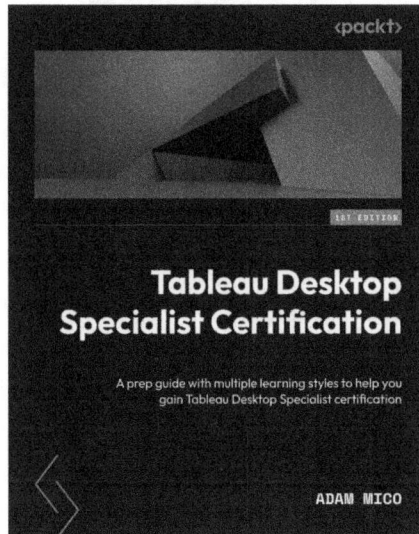

Tableau Desktop Specialist Certification

Adam Mico

ISBN: 978-1-80181-013-5

- Understand how to add data to the application
- Explore data for insights in Tableau
- Discover what charts to use when visualizing for audiences
- Understand functions, calculations and the basics of parameters
- Work with dimensions, measures and their variations
- Contextualize a visualization with marks
- Share insights and focus on editing a Tableau visualization

Share Your Thoughts

Now you've finished *Tableau Certified Data Analyst Certification Guide*, we'd love to hear your thoughts! Scan the QR code below to go straight to the Amazon review page for this book and share your feedback or leave a review on the site that you purchased it from.

https://packt.link/r/1803243465

Your review is important to us and the tech community and will help us make sure we're delivering excellent quality content.

Download a Free PDF Copy of This Book

Thanks for purchasing this book!

Do you like to read on the go but are unable to carry your print books everywhere?

Is your eBook purchase not compatible with the device of your choice?

Don't worry, now with every Packt book you get a DRM-free PDF version of that book at no cost.

Read anywhere, any place, on any device. Search, copy, and paste code from your favorite technical books directly into your application.

The perks don't stop there, you can get exclusive access to discounts, newsletters, and great free content in your inbox daily.

Follow these simple steps to get the benefits:

1. Scan the QR code or visit the link below:

https://packt.link/free-ebook/9781803243467

2. Submit your proof of purchase.
3. That's it! We'll send your free PDF and other benefits to your email directly.

www.ingramcontent.com/pod-product-compliance
Lightning Source LLC
Chambersburg PA
CBHW072008230326
41598CB00082B/6870